The Chromosomal Imbalance Theory of Cancer

Autocatalyzed Progression of Aneuploidy *is* Carcinogenesis

The Chromosomal Imbalance Theory of Cancer

Autocatalyzed Progression of Aneuploidy *is* Carcinogenesis

David Rasnick

Oakland, CA

USA

CRC Press
Taylor & Francis Group
an **informa** business
www.crcpress.com

6000 Broken Sound Parkway, NW
Suite 300, Boca Raton, FL 33487
270 Madison Avenue
New York, NY 10016
2 Park Square, Milton Park
Abingdon, Oxon OX14 4RN, UK

Science Publishers
Jersey, British Isles
Enfield, New Hampshire

Published by Science Publishers, an imprint of Edenbridge Ltd

- St. Helier, Jersey, British Channel Islands
- P.O. Box 699, Enfield, NH 03748, USA

E-mail: *info@scipub.net*   Website: *www.scipub.net*

Marketed and distributed by:

	6000 Broken Sound Parkway, NW Suite 300, Boca Raton, FL 33487
CRC Press	270 Madison Avenue New York, NY 10016
Taylor & Francis Group an **informa** business www.crcpress.com	2 Park Square, Milton Park Abingdon, Oxon OX14 4RN, UK

Copyright reserved © 2012

ISBN: 978-1-57808-737-2

Library of Congress Cataloging-in-Publication Data

Rasnick, David William, 1948-
  The chromosomal imbalance theory of cancer : autocatalyzed
progression of aneuploidy is carcinogenesis / David Rasnick.
      p. cm.
  Includes bibliographical references and index.
  ISBN 978-1-57808-737-2 (hardcover : alk. paper)
  1. Carcinogenesis. 2. Cancer--Genetic aspects. 3. Aneuploidy.
I. Title.
  RC268.5.R37 2012
  616.99'4071--dc23

                                                          2011038881

Printed in the United States of America

Preface

Tumors destroy man in an unique and appalling way, as flesh of his own flesh, which has somehow been rendered proliferative, rampant, predatory, and ungovernable. They are the most concrete and formidable of human maladies, yet despite more than 70 years of experimental study they remain the least understood.

(Rous 1967)

The broadly held conviction among researchers is that cancer ultimately results from an abnormality of the genome. The two principal competing theories on the nature of that abnormality is the subject of this book: Molecular medicine's search for the "material" cause of cancer in the form of gene mutations, and the chromosomal imbalance explanation that cancer results from global alterations in the dynamical relationships among all the genetic and metabolic activities of a cell independent of gene mutations.

In 1969, President Nixon proposed to reduce the budget of the National Cancer Institute (NCI). However, faced with the magnitude of the cancer problem, plus other political considerations, Nixon reversed himself embracing as his own the National Cancer Act sponsored by Senators Kennedy and Rogers and declared a national "war on cancer" in 1971 (Rettig 2006). Planners of this war predicted that technology would conquer cancer as it had conquered space and molecular biology would lead the way.

In 1986, John Bailar and Elaine Smith of the Harvard School of Public Health assessed the overall progress against cancer during the years 1950–1982. In the United States, these years were associated with increases in the number of deaths from cancer, in the crude cancer-related mortality rate, in the age-adjusted mortality rate, and in both the crude and the age-adjusted incidence rates, whereas reported survival rates (crude and relative) for cancer patients also increased (Bailar and Smith 1986). Notwithstanding progress on minor fronts, they concluded we are losing the war against cancer.

Eleven years later, Bailar and Gornik took another look at how the campaign was going and declared the war against cancer is far from over (Bailar and Gornik 1997). "Will we at some future time do better in the war against cancer?" the authors asked. "The present optimism about new therapeutic approaches rooted in molecular medicine may turn out to be justified, but the arguments are similar in tone and rhetoric to those of decades past about chemotherapy, tumor virology, immunology, and other approaches. In our view, prudence requires a skeptical view of the tacit assumption that marvelous new treatments for cancer are just waiting to be discovered."

In 2004, three federal reports (The CDC's *Morbidity and Mortality Report*, June 25, The Annual Report to the Nation on the Status of Cancer, published in *Cancer*, July 1, and "Living Beyond Cancer: Finding a New Balance" issued by the President's Cancer Panel in early June) said the number of cancer cases in the United States had reached a new high, and more people are alive after a diagnosis of cancer than ever before (Twombly 2004). It was not clear exactly what that declaration meant, however. Some took this to mean there had been marked progress in the treatment of cancer. Others were quick to question the implied widespread treatment success, saying the numbers are inflated by increased detection of non-lethal cancers by screening and there was no information on the quality of life. Even Julia Rowland, director of the NCI's Office of Cancer Survivorship said, "The effect of including those cancers in the data pool is that 5-year survival rates increase because more people who may never have otherwise known they had cancer are now considered survivors, thereby masking the more important question of whether progress has been made in treating advanced solid tumors."

John Bailar, professor emeritus of health studies at the University of Chicago agreed. He pointed out that the reports by the CDC and the President's Cancer Panel directly compared "survival" between two different time frames decades apart. He said that made no sense given the potential for over-diagnosis by increased screening. Even more recently, a 2005 article (Leaf 2004) and two books (Epstein 2005, Faguet 2005) pulled few punches criticizing the paltry progress and dashed hopes in the war on cancer.

In an editorial titled "Our Contribution to the Public Fear of Cancer", Bernard Strauss said, "the scientific community has managed to confuse the public about the causes of cancer and to add to an almost irrational fear of the disease. The only way to allay this fear is to development effective treatment and to understand how cancer develops... . The public's responses to discussions of cancer are reminiscent of societies responses to the threat of epidemics before the nature of infectious disease was understood" (Strauss 1998).

What is the public to make of the confusion caused by the experts themselves? The public's dread of cancer and the fear of plague in the Middle Ages have this in common: no rational explanation for the disease and no way to combat it. But what makes cancer so intractable and mysterious, the biological equivalent of Fermat's last theorem? The answer lies in the way scientists and clinicians have been looking at the problem. Most cancer researchers think they already know the basic cause of cancer: genetic mutations in specific genes (Strauss 1998). However, the gene mutation hypothesis has not led to an understanding of even the most basic questions of how cancer starts and progresses. For example, in a commentary in the *Proceedings of the National Academy of Sciences*, Boland and Ricciardiello asked: "How many mutations does it take to make a tumor?" (Boland and Ricciardiello 1999). The answer was apparently 11,000 (Stoler et al. 1999). Boland and Ricciardiello rightly asked how does this result fit with central concepts such as clonal expansion and multi-step carcinogenesis? Indeed, questions that go to the heart of the mutation theory, which currently says only 4–6 mutations (Hahn and Weinberg 2002b) are needed to cause cancer.

If the current doctrine that cancer is caused by gene mutations was on the right track, the confusion and debate among cancer experts should have diminished in recent years instead of accelerating. Furthermore, cancer statistics should by now show obvious signs of progress but they don't. The worsening situation is leading some cancer researchers to look for an escape from the quagmire of mutation theory. What is needed is a new, more productive way to think about cancer.

The solution one comes up with depends strongly on how one looks at the problem. To see this, consider your favorite puzzle or even better, a well executed magic trick. A world-class magician produces surprise and delight by negating everyday experience and shattering the rules of causality. The *magic* in the magic trick is to make the audience look at the trick in such a way as to make it appear incomprehensible, unfathomable, impenetrable, baffling, perplexing, mystifying, bewildering—how cancer appears today. However, looking at the same magic trick in a different way (the way another magician would) reveals it to be completely consistent with the logic of how things happen. Once the trick is revealed, the magic disappears and the rational world is restored. By looking at the cancer problem in a different way it is possible to lift the shroud concealing the unifying simplicity behind cancer.

Interest in cancer cytogenetics influenced human cytogenetics much more profoundly than is currently appreciated. For example, the main goal behind the study that eventually led to the description of the correct chromosome number in man was to identify what distinguished a cancer cell (Tjio and Levan 1956). The motivation was not primarily an interest in the normal chromosome constitution, which at that time had no obvious implications, but the hope that such knowledge would help answer the basic question of whether chromosome changes lay behind the transformation of a normal cell to cancer (Heim and Mitelman 2009).

Normal human cells turned out to have 23 different chromosomes that come in pairs, half from each parent, to yield a total of 46 chromosomes. Such cells are said to be "diploid." Cells found in solid tumors, on the other hand, typically have 60–90 chromosomes

(Shackney et al. 1995a). Their ploidy is "not good," in other words, and the Greek term is "aneuploid." It is a word you will have a hard time finding in the cancer chapters of the leading textbooks of biology.

Recall that the genes (of which there may be 25,000 or so in humans (Collins et al. 2004)) are strung along the chromosomes, so that each chromosome contains thousands of genes. Any cell with a chromosome number different from 46 and not an exact multiple of 23, or with an abnormal complement of chromosomes that add up to 46, is an aneuploid cell. Thus, aneuploid cells contain an imbalance in the complement of genes and chromosomes compared to the normal or "diploid" cell. This imbalance in the chromosomes leads to a wide variety of problems, one of which is cancer.

Another problem caused by aneuploidy that is familiar to most people is Down syndrome. This results when a baby is born with three copies of chromosome 21 instead of the normal two. Just one extra copy of the smallest chromosome, with its thousand or so normal genes, is sufficient to cause the syndrome (Shapiro 1983). Most Down fetuses are spontaneously aborted. Nonetheless, the imbalance is small enough (47 chromosomes) to permit occasional live births. The level of aneuploidy is therefore far below the threshold of 60–90 chromosomes found in invasive cancer, but it gives these patients a head start toward developing the same cancers that normal people get (Hill et al. 2003). Down syndrome patients have up to a 30-fold increased risk of leukemia, for example, compared to the general population (Patja et al. 2006, Shen et al. 1995, Zipursky et al. 1994).

There is one important difference between the small chromosome imbalance found in Down syndrome, and the more pronounced aneuploidy of cancer cells. With Down syndrome, the defect occurs in the germ line and so the chromosomal error is present in every cell in the body. But the defect that gives rise to the unbalanced complement of chromosomes in cancer cells is "somatic". That is, it occurs in a particular cell after the body is formed. In the course of life, cells constantly divide by a process called mitosis. When errors in mitosis occur, as they often do, the possibility exists that a daughter cell will be aneuploid.

Aneuploidy destabilizes a dividing cell in much the same way that a dent disrupts the symmetry of a wheel, causing ever-greater distortions with each revolution. As aneuploid cells divide, their genomes become increasingly disorganized to the point where most of these cells stop dividing and many die. But rarely, and disastrously, an aneuploid cell with the right number and combination of chromosomes wins the genetic lottery and keeps right on going. Then it has become a cancer cell.

Cells with a normal number of chromosomes are intrinsically stable and not prone to transformation into cancer. What, therefore, causes normal cells to become aneuploid? That is a hotly contested question. It is known, however, that radioactive particles striking the nucleus or cytoplasm either kill or damage a cell. When the damaged cell then divides by mitosis, an error may arise leading to chromosomal imbalance. In short, radiation can cause aneuploidy. And certain chemicals, such as tars, also give rise to aneuploid cells. Tars and radiation sources are known carcinogens. In fact, all carcinogens that have been examined do cause aneuploidy.

That is a strong argument for the aneuploidy theory of cancer, but in order to understand the controversy one must understand the alternative theory. Everyone has heard of it because it is in the newspapers all the time. It is the gene mutation theory of cancer. According to this theory, certain genes, when they are mutated, turn a normal cell into a cancer cell. This theory has endured since the 1970s, and more than one Nobel Prize has been awarded to researchers who have made claims about it. One prize-winner was the former director of the National Institutes of Health, Harold Varmus. According to some researchers, the mutation of just one, or perhaps several genes, may be sufficient to transform a normal cell into a cancer cell.

In contrast, aneuploidy disrupts the normal balance and interactions of many thousands of genes, because just one chromosome typically contains thousands of genes. And a cancer cell may have several copies of a given chromosome. For this reason alone, aneuploidy is far more devastating to the life of a cell than a small handful of gene mutations.

The fundamental difference between the chromosomal imbalance theory and the reigning gene mutation theory may be put this way. If the whole genome is a biological dictionary, divided into volumes called chromosomes, then the life of a cell is a Shakespearean drama. If one were to misspell a word here and there, in *Hamlet* for example, such "mutations" would be irrelevant to the vast majority of readers, or theater-goers. A multicellular organism is at least as resistant to "gene mutations" as a Shakespeare play.

On the other hand, without "mutating" a single word, one could transform the script of *Hamlet* into a legal document, a love letter, a declaration of independence, or more likely gibberish, by simply shifting and shuffling, copying and deleting numerous individual words, sentences and whole paragraphs. That is the literary equivalent of what aneuploidy does. The most efficient means of rewriting a cell's script is the wholesale shifting and shuffling of the genes, which aneuploidy or chromosomal imbalance accomplishes admirably.

Aneuploidy is known to be an efficient mechanism for altering the properties of cells, and it is also conceded that aneuploid cells are found in virtually all solid tumors. Bert Vogelstein of Johns Hopkins University has said that "at least 90 percent of human cancers are aneuploid." The true figure is 100 percent since there is not one confirmed diploid cancer (Section 4.4.4).

Nonetheless, the presence of mutations in a handful of genes continues to be viewed as a significant, even a causal factor in carcinogenesis, even though any given mutated gene is found in only a minority of cancers. Cells with mutated genes can indeed be found in cancerous as well as normal cells, but it is becoming increasingly clear the vast majority of mutations are innocuous. Hence they are readily accommodated during the expansion of barely viable aneuploid cells as they compete for survival with their more viable chromosomally balanced counterparts. The current emphasis in cancer research on the search for mutant genes in a perpetual background of aneuploidy is a classic example of not seeing the forest for the trees.

Thomas Kuhn remarked that the great theoretical advances of Copernicus, Newton, Lavoisier, and Einstein had less to do with definitive experiments than with looking at old data from a new perspective. Sufficient (indeed overwhelming) evidence is already at hand to convict aneuploidy of the crime of cancer and release gene mutations from custody (Aldaz et al. 1987, Aldaz et al. 1988a, Aldaz et al. 1988b, Brinkley and Goepfert 1998, Duesberg et al. 1998, Duesberg 1999, Duesberg et al. 2000a, Duesberg et al. 2000b, Duesberg et al. 2000c, Duesberg et al. 2001a, Duesberg et al. 2001b, Duesberg and Li 2003, Duesberg 2003, Duesberg et al. 2004a, Duesberg et al. 2004b, Duesberg et al. 2006, Fabarius et al. 2003, Heng et al. 2006b, Klein et al. 2010, Li et al. 1997, Li et al. 2000, Li et al. 2009, Liu et al. 1998, Rasnick and Duesberg 1999, Rasnick and Duesberg 2000, Rasnick 2000, Reisman et al. 1964a, Reisman et al. 1964b, Ye et al. 2009). Nevertheless, the gene mutation theorists, when faced with the undeniable evidence that aneuploidy is necessary for cancer, have adopted a fall-back position. They argue that gene mutations must initiate the aneuploidy (Sen 2000), or as the *Scientific American* reported, referring to a researcher in Vogelstein's lab, "[Christoph] Lengauer insists aneuploidy must be a consequence of gene mutations" (Gibbs 2001). There would be no need to "insist" if there were proof that gene mutations really do cause aneuploidy and cancer.

What would gravely weaken the aneuploidy theory would be confirmed cases of diploid cancer (in which the tumor cells have balanced chromosomes), and with the culprit genes found lurking in every cell. That would go a long way toward proving the gene mutation theory. But where has that been demonstrated? It would be a front-page story. The truth is that researchers have not yet produced any convincing examples of diploid cancer.

In fact, the evidence is going the other way. There is a growing list of carcinogens that do not mutate genes at all (Section 4.1.4). In addition, there are no cancer-specific gene mutations (Section 4.4.2). Even tumors of a single organ rarely have uniform genetic alterations (Section 4.4.3). And, in a rebuttal that should be decisive, no genes have yet been isolated from cancers that can transform normal human or animal cells into cancer cells (Section 4.4.4). Moreover, the latent periods between the application of a carcinogen and the appearance of cancer are exceedingly long, ranging from many months to decades, in contrast the effects of mutation are instantaneous (Section 4.4.5).

The goal of billions of dollars and decades of research was to come up with a clear and simple statement of how cancer genes cause or promote cancer. This was certainly the hope and expectation of most cancer researchers. One of the hallmarks of a bad theory is when its evolution becomes so complex and confused that experts in the field have difficulty explaining it. Thomas Ried, a major researcher at the National Cancer Institute in Bethesday, recently labored to...

> *speculate that the activation of specific oncogenes, and the inactivation of tumor suppressor genes act in concert with the deregulation of genes as a consequence of low-level copy number changes that provide the metabolic infrastructure for increased proliferation. One of the challenges in understanding the genome mutations in carcinomas will be to elucidate whether the presence of a tumor suppressor gene on frequently lost chromosomes, or the presence of an oncogene on frequently gained chromosomes is sufficient to fully explain the reason for the defining and recurrent patterns of genomic imbalances. In other words, we will need means to experimentally dissect the relative contribution of specific oncogene activation vis-a-vis the global transcriptional deregulation imposed by chromosome-wide copy number changes. Only then will we be in a position to truly verify or falsify Boveri's central statement, i.e., the dominant role of inhibiting and promoting chromosomes that formed the basis for his chromosome theory of cancer.*
>
> (Ried 2009)

The conceptual barriers to accepting aneuploidy as the cause of cancer are not trivial but they shrink in comparison with the political and sociological obstacles. US taxpayers have forked over hundreds of billions of dollars in the war on cancer only to find that after 40 years of battling viruses, "oncogenes", and "tumor suppressor" genes we are losing the war (Epstein 1998). But it is a one-front war with almost no resources devoted to alternative approaches. In spite of a century of evidence implicating aneuploidy as the cause of cancer, a leading researcher guesses that, "If you were to poll researchers ... 95 percent would say that the accumulation of mutations [to key genes] causes cancer" (Gibbs 2001).

The biotech industry has bet heavily on cancer diagnostics and therapeutics based entirely on the gene mutation theory. The highly publicized sequencing of the human genome, the commercialization of diagnostic tests for cancer genes (Arnold 2001, Hanna et al. 2001, Wagner et al. 2000), and the hype about Gleevec being "at the forefront of a new wave of cancer treatments [that] differs from other existing chemotherapies because it affects a protein that directly causes cancer" (McCormick 2001) make it even more difficult for researchers to consider the possibility that mutant genes may not cause cancer after all.

Max Planck said that, "A new scientific truth does not triumph by convincing its opponents and making them see the light, but rather because its opponents eventually die, and a new generation grows up that is familiar with it" (Planck 1949). It is encouraging to see that a new generation of cancer researchers are more inclined to accept aneuploidy as an alternative to gene mutation.

Chromosomal imbalance theory shows how gene mutations are not powerful enough to cause cancer (Section 5.4). It explains how cancer is initiated (Chapter 5) and why progression takes years to decades (Section 6.1.3). It explains the global or macroscopic characteristics that readily identify cancer: anaplasia, autonomous growth, metastasis, abnormal cell morphology, DNA indices ranging from 0.5 to over 2, genetic instability, and the high levels of membrane-bound and secreted proteins responsible for invasiveness and loss of contact inhibition (Chapters 5 & 6). It explains the common failure of chemotherapy (Section 7.3) and why cancer cells often become drug resistant even to drugs they were never exposed (Sections 5.3.5 & 6.2.4). It provides objective, quantitative measures for the detection of cancer and monitoring its progression (Section 7.2). It suggests non-toxic strategies of cancer therapy and prevention (Section 7.3). The chromosomal imbalance theory is the most comprehensive, productive, and satisfying explanation of carcinogenesis. In short: *The Autocatalyzed Progression of Aneuploidy is Carcinogenesis.*

David Rasnick

Contents

1

Introducing Cancer

What can be the nature of the generality of neoplastic changes, the reasons for their persistence, their irreversibility, and the discontinuous, steplike alterations that they frequently undergo?

(Rous 1967)

I have learned that there is a vast and deadly gap between the reality of cancer, which strikes human beings, and the theory of cancer, which thousands of researchers are using in their search for a cure.

(Dermer 1994)

…benign tumors which, having all the fundamentals of neoplasia "without the frills," may yield the more crucial information about its essential nature.

(Foulds 1954)

The origin and nature of cancer has been one of the great enigmas since the ancient Egyptians and Greeks. The central paradox is that tumors are us and yet not us. Paleopathologic findings and examination of ancient mummies indicate that cancer occurred in prehistoric time. According to Steven Hajdu, the first written description of cancer is found in the Edwin Smith Papyrus (3000 BC) and the Ebers Papyrus (1500 BC) (Hajdu 2004). The Edwin Smith Papyrus contains the earliest description of breast cancer, with the conclusion that there is no treatment. In the Ebers Papyrus, enlarged thyroids, polyps, and tumors of the skin, pharynx, stomach, rectum, and uterus are described. Although cancer was not a prevalent disease in antiquity (because most people did not live to old age), it is of interest that in ancient writings breast cancer is mentioned far more often than any other malignancy.

According to Hajdu, Hippocrates (460–375 BC), a native of Greece and contemporary of Socrates and Plato, was a skillful diagnostician (Hajdu 2004). Hippocrates classified diseases according to the principal symptoms and came to the conclusion that all diseases, including cancer, originate from natural causes. Hippocrates noted that growing tumors occur mostly in adults and the growths reminded him of a moving crab (*karakinos* in Greek), which led to the modern term "cancer" to indicate a malignant tumor. He recognized cancers of the skin, mouth, breast, and stomach. He knew about anorectal condylomas and polyps and recommended examination with a speculum if they were higher up in the colon. Hippocrates described breast carcinoma with bloody discharge from the nipple and called attention to the danger of bloody ascites. He observed hard slow-growing tumors of the cervix, which were associated with bleeding, emaciation, edema, and caused death. He mentioned superficial and deep-seated tumors in the armpit, groin, and thigh in older people. Hippocrates summed up his recommendation for treatment by writing that tumors that are not cured by medicine are cured by iron (knife), those that are not cured by iron are cured by fire (cautery), and those that are not cured by fire are incurable. For occult or deep-seated tumors, he advised not to use any treatment because if treated, the patient would die quickly. If not treated, the patient could survive for an extended period.

Hajdu traced the emigration of Greek physicians to Rome after Greece became part of the Empire in 146 BC (Hajdu 2004). Celsus, a native of Greece, received most of his medical education at the best medical school at the time, which was located in Alexandria, Egypt. Celsus was a proud citizen of his adopted Rome and made Latin the language of medicine for the first time. Celsus continued the Hippocratic tradition by comparing cancer with a crab, because it adheres to surrounding structures in a manner similar to how the crab holds on to anything in its claws. Celsus distinguished several varieties of cancer. He knew about superficial cancers of the face, mouth, throat, and penis but he also mentioned cancers of the liver, spleen, and visceral organs such as the colon. In *De Medicina*, Celsus introduced the first classification for breast carcinoma.

He recommended aggressive surgical therapy but he believed that only at the early stage could these tumors be removed. He cautioned that even after excision, and when a well-healed scar was formed, breast carcinomas may recur with swelling in the armpit and cause death by spreading into the body.

Galen was a theorist and a prolific writer but, despite the fact that he practiced in Rome and his spoken language was Latin, he used Greek in his writings. He penned some 500 medical papers, including 100 notes concerning tumors and cancer. Although he also compared cancer with the crab and advised surgery by cutting into healthy tissue around the border of the tumor, he accepted the Roman prejudice against surgical procedures. He believed that the best surgeon was the one who operated only as a last resort. Galen held that cancer was a constitutional disease that afflicts the sick, particularly those with a melancholic disposition. After the fall of the Roman Empire in AD 476, Celsus, the first Roman physician of excellence, was forgotten but Galen's rigid humoral theory and authoritative reasoning that left no questions unanswered were promoted. His thinking was well suited to Christian theology as well as Byzantine and Arabic teachings. Galen's canonical theories obstructed progress in medicine and delayed advances in understanding cancer until the end of the 18th century (Hajdu 2004).

The reports of Agricola (circa 1494–1555) and Paracelsus (circa 1493–1541), who described the disease of Schneeberg miners, are sometimes quoted as the first etiological observations on human cancer (Tomatis and Huff 2002). They were probably the first to provide an accurate description of an occupational disease, but they did not actually mention the words cancer or tumor. It was only in the middle of the 19th Century that miners' disease was actually considered to be neoplastic. Described initially as lymphosarcoma, it was finally diagnosed as bronchogenic carcinoma by Harting and Hesse in 1869 (Shimkin 1979).

Among the first reports that environmental agents could cause human cancer were those of John Hill, who reported that nasal tumors were associated with the use of tobacco snuff (Hill 1761), and of Percival Pott, who noted the causality of scrotal cancer in chimney sweeps (Melicow 1975). Thus, tobacco and soot became the first two leading causes of human cancer. The next reports of environmental carcinogens came about a century later and concerned the occurrence of skin cancer after therapeutic use of arsenic in Fowler's solution (Hutchinson 1888) and the high incidence of urinary bladder tumors in aniline dye workers (Dietrich and Dietrich 2001). The International Labour Office published a report in 1921 in which certain aromatic amines were labeled as carcinogenic to exposed workers (Tomatis and Huff 2002). The causal relationship between pulmonary cancer and occupational exposure of miners to high concentrations of radon was firmly established in the early 1940s (Schuttmann 1993). Several industrial production processes were subsequently identified as sources of exposure to carcinogens, and numerous chemicals and chemical mixtures were recognized as causative agents of human cancer (Tomatis et al. 1989).

Ionizing radiation may have been the first carcinogen encountered by the human species. Ionizing radiation was recognized as a cause of human tumors seven years after initiation of its artificial use by Roentgen in 1895 (Frieben 1902, Sick 1902). That was an exceptionally short time in comparison with the much longer periods—in certain instances several decades—before which certain chemicals that had been introduced into the environment were recognized as human carcinogens. Over 40 years passed before the natural radioactivity present in uranium mines was accepted as carcinogenic (Schuttmann 1993).

These environmental causes of cancer were called exogenous. Towards the end of the 19th Century, endogenous or heritable (meaning genetic) causes of cancer were hypothesized. The classification of the causes of cancer into these two broad categories (exogenous and endogenous) had the practical effect of concentrating funding of those research efforts that offered the greatest possibility of success. For a long time, that meant research was focused on the identification of exogenous or environmental carcinogens. Research on the endogenous component of carcinogenesis eventually gained importance and gradually took over when the methods of molecular biology were applied to cancer research later in the 20th Century. The research on viruses played an indirect but important role in this shift of interest.

In ancient times, cancer was not considered to be transmissible. However, between the seventeenth and eighteenth centuries, cancer was believed to be a contagious disease. Special hospitals were built in some European countries to isolate patients with cancer, almost in the same way as was done with patients with leprosy (Tomatis and Huff 2002). This period did not last long and the conviction that cancer was not contagious again prevailed since a microorganism could not be related to cancer. This may at least partly explain the little attention paid to the experiments of Payton Rous at the beginning of the 20th Century on the role of viruses in the origin of cancer (Rous 1911).

Nevertheless, the hypothesis of a parasitic origin of cancer had a short-lived celebrity in 1926, when the Danish scientist Johannes Fibiger received the Nobel Prize for claiming that gastric cancer in rats was caused by the nematode *Spiroptera carcinoma* (Fibiger 1926). However, it was later shown that Fibiger's results were not reproducible and, instead, it was decided a deficiency of vitamin A had caused cancer in his rodents (Petithory et al. 1997, Stolt et al. 2004). Perhaps embarrassed by Fibiger's Nobel Prize, it was forty years before the Nobel Foundation again offered its honors for cancer research. The hypothesis of a viral origin for human cancer never completely disappeared, however. In the first half of the 20th Century, the viral theory gained support in France and the Soviet Union (Shimkin 1979).

The rivalry between proponents of chemical and viral carcinogenesis was based in part on different schools of thought but was also related to competition for research funds. In general, scientists in the field of viral carcinogenesis, and later of molecular biology, considered the scientists involved in chemical carcinogenesis were old-fashioned, while those working in chemical carcinogenesis looked with suspicion at the viral oncologists (Tomatis and Huff 2002). Chemical carcinogenesis prevailed, in terms of funding and popularity, until the mid-1960s and early 1970s. Then came President Nixon's declaration of war on cancer, after which cancer virology became fashionable, highly advertised, and received most of the funding. Unfortunately, the viral approach did not produce the hoped-for solution to the cancer riddle. But it did contribute to the development of scientists with new skills and interests, who became essential to the rapid development of molecular biology and molecular genetics. This was probably the most consequential (albeit unintended) result from Nixon's war on cancer. Even the strongest traditional disciplines, such as biochemistry and pathology (and more recently epidemiology), have become molecular. Cancer research has become almost exclusively an exercise in characterizing and manipulating genes.

Cancer facts

Cancer is defined as any malignant growth or tumor caused by abnormal and uncontrolled cell division; it may spread to other parts of the body through the lymphatic system or the blood stream. The hundreds of different types of cancer are distinguishable in their details yet they all display the global or macroscopic characteristics that a theory of cancer must explain: anaplasia (a change in the structure and orientation of cells, characterized by a loss of differentiation and reversion to a more primitive form), autonomous growth, metastasis, abnormal cell morphology, DNA indices (amount of DNA in cancer cells divided by the amount in normal cells) ranging from 0.5 to over 2, genetic instability, the high levels of membrane-bound and secreted proteins responsible for invasiveness and loss of contact inhibition, multi-drug resistance, and the exceedingly long times of up to decades from carcinogen exposure to the appearance of cancer.

There is no clear evidence of cancer in living plants, and cancer seems to be absent in modern invertebrates, excluding some experimental conditions (Capasso 2005). The only solid evidence for true neoplastic diseases is in vertebrate animals, excluding amphibians. Neoplasms in birds are relatively common but are strictly limited to captive animals. In wild bird populations neoplasms seem extremely rare if present. While cancer is also rare in wild mammals, the prevalence of neoplasms in modern humans is particularly high and similar to that in domestic dogs and captive birds (Capasso 2005).

As we have seen, almost all types of modern human neoplastic diseases have been documented in ancient human remains. However, the prevalence of cancer has increased tremendously over the past century. For example, in Germany the mortality due to cancer was only 3.3% in 1900 and climbed to over 20% in 1970 (Capasso 2005). Today, about half of all men and one-third of all women develop cancer and about 20% of all deaths are due to cancer (Parker et al. 1996). This is an impressive increase and seems to demonstrate that the increase in cancer prevalence is a recent biological event.

The phenomenal growth of cancer has been linked to the aging of modern populations. Indeed, over the past century, especially in recent decades and in developed countries, life expectancy has steadily increased from about 30–40 years to 70–80 years (Capasso 2005). This fact is certainly a major factor behind the increase in cancer prevalence over the past century because cancer is clearly an age-progressive disease whose prevalence ranges from about 1.8% for those under 39 years old to 27% among those 60–79 years old (Parker et al. 1996). In the United States in 1996, more than 87% of cancer deaths occurred in subjects aged 55 years and older. The aging of populations increases the time of exposure of each individual to environmental carcinogens.

However, the aging of the human population is not sufficient to explain the tremendous increase in the prevalence of cancer around the world. Geography is now recognized as a risk factor for cancer, relatively independent of the life expectancy characteristics of individual countries (Parkin et al. 2001). The past century substantially added to the burden of natural carcinogens. Nuclear testing begun in the 1950s increased environmental radioactivity. The exponential growth of synthetic carcinogens is concentrated in urban and metropolitan habitats.

These considerations led Luigi Capasso to conclude: "The impressive increase in cancer prevalence documented in human populations over the last century is associated with modern man. It is a completely new phenomenon and has no precedents in the history of animals on the Earth. The high prevalence of cancer contributes to limiting the increase in life expectancy, and seems to be associated with the modern lifestyle. This lifestyle is characterized by living in a completely artificial environment (i.e., a prevalently indoor and metropolitan life in an environment in which we undergo prolonged exposure to environmental carcinogens associated with an increase in carcinogenic pollution). The high prevalence of cancer in vertebrates that share this new lifestyle with us in our almost completely artificial environments (i.e., domestic dogs and birds) seems to confirm this picture" (Capasso 2005).

Cancer requires many cell divisions

Cancer researchers have divided the process of carcinogenesis into to the initiation and promotion phases. This is because of the many years up to decades between exposure to carcinogen and appearance of cancer. John Cairns said, "the distinction between early and late steps in carcinogenesis is surely worth making if only to remind us that we really have no idea what is going on during the long interval that usually elapses between the initial stimulus and appearance of cancer" (Cairns 1978).

The initiation step is understood to be a consequence of exposure to carcinogen but the tissues and cells show no signs of malignancy. Promotion is the ability of carcinogen-treated cells to divide sufficiently to progress to frank cancer, as hinted at by several observations. Croton oil (a blistering agent and purgative extracted from the seeds of an Indian tree) has long been used in experiments in animals to promote the generation of cancer after first treating with various carcinogens. Croton oil by itself is not a carcinogen. When croton oil is rubbed on the skin it produces a local reddening and thickening and temporarily increases the rate of cell proliferation (Berenblum and Shubik 1947).

For the production of cancer in internal organs, certain special methods of promotion have been found: for example, partial excision of the liver forces the remaining liver cells to multiply and compensate for the loss, and this will promote liver cancer in mice that have been fed an initiating carcinogen. Similarly, cancers of the thyroid (Hall 1948) and ovary (Furth and Sobel 1947) were provoked by appropriate hormones. Leukemia can be produced in rats treated with X-rays followed by repeated bleeding so that cells in the bone marrow have to increase their extent or rate of multiplication (Gong 1971).

All these procedures have in common increased cell multiplication. The unanswered question for Cairns and others was why such provoked multiplication should be dangerous to tissues like skin and marrow, where cells are multiplying the whole time even in the absence of an artificial stimulus? For Cairns, no plausible explanation has emerged from the mutation theory since promoters by definition don't cause cancer. Had Cairns and others paid more attention to the well-known fact at the time that normal cells have exactly the balanced number of 46 chromosomes and cancer cells do not, the answer may have come sooner.

The rate of multiplication of cells in the body is normally closely controlled, and precisely matches their rate of loss. In skin, the deepest cells divide just fast enough to replace the cells that are being shed continuously from the surface, so that the total population of cells remains constant. Cairns points out, "It is a common mistake to assume that cancer cells divide faster than the normal cells from which they were derived" (Cairns 1978). The fact is, most cancer cells divide at about or below the rate of normal cells. As explained in detail in Chapters 5 & 6, carcinogens produce chromosomally unbalanced (aneuploid) cells, which are always damaged cells that divide with difficulty, if at all. However, if the damaged aneuploid cells can be encouraged or forced to divide, eventually a population of aneuploid cells may evolve to the point where enough survive to become a self-propagating cancer. Cancer cells are able to increase in number because a greater proportion continue dividing than is normally allowed (e.g., skin) or don't stop dividing in tissues where they should (e.g., liver) (Steel 1972).

It is widely assumed that tumor cells grow exponentially (Shackney 1993). Exponential growth agrees with the unlimited proliferative activity of tumor cells recorded in early, mainly *in vitro*, studies. However, it is generally recognized that the assumption of exponential growth is applicable to virtually no solid tumor growing *in vivo* (Shackney et al. 1978). For example, there is an evident discrepancy between the exponential tumor growth theory and experimental data obtained from tumor cells growing *in vivo*, where tumor doubling times have been found to greatly exceed cell cycle times (Brú 1998). Brú et al. demonstrated that actively proliferating cells in culture as well as tumors in animals are restricted to the colony or tumor border and inhibited in the innermost areas (Brú et al. 2003).

The restriction of cell proliferation to the tumor contour mathematically implies a linear growth rate. Scaling techniques used to analyze the fractal nature of cell colonies growing *in vitro*, and of tumors developing *in vivo*, showed them to exhibit exactly the same growth dynamics independent of cell type (Brú et al. 2003). These results invalidate the hypothesis that the main mechanism responsible for tumor growth is nutrient competition among cells. These issues are of great importance since both radiotherapy and chemotherapy are entirely based on the kinetics of cell growth.

Aneuploidy is more pronounced in advance solid tumors than in earlier stages (Aldaz et al. 1987, Aldaz et al. 1988a, Caratero et al. 1990, Tomita 1995). This is consistent with the theory that the level of aneuploidy increases with the number of cell divisions as described in Chapters 5 & 6. If an exponential growth regime is assumed, each cell must undergo 32 divisions to form a 2 cm^3 tumor (~10^9 cells). However, in a linear growth regime, the number of divisions by cells on the surface would be about 30 times greater than at the center (Brú et al. 2003). Naturally, this leads to a higher frequency of genetic abnormalities in cells at the growing tumor border. Metastases are generated from the most malignant cells from the border of a primary tumor, which is consistent with the observation that metastatic cells are more aneuploid than those of the primary tumor (Futakawa et al. 1997).

A linear growth regime provides a much better explanation of this than does exponential growth (Brú et al. 2003).

One of the important clinical consequences of linear growth is that it explains the discrepancy between anatomic-pathological analysis of biopsies and the diagnosis of many cancers. The doctor who performs the biopsy usually takes a sample from the center of the tumor to be sure that what is taken corresponds to the lesion. But if growth is linear, and the malignancy of cells increases along the tumor radius, such a biopsy would always take less malignant cells and might lead to a diagnostic error (Liberman et al. 2000).

The apparently obvious criterion that cancer cells must survive in order to cause disease is less-well considered because of the belief cancer cells are immortal, possessing properties superior to their normal precursors, and bent on our destruction. As the story unfolds, it will become clear that cancer cells are damaged cells, spontaneously die at high rates, and are inferior in most respects to normal cells. As Peyton Rous said, "That cancer cells are often sick cells and die young is known to every pathologist" (Rous and Kidd 1941). Leslie Foulds emphasized the point saying it is only the "successful" tumors that attract attention; the "unsuccessful" ones escape notice (Foulds 1954). Of course, it is the "successful" tumors we will be considering in the next few chapters. Chapter 7 discusses the massive, spontaneous destruction of cancer cells in search of clues to spontaneous remission and non-toxic cancer therapies.

The long journey to the modern chromosomal imbalance (aneuploidy) theory of cancer began with David Hansemann and Theodor Boveri (Chapter 2).

2

Boveri's Theory of Cancer was Ahead of its Time

Of the many biological monographs that appeared in Germany at the end of the nineteenth century and the beginning of the twentieth, very few are now remembered. Theodor Boveri's Concerning the Origin of Malignant Tumours *is a remarkable exception. It is regularly quoted, misquoted and quoted out of context in our present day cancer research literature. That it should be quoted is in itself unsurprising, for it is one of the most prescient theoretical statements that the history of cancer research has produced. But why it should be so often misquoted is not so immediately obvious.*

(Harris 2007)

Almost the first thing researchers noticed from the time they began looking at cancer cells under the microscope was that they had excess chromosomes. One of the pioneers was David Hansemann who observed asymmetric mitoses in all the epithelial cancers he examined (Hansemann 1890). He seems to have been the first to connect cancer to an imbalance in the number of chromosomes. Hansemann captured the essence of cancer as, "a process carrying the cell in some entirely new direction—a direction, moreover, which is not the same in all tumors, nor even constant in the same tumor.... . The anaplastic cell then is one in which, through some unknown agency, a progressive disorganization of the mitotic process occurs, which in turn results in the production of cells that are undifferentiated in the sense that those functions last to be acquired, most highly specialized...are more or less lost; but redifferentiated in the sense that the cancer cell is not at all an embryonic cell, but is a new biologic entity, differing from any cell present at any time in normal ontogenesis. But...this entity displays no characters absolutely and completely lacking in the mother cell... . Its changed behavior depends on exaltation of some qualities, and depression of others, all at least potentially present in the mother cell." (Translated by Whitman (Whitman 1919).) Hansemann's "unknown agency" has turned out to be the relentless randomization of the genome caused by aneuploidy, the subject of this book.

But it was Theodor Boveri who offered the first coherent theory (including a mechanism) of how chromosomal imbalance (aneuploidy) leads to cancer and produces all its characteristic properties (Boveri 1914). Nowadays, Boveri's field of research would go by the name developmental genetics. He and Walter Sutton (Sutton 1903) were the first to consider the chromosomes to be the matter of heredity, and Boveri may have been the first to present clear experimental proof for this assumption using sea urchin eggs (Wolf 1974).

To summarize, Boveri produced fragmented sea urchin eggs by shaking them. This resulted in some cell fragments that included the nucleus and others that did not. After fertilization, the fragments without the egg nucleus developed into normal pluteus larvae, although they contained only the male set of chromosomes, or half the normal complement of 18 chromosomes. Occasionally, an unfertilized egg also developed into a pluteus which possessed only the maternal half of the complete chromosome set. From this experiment it was shown that the nuclei of the male and female gametes contained homologous information. Thus the genetic material was in the nucleus and the chromosomes were the matter of heredity preserved from one cell generation to the next.

Subsequently, Boveri demonstrated in a series of experiments using multipolar mitoses that the individual chromosomes were functionally distinct elements, confirming the speculation of Sutton (Sutton 1903). If a sea urchin egg was fertilized with two sperm, a tetrapolar spindle was formed during the first cell division. Boveri had discovered earlier that the centrosome was introduced into the egg cytoplasm by the sperm. Thus, after double fertilization, the two centrosomes divided to give rise to a tetrapolar spindle apparatus that had to deal with three haploid sets of chromosomes. As a consequence of this imbalance, the chromosomes were distributed unequally to daughter cells in almost all instances. The majority of the embryos developed normally until gastrulation but at that stage they died off.

In another experiment, Boveri produced embryos with tripolar spindle apparatus. In this case, a regular distribution of the chromosomes formed after double fertilization should occur more frequently than in the tetrapolar cells. The resulting embryos proceeded to various stages of development, however, a number of abnormalities occurred. The abnormally developed embryos were illuminating. In many cases the abnormalities were restricted to certain segments of the embryo, while other segments were not affected. Boveri even called some of the abnormal embryos tumors. From these experiments with dispermic sea urchin eggs, Boveri concluded that the individual chromosomes were endowed with different qualities. Thus, the chromosome theory of inheritance was settled. The uniqueness of individual chromosomes was to constitute one of the fundamentals of his theory of malignancy.

In his first publication on dispermic sea urchin eggs in 1902, Boveri had conceived of the idea that malignant tumors could be due to an abnormal chromosome constitution, originating, for example, as the consequence of a multipolar mitosis but cancer had not been the subject of that work (Boveri 1902/1964). Since the possible connection between abnormal mitoses and malignant tumors had been discussed now and then in the literature, but time and again rejected, Boveri decided to deal with the problem of cancer in his famous monograph of 1914 (Boveri 1914, Harris 2007). "[T]he essential element in my hypothesis," Boveri said, "is not the abnormal mitoses, but a particular abnormal composition of the chromatin, irrespective of how it arises. Any event that produces this chromatin composition would eventually generate a malignant tumor and can be considered tumorigenic. Each process which brings about this constitution would result in the origin of a malignant tumor: a disorder in certain chromosomes produced by a hereditary condition; destruction of chromosomes by intracellular parasites; damage of particular chromosomes by external agents that spare others; and other factors [X-rays and chemicals]" (Harris 2007). He discussed in some detail how abnormal numbers of centrosomes could lead to an imbalance in the number of chromosomes during cell division. Recent experimental evidence confirms Boveri's prediction regarding the role centrosomes play in the generation and progression of aneuploidy (Ganem et al. 2009, Silkworth et al. 2009).

Boveri's theory predicted that cancer results from a single cell that has acquired an abnormal chromosome constitution. In other words, he predicted the clonal origin of cancer. "However long it might take to prove this idea," he continued, "I believe there is no escaping it; and I am convinced any theory of malignancy that does not take account of its unicellular origin is doomed" (Harris 2007).

"The primordial tumorigenic cell, as I [Boveri] propose to call it in what follows, is, according to my hypothesis, a cell that harbors a specific faulty assembly of chromosomes as a consequence of an abnormal event. This is the main cause of the propensity for unrestricted proliferation that the primordial cell passes to its progeny so long as these continue to multiply by normal mitotic binary fission. But all the other abnormal properties that the tumor cell exhibits are also determined by the abnormal chromosome constitution of the primordial cell, and these properties will also be inherited by all the progeny of this cell so long as subsequent cell division takes place by normal bipolar mitosis".

It is well known that a tumor cell has an abnormal metabolism. According to Boveri, since "the individual chromosomes have different qualities, chromosome aberrations will result in deviant metabolic functions. If, therefore, certain chromosomes are missing and others are present in abundance, certain substances will be produced also in abundance, and there will be a deficiency in others" (Wolf 1974).

Boveri's use of the word specific has caused no end of confusion to this day about what he meant. Specific has led many researchers in pursuit of well-defined abnormal combinations of chromosomes as specific causes of cancer. The functionally distinct qualities of individual chromosomes implies that a specific assortment of chromosomes should produce a characteristic phenotype. As a developmental biologist, Boveri knew better than most that phenotype is a moving target, which is especially true of cancers. (See Sections 5.3.7 & 6.1.6 for an updated discussion of specific aneusomies.) This led him to hypothesize the co-evolution of karyotype and phenotype: "But it is also conceivable that the creation of certain abnormal chromosome combinations so perturbs the equilibrium within the nucleus that particular chromosomes go on changing under the influence of changes in other chromosomes. One group of chromosomes might eventually preponderate and perhaps even suppress the activity of others. It is therefore understandable that a malignant tumor that is at first closely similar to its tissue of origin progressively becomes less so and eventually becomes completely unrecognizable" (Harris 2007). This is equivalent to the autocatalyzed progression of aneuploidy discussed in Sections 5.3 & 5.3.11.

In Boveri's time X-rays and certain chemicals were known to cause aneuploidy. Boveri said the interval between the time of the insult and the origin of a tumor may be explained by the assumption that the cancer-causing agent first interferes with the process of cell division, producing an aneuploid cell. In the second step, the aneuploid cell must be stimulated to divide further, producing daughter aneuploid cells. In heavily proliferating tissues, the risk of a tumor is increased.

Boveri pointed out that a natural consequence of his aneuploidy theory was that the risk of tumors would increase with age since in aging cells the process of cell division is more frequently disturbed (Wolf 1974). In addition, enough time has elapsed in an older organism for many cell divisions to have occurred. Boveri even predicted tumors that had the correct number of chromosomes but with an abnormal complement—the so-called pseudo-diploid cancers. The essentials of Boveri's chromosomal imbalance theory of cancer are as valid today as in 1914.

An overwhelming body of evidence has established an inseparable bond between cancer and aneuploidy (Atkin and

Baker 1966, Atkin and Baker 1990, Gebhart and Liehr 2000, Heim and Mitelman 1995a, Koller 1964, Koller 1972, Mertens et al. 1997, Mitelman 1994, Sandberg 1990). By 1969, Albert Levan was confident enough to say that, "there is safe evidence that carcinogenesis, as well as all stages of malignancy, is accompanied by chromosomal irregularities..." (Levan 1969). But he went on to add that, "nothing is known, however, as to the significance of these chromosome irregularities in relation to the carcinogenic transformation." In other words, he raised the question: is chromosomal imbalance (aneuploidy) a cause or consequence of cancer? Variations of Levan's declaration of ignorance regularly appear. The first paragraph of a 2009 paper, for example, says: "[V]ery little is known about the oncogenic role, if any, of numerical chromosomal aberrations, *aneuploidy*, which are the most common abnormalities in cancer" (Ganmore et al. 2009).

While leaving the question open, Levan acknowledged that aneuploidy satisfies at least one requirement of a cause: "Chromosome variation is an integrated part of tumor development from the earliest beginning of carcinogenesis to the latest progressive stages. Even before any malignancy has started chromosome variation in a normal tissue is generally associated with an increased tendency to cancer" (Levan 1969).

2.1 ANEUPLOIDY THEORY "GOT LOST"

Boveri's monograph was published in Germany in 1914, the year that marked the outbreak of the First World War. He died a year later. His views on cancer had met a chilly response from the medical community, but it was not this that determined the silence that followed the appearance of his monograph. Its fate was mainly determined by the fact that the First World War and its ruinous aftermath for Germany effectively closed down biological research there for more than a decade. Exploration of his ideas was no doubt also delayed by the inadequacy of the tumour chromosome preparations that could be made at the time.

(Harris 2007)

By the mid 20th Century, the karyotypic abnormalities in both human and animal cancers seemed to be of two essentially different kinds: nonrandom changes preferentially involving particular chromosomes and a frequently more massive random or background variation affecting all chromosomes. To differentiate between the two could be exceedingly difficult, however. In spite of painstaking efforts, little progress was made in cancer cytogenetics during this period (Heim and Mitelman 2009). To this day, cytogeneticists continue to focus on the minority of "nonrandom" aneusomies often present in individual cancers from the same tissue of origin (Fabarius et al. 2002, Heim and Mitelman 2009), because "no completely specific primary or secondary karyotypic abnormality has been identified" (Hoglund et al. 2001b).

More recently, the detection of aneuploidies by the technique of comparative genomic hybridization has revealed that high percentages of human cancers from the same tissue-of-origin have similar, but again not identical aneusomies (Albertson et al. 2003, Baudis 2007, Gebhart and Liehr 2000, Jiang et al. 1998, Kallioniemi et al. 1994a, Mahlamaki et al. 1997, Richter et al. 1998, Richter et al. 2003). In view of this, many researchers suggested that these common aneusomies encode common genes that are necessary for carcinogenesis (Albertson et al. 2003, Baudis 2007, Gebhart and Liehr 2000, Hoglund et al. 2001b, Jiang et al. 1998, Kallioniemi et al. 1994a, Mahlamaki et al. 1997, Richter et al. 1998, Richter et al. 2003).

Some, favoring the importance of oncogenes and tumor suppressors (discussed in Chapters 3 & 4), have argued against a role for aneuploidy as a driving force in tumorigenesis (Zimonjic et al. 2001). Others have argued that aneuploidy is only a benign side effect of transformation (Hede 2005). Still others have suggested that aneuploidy promotes tumor progression but not initiation (Johansson et al. 1996).

In 1974, Susmo Ohno succinctly captured the aneuploidy-cancer dilemma: "It is tempting to be satisfied with a sweeping statement such as 'genetic imbalance (aneuploidy) is either the prerequisite or the cause of malignancy,' inasmuch as conditions that enhance production of aneuploid somatic cells appear also to predispose to the development of malignant tumors, and since most, if not all, malignant cells are aneuploid. The trouble with such a statement is that neither the reader nor the maker of the statement has any idea of what it means precisely" (Ohno 1974).

Unable to resolve the dilemma of how to rank the importance of aneuploidy, Barrett et al. said in 1983: "It is very difficult to rule out aneuploidy as a result of, rather than a cause of, neoplastic development. However, we feel that the repeated findings of nonrandom aneuploidy in a variety of preneoplastic and neoplastic lesions induced by a variety of diverse carcinogenic insults (chemical, viral, hormonal, and foreign body) indicate that these changes are important in carcinogenesis. Since many carcinogens can induce aneuploidy, it seems important to consider the role of induced aneuploidy in carcinogenesis. Klein has stressed that in murine T-cell leukemia, a common cytogenetic change (trisomy 15) is found with diverse initiating agents (spontaneous, viral, and chemical) (Klein 1979). Since all these agents can induce aneuploidy, the role of aneuploidy in the initial phases of tumor development should be further studied. Whether aneuploidy induction by chemicals is the primary event in neoplastic development, or only facilitates the progression of spontaneous or induced changes, does not diminish its importance in neoplastic progression" (Tsutsui et al. 1983).

Unmindful of Boveri's hypothesis of co-evolving karyotype and phenotype, it was the absence of cancer-specific karyotypes that demolished enthusiasm for aneuploidy as a cause of cancer. Peyton Rous, the discoverer of the Rous sarcoma virus, summed up the situation saying: "Persistent search has been made, ever since Boveri's time, for chromosome alterations which might prove characteristic of the neoplastic state—all to no purpose" (Rous 1959). Thirty-six years later, Harris said a search for cancer-specific karyotypes "utterly failed to identify any specific chromosomal change that might plausibly be supposed to have a direct causative role in the generation of a tumour" (Harris 1995).

The discovery of the double helical structure of DNA by Watson and Crick in 1953 augured the molecularization of biology (Watson and Crick 1953). The trend toward the molecular, and the idea current in the 1970s that viruses may be a major cause of cancer, pushed Boveri's theory further into the background. According to an editorial in Science in 1999, "Over the following decades, however, [Boveri's] idea got lost, as researchers concentrated on understanding the specific gene malfunctions that lead to cancer" (Pennisi 1999). The idea got lost so completely that it is now no longer mentioned in the cancer chapters of leading textbooks of biology (Alberts et al. 1994, Lewin 1994, Lodish et al. 1999, Watson et al. 1987). As a result scientists studying aneuploidy now compare their work to "resurrection" (Brinkley and Goepfert 1998).

Even cytogeneticists have disregarded the aneuploidy hypothesis in favor of gene mutation. For example, Nowell wrote in an influential article in Science in 1976, "It is certainly clear that visible alterations in chromosome structure are not essential to the initial change" (Nowell 1976). Twenty years later Mitelman et al. wrote, "We propose that unbalanced primary changes [aneuploidy], in fact, are secondary, the primary being submicroscopic. There are no unbalanced primary changes, only secondary imbalances masquerading as primary" (Johansson et al. 1996). Aneuploidy, if considered at all, is now viewed as just one of several mechanisms that alter the dosage of hypothetical oncogenes and tumor suppressor genes (Cahill et al. 1999, Orr-Weaver and Weinberg 1998). For example, Mitelman et al. stated in 1997, "Obviously, the pathogenetically important outcome of cytogenetically identified gains or losses of chromosomal material may simply be ascribed to amplification or deletion of single oncogenes or tumor suppressor genes..." (Mitelman et al. 1997a).

The absence of defined cancer-specific karyotypes remains the biggest conceptual barrier preventing most oncogene researchers from considering aneuploidy as the cause of cancer. This is largely due to an ingrained belief in specific genetic causes of cancer, modeled after the hard-won lesson that particular bacteria and viruses were the specific causes of infectious diseases. The pharmaceutical development of a manageable number of specific and highly effective drugs against cancer, analogous to antibiotics for infectious diseases, requires an equally manageable number of well-defined targets. The tsunami of 11,000 gene mutations per cancer cell reported by Stoler et al. obliterated the idea of a manageable number of targets (Stoler et al. 1999).

3

Genesis of "The Enemy Within"?

We now believe that we know the 'why for these happenings,' and it takes a remarkable form. Two sorts of genes govern the proliferation of our cells: proto-oncogenes, which serve as accelerators to activate the genes of the cell; and tumor suppressor genes, which serve as brakes to retard the growth of cells. Jam an accelerator or remove a brake, and the cell may be unleashed to relentless proliferation.... . It now appears that most if not all forms of cancer may contain damage to one or another specific proto-oncogene. In each instance, the damage somehow unleashes the gene or its protein product, driving the cell to relentless proliferation. A genetic engine for the cancer cell has been found.

(Bishop 1995)

Cancer has been generally described as a stepwise progression driven by a series of somatic gene mutations occurring in a linear pattern of cancer evolution (Fearon and Vogelstein 1990, Jackson and Loeb 1998, Nowell 1976). This prevailing assumption led to a major effort to identify mutated genes and their defined molecular pathways, as well as an attempt to establish recurrent genetic patterns of cancer progression. What is meant by somatic mutation has evolved over the years. Originally it was an umbrella concept covering variation in the number of whole chromosomes, chromosomal rearrangements, duplication or elimination of a part of individual chromosomes, and finally point mutations. Point mutations are changes in any of the four nucleotides of specific genes along a chromosome. Thus, for several decades aneuploidy (chromosomal imbalance) was on equal footing with point mutation. As noted above, the molecularization of biology led to the seductive idea that cancer could be explained and understood as mutations in the smallest genetic units—i.e., damage to individual genes. The massive chromosomal aberrations observed since Hansemann's time were now thought irrelevant and cursed as a nuisance complicating the experimental work of decoding the "real" genetics of cancer.

The point mutation juggernaut was cemented in 1989 when the Nobel Prize in Physiology or Medicine was awarded to J. Michael Bishop and Harold E. Varmus "for their discovery of the cellular origin of retroviral oncogenes." Erling Norrby of the Karolinska Institute encapsulated the gene mutation theory of cancer in his presentation speech (Norrby 1989):

> *Michael Bishop and Harold Varmus, and their collaborators managed to develop a molecular probe, which could selectively identify the tumor-inducing gene in the Rous virus. By use of this probe it was demonstrated that the critical gene was present in normal cells from all species. To their own great surprise and that of the scientific community, they had to draw the conclusion that the tumor-inducing gene in the Rous virus was of cellular origin. Does this mean that we are carrying cancer genes in our cells? Obviously not. However, in our cells there is a family of probably several hundred genes which are old in evolutionary terms and which control the normal growth and division of cells. Disruptions in the functioning of one or more of these genes can cause one cell to slip out of the network of growth control. The cell runs amok and a tumor may be the result... . Since the family of growth-controlling genes, of which we have now identified more than 60, was demonstrated in tumor cells, they were given the somewhat illogical name oncogenes. This name derives from the Greek term* onkos *meaning tumor... . Cancer originates in disturbances in the genetic material of cells. Yet in most cases a single disturbance is not sufficient, but instead an accumulation of several critical injuries is required. This is the reason why cancers usually occur relatively late in life. Abnormally functioning oncogenes have now been demonstrated in many types of tumors in man. For the first time we are beginning to comprehend the complicated mechanisms behind the development of this group of diseases. New opportunities for diagnosis and treatment of various forms of cancer are now becoming available.*

In 1995, Bishop described how five major themes converged to construct "the intellectual edifice that now houses cancer research" (Bishop 1995):

1. *Cancer is Clonal*

"The first and primeval theme is that the phenotype of neoplastic cells is heritable, no matter what its cause. One cancer cell begets two others and so on *ad infinitum*, almost without exception... . The clonality of tumors has become an article of faith among most students of cancer. But there is an irony here, because the heritable nature of the neoplastic phenotype bespeaks a genetic stability that is illusory. In reality, we now know that an expanding clone of potential cancer cells diversifies with startling rapidity, a rapidity that is essential to the success of tumorigenesis. Every tumor represents the outcome of an individual experiment in cellular evolution, fueled by a genetic instability that we have brought to view only recently, and overseen by relentless selection for advantageous cellular properties. If cancer begins as a single cell that eventually progresses to a full blown malignancy, what initiates the deadly sequence? For at least two centuries, the focus has been on external causes. Only in the past two decades, however, have we convinced ourselves that those causes might be united by a common mechanism: the mutation of DNA, the second of our converging themes".

2. *Mutation of DNA*

"In any poll of modern molecular biologists, you could probably get a rousing majority for the view that carcinogens act by affecting DNA. But still, there are lingering difficulties. It all began promisingly enough, with the development by Ames and others of tests that allowed the detection of mutagenicity in cultures of bacteria or vertebrate cells. At first, there appeared to be a direct correlation between relative mutagenicity in these short term tests and relative carcinogenicity in rodent models. But now we know that perhaps half of all the substances that are carcinogenic in rodent tests do not score as mutagens in the short term tests, a discrepancy that remains inadequately explained. One standing rationalization is that nonmutagenic carcinogens are cytotoxic, and this leads to compensatory cellular proliferation that favors intrinsic mutagenesis. These bothersome ambiguities have caused no end of strife in the efforts to identify and regulate the use of carcinogenic agents. And they illustrate why efforts to study the mechanisms of external carcinogenesis have contributed only tangentially to the search for the inner malady of cancer cells".

3. *Abnormal Chromosomes*

"By the end of 1989, editors at the late, lamented journal *Cancer Cells* were able to compile a list of 114 chromosomal aberrations associated in a specific manner with cancer. And a recent review in *Nature* listed 52 chromosomal translocations whose breakpoints have received molecular attention sufficient to identify the affected gene or genes. Thus, cytogenetics has succeeded in pointing to the genetic apparatus as the ailing organ of the cancer cell. But the implication that malfunctioning genes might propel neoplastic growth could not be tested with microscopy. Instead, it was the study of tumor viruses that produced the first explicit example of cancer genes—the fourth of our converging themes".

4. *Oncogenes*

"Steven Martin used Rous sarcoma virus to isolate mutants that were temperature-sensitive solely for transformation (Martin 1970). By this means, he was able to show that a viral gene was required for both the initiation and maintenance of transformation, yet made no contribution to viral replication. The gene was soon dubbed v-*src* and the generic term 'oncogene' also came into general use. Second, Peter Duesberg and Peter Vogt reported the first of their studies showing that the transforming ability of Rous sarcoma virus could be lost by deletion, and that the deletions provided a physical definition of the oncogene v-*src*. Then, in a spirited race that ended in a dead heat, Duesberg and his colleagues on the one hand, Weissman and his on the other, mapped v-*src* to its location on the genome of Rous sarcoma virus. But do not let the term 'map' mislead the young among you with a false image of restriction enzymes and Southern blots: this was life before recombinant DNA: the experiments were models of imagination and technical tours de force".

"The discovery that the virus of Peyton Rous uses a gene to elicit cancer brought clarity to what had been a muddled business. There had been hints before that the elemental secrets of cancer might lay hidden in the genetic dowry of cells. But here in Rous sarcoma virus was an explicit example of a gene that can switch a cell from normal to cancerous growth. Now a more ambitious question arose. Might the cell itself have such genes? … First, there was an evolutionary puzzle. The genome of Rous sarcoma virus has three genes devoted to viral replication. But v-*src* is not one of these: it is fully dispensable, without harm to the virus. Why then is it there? Second, there was the Oncogene Hypothesis from Robert Huebner and George Todaro (Huebner and Todaro 1969), which attributed all cancer to the activation of oncogenes, implanted into vertebrate germ lines by retroviral infection eons ago. We thought we might get some purchase on both of these issues by looking for a counterpart of v-*src* in normal cells. It required the better part of four years, but in the end, we reached the conclusion that vertebrate cells do indeed carry a version of *src*".

"But the vertebrate *src* proved to be a normal cellular gene, not a retroviral intruder. The results stood the Oncogene Hypothesis on its head. Eventually, it became clear that *src* is indeed a normal cellular gene (c-*src*), grafted into the virus of Peyton Rous by an accident of nature we call 'transduction.' So the virus is an inadvertent pirate; the booty is a cellular gene with the potential to become a cancer gene when mutated in certain ways… . The need arose for a generic term to describe the cellular progenitors of retroviral oncogenes. The first to find general usage was 'cellular oncogene.' But some investigators were uncomfortable with this term, because of its unwarranted implication that the cellular genes might be tumorigenic in their native state… . Proto-oncogene has come into general use, as the colloquial counterpoise to oncogene. It has also become an embarrassment, because the precise connotation of the word is that of prototype rather than progenitor, not far removed from the offensive connotation of the term 'cellular oncogene' ".

"Any influence that can damage a proto-oncogene might give rise to an oncogene, even if the damage occurred without the gene ever leaving the cell, without the gene ever encountering a virus. In this view, proto-oncogenes become precursors to cancer genes within our cells, and *damage to genes becomes the underpinning of all cancers...* . In 1979, Shih and Weinberg (Shih et al. 1979), motivated by the precedents of tumor virology, used DNA-mediated gene transfer to demonstrate the presence of a biologically activate oncogene in chemically transformed cells (Shilo and Weinberg 1981). That finding led, in turn, to the discovery that many human tumors harbor mutant versions of *ras* proto-oncogenes—once again, culprits already familiar to us from the study of retroviruses. As the exploration of genetic damage in tumor cells proceeded, the repertoire of proto-oncogenes grew. The sum now exceeds one hundred. And as the tally grew, so did the tie between proto-oncogenes and cancer. It now appears that most if not all forms of cancer may contain damage to one or another specific proto-oncogene. In each instance, the damage somehow unleashes the gene or its protein product, driving the cell to relentless proliferation. A genetic engine for the cancer cell has been found".

5. *Tumor Suppressor Genes*

"The route to the isolation of tumor suppressor genes lay through the study of inherited cancer—the last of our convergent themes. ... Alfred Knudson published the first of his seminal papers that formulated how many genetic events might be involved in the genesis of retinoblastoma [he proposed two] (Knudson 1971), how those events might represent independent mutations, and how inheritance of one of these mutations would explain the increased susceptibility to retinoblastoma in affected families... . The story reached its first climax in 1986, with the isolation of the retinoblastoma gene, RB1. After decades as chimera, tumor suppressor genes had become a physical reality. The uncovering of RB1 and its role in tumorigenesis dramatized the way in which epidemiology, population genetics, cytogenetics, and molecular biology can be joined into an integrated assault on cancer. And it was a harbinger of much more to come. Evidence has now been obtained for recessive genetic lesions in most if not all forms of human cancer. At least a dozen of the affected genes have been identified directly by the use of recombinant DNA, and there are probably many more to come... . And in continuation of the precedent established by retinoblastoma, most other familial cancers are apparently based on recessive, inherited deficiencies of tumor suppressor genes. Indeed, the familial syndromes have remained major points of departure in the isolation of additional tumor suppressor genes".

This was a neat, logical, and convincing story to most cancer researchers, which Bishop aptly characterized as the genetic paradigm that dominates cancer today. Thus, dominant gain-of-function mutations in proto-oncogenes directly cause cancer, while recessive loss-of-function mutations in tumor suppressor genes, which "guard" against cancer, allow oncogenes to do their worst.

In 2000, Hanahan and Weinberg listed six hallmarks of cancer: 1) sustained proliferative signaling, 2) evading growth suppressors, 3) resisting cell death, 4) replicative immortality, 5) sustained angiogenesis, 6) tissue invasion and metastasis (Hanahan and Weinberg 2000). In 2011, these authors added two emerging hallmarks: 7) deregulating cellular energetics, 8) avoiding immune destruction; and two enabling characteristics: 9) genome instability, 10) tumor-promoting inflammation (Hanahan and Weinberg 2011). "However," Ashworth et al. point out, "numerous questions remain about how mutations in DNA lead to the acquisition of these traits" (Anderson et al. 2011).

Bishop correctly stated, every tumor represents the outcome of an individual experiment in cellular evolution, fueled by a genetic instability and overseen by relentless selection for advantageous cellular properties, which is consistent with virtually all theories of cancer, including aneuploidy. He forthrightly acknowledged the paradox that at least half of all known carcinogens are not mutagenic. Believing damaged genes must cause cancer, he extracted from the massive chromosomal irregularities of cancer cells the few genes severed at breakpoints (cleavage sites) of chromosomes as being significant.

But what really paved the road to oncogenes was the discovery that certain viruses could rapidly transform normal cells *in vitro*, but only when they contained certain genes pilfered from host cells. This led Weinberg to conclude, "By infecting cells growing in monolayer culture [with oncogenic viruses], an investigator could recapitulate a process that hitherto was thought to occur only in the matrix of a living tissue. With one blow, the process of malignant transformation was demystified" (Weinberg 1989). The demystification was fleeting, however.

As noted above, the special cancer genes discovered by Bishop and Varmus turned out to be normal animal or human genes, which implied everyone should have cancer. To overcome this problem it was proposed that mutations in these normal genes were required to turn them into oncogenes. (That solution, of course, introduced another problem: how did the normal genes without mutations transform virus-infected cells? Section 4.3 provides the answer.) Bishop said the damaged caused by mutation "somehow unleashes the gene or its protein product, driving the cell to relentless proliferation." The "somehow" remains unknown to this day.

3.1 CLONAL CANCER

The development of human cancer is widely thought to entail a series of events that causes a progressively more malignant phenotype. Nowell first proposed the clonal evolution of tumor cell populations to explain how malignant tumors arise over time (Nowell 1976). This hypothesis predicts that tumor cells of the ultimate stage will carry each of the events, cells of the penultimate stage will carry each of the events less the last one, and so on (Cavenee et al. 1989). A clonal population of cells is defined as those cells arising from the mitotic division of a single somatic cell (Secker-Walker 1985).

The most compelling evidence for monoclonality comes from analyses of the gene products of malignant cells, and these may be located on the surface, within the cell or secreted. "With tumors of B-cell lineage, monoclonality is most easily established by the fact that all cells synthesize the same immunoglobulin. The demonstration of unique DNA rearrangements of the antigen receptors has proved the monoclonality of most T-cell malignancies so far studied. For non-lymphoid cancers, studies of clonality have been based on the mosaicism which exists in normal tissues of women heterozygous for the two allelic forms of the enzyme glucose-6-phosphate dehydrogenase, the gene for which is on the X-chromosome. The cells from tumors of such women are usually—though not invariably—composed predominantly of one of the alleles and therefore monoclonal" (Alexander 1985).

The assessment of clonality may be dependent on the technique used. For example, the investigation of two cell populations may suggest two independent clones by immunoglobulin gene analysis but a single clone by X-linked DNA polymorphism analysis, reflecting the earlier occurrence in development of X-chromosome inactivation (Wainscoat and Fey 1990). A further consideration which complicates the analysis of clonality is that cancer cells constituting a single clone are not genetically identical (Lengauer et al. 1998, Levan and Biesele 1958, Rasnick and Duesberg 1999) since clonal evolution is inevitable within such populations of cells (Nowell 1976). Karyotypic heterogeneity of cancer cells is not incompatible with the concept that they all derive from a single transformed cell. The heterogeneity of the cancer cells comprising a tumor stems from karyotypic instability, which is perhaps the most characteristic difference between cancer and normal cells (Breivik and Gaudernack 1999).

Alexander concluded in 1985, the evidence taken together strongly suggests that many, if not most, cases of human cancer are monoclonal. However, in light of other evidence, the clonality of cancer posed a problem. He pointed out that: "In view of the large numbers of cells which are at risk in an adult organism and the relative rarity of cancer, mutations resulting in transformation would be expected to be extremely infrequent. Yet *in vitro* transformation of mammalian cells into a phenotype capable of growing as malignant tumours when transplanted into animals occurs remarkably readily and in tissue culture the transformation of normal cells to ones which exhibit malignant characteristics is far from being very rare or infrequent. Because of the ease of dosimetry this is most readily demonstrated for the carcinogenic effect of ionising radiations (Borek 1982) although the same applies to chemical carcinogens or indeed 'spontaneous' transformation. One Gray of X-rays given to embryonic cells in tissue culture causes one cell in 10^4 to be transformed and after clonal expansion such a transformed cell will grow as a tumour *in vivo*. Yet clearly the carcinogenicity of X-rays for intact animals is many orders of magnitude less than would follow from the induction at the rate observed *in vitro* of a single malignant cell when one considers the number of cells capable of being transformed" (Alexander 1985).

Alexander proposed "A way out of the conflict between the ease of cell transformation *in vitro* and the rarity of tumours *in vivo* is to abandon the concept that tumours arise from a single cell. The finding of monoclonality in clinically detectable cancer and leukaemia when more than 10^{10} cancer cells are present does not mean that initially the cancer arose from a single cell. For example, there is growing evidence for multiclonality in a wide range of human tumors (Helmbold et al. 2005), nevertheless, when laser capture microdissection is used to excise distinct foci from polyclonal lesions, each individual focus was found to be monoclonal (Wu et al. 2003). Initially the malignant proliferation could be polyclonal and monoclonality could be a late event due to selection of cells from the different clones" (Alexander 1985). Indeed, in chemically induced sarcomas of mice Woodruff et al. documented instances in which an originally polyclonal tumor progressively became monoclonal (Woodruff et al. 1982).

Alexander proposed the concept that the initiation of tumor growth *in vivo* requires the participation of several independently transformed cells and that it is only when a minimum number of transformed cells come together that they create a micro-environment which permits their unlimited proliferation and the production of a malignant lesion. This model would account for the finding that tumors arise very much less frequently *in vivo* than would correspond to the occurrence of transformation at the cellular level *in vitro* of a culture system containing a comparable number of cells. But, he asked, why should a single transformed cell be competent to grow as a clone *in vitro* whereas *in vivo* in the tissue in which it originates it does not proliferate? He said, this situation is not as absurd as it at first appears because in an animal a cell that has undergone transformation to malignancy has to grow in the environment of extra cellular fluid which in composition resembles plasma whereas in cell culture clonal growth from a single cell occurs in serum (Alexander 1985). A more compelling explanation is the organism, as an integrated whole, with its ability to heal and repair, is consequently much more resistant to transformation to malignancy than cells in culture without the organism's resources and advantages.

Alexander's analysis of clonality of cancer led him to a very interesting hypothesis to explain the well known fact that metastases appear predominantly in specific tissues and organs. While tumor emboli consisting of more than one cancer cell give rise more frequently than single cells to lung metastases, there can be little doubt that single cells are capable of causing blood borne metastasis, especially in organs other than the lung to which they must have gained access via the arterial circulation. However, one of the most striking aspects of the metastatic process is the peculiarity of the relative frequencies of metastases in different organs. Clinical postmortem studies have shown that cancer cells that have passed beyond the lung into the arterial circulation grow selectively in certain organs (Willis 1967). In experimental animals, organ preference can be demonstrated by injecting cancer cells into the left ventricle (so as to avoid the filtering effect of the lung which arises if cells are given intravenously) whence they are distributed via the arterial circulation to all of the organs. Several investigations had shown that following this procedure few, if any, metastases occurred in gut and muscle which received the majority of the blood, but occurred instead in adrenal, bone, ovary and other organs that took only a small fraction of the cardiac output (Alexander 1985).

Alexander et al. made a detailed study of the initial distribution, trapping, cell death and eventual incidence of metastases for three histologically different rat tumors following intracardiac injection of their cells (Murphy et al. 1985). In their studies the proportion of the cells arrested in different organs paralleled the blood flow to the organs (i.e., the cells went where the blood went) but the probability that a cancer cell deposited in an organ caused a macroscopic metastasis varied very widely between different organs. They found one out of ten cells trapped in the adrenal caused a metastasis, whereas in skeletal muscle the figure was one in 10^5. This organ preference did not have an immunological basis as the same distribution was seen in genetically athymic (Nu/Nu) rats and rats immunosuppressed with cyclosporin A. Alexander et al. speculated that an isolated cancer cell was not capable of autonomous growth unless it finds itself in a tissue capable of supplying it with growth factors which act like transforming growth factor (TGF), or which potentiate TGF (Alexander et al. 1985). "Once growth has started it will be self sustaining since a cluster of cancer cells will ensure the necessary concentration of TGF in the fluid around the metastasis".

The existence of dormant metastases in organs distant from the primary tumor could be similarly explained. In animal models, the presence in the lung of dormant cancer cells, which stemmed from blood borne spread from a distant primary tumor, could be demonstrated by transplantation. In the lung, the cells do not grow but when a cell suspension from the lung taken from animals, from which the "primary" had been surgically removed a week previously, is injected into the peritoneal cavity, then tumors indistinguishable from the "primary" grew out (Alexander 1983).

Alexander said the concept that more than one cell needs to undergo transformation before a tumor can develop is in some ways a re-expression of theories which saw cancer as a generalized tissue disorder. The evidence for this is compelling for bladder carcinoma and attention had been drawn to this old concept by Rubin in a critical analysis of the role of mutational events in carcinogenesis (Rubin 1984). Alexander concluded by saying that discoveries in the field of polypeptide growth factors and in particular their constitutive synthesis by malignant cells provided a biological framework in which the clonal growth of malignant cells *in vitro* can be reconciled with a hypothesis that in general tumors occurring in animals are not clonal in origin, but that their genesis requires the interaction and co-operation of several transformed cells. The monoclonality of macroscopic tumors need not reflect a clonal state at early stages of tumor development, so much as the cumulative effect of selective pressures upon polyclonal populations during active growth.

3.2   TUMORIGENIC RETROVIRUSES

Save in the case of the relatively small, though slowly increasing, number of viruses, no inkling has been obtained of what happens when a cell becomes neoplastic, nor of how its power is passed on when it divides. Man must and will find out.

(Rous 1967)

Retroviruses have set back cancer research many decades.
(Peter Duesberg 2010, personal communication)

Peyton Rous discovered the viral etiology of a chicken sarcoma in 1911 through his interest in tumor transplantability to new hosts by a filtrate (Rous 1911). He said, "The behaviour of the new growth has been throughout that of a true neoplasm, for which reason the fact of its transmission by means of a cell-free filtrate assumes exceptional importance." He fully realized from the outset that this was "a unique and significant finding." He also realized that the significance of the discovery depended on the true nature of the induced growth. As an experienced pathologist he confidently stated: "The [pathological] picture [of the growth] does not in the least suggest a granuloma... . [I]t exhibits to a special degree, not merely a few, but all those features by which the malignant neoplasms are characterized." But Ponten took a different view. He said, "the [early] oncogenic property of Rous sarcoma virus was more like that of an agent inducing benign overgrowth than one which gives autonomous malignant neoplasms" (Ponten 1976).

Rous's discovery had little immediate influence at the time because scientists were not prepared to think of viruses as agents of cancer (Dulbecco 1976). It took more than four decades following Rous's discovery before viruses were taken seriously by cancer researchers. The situation changed radically in the 1950s when the isolation of murine leukemia virus by Ludwik Gross (Gross 1951, Gross 1957) catalyzed the field, winning young converts to the expanding search for tumor viruses. Several of the viruses isolated during this period became important model systems, actively studied at the cellular and molecular levels to this day. The Friend and Rauscher murine leukemia viruses provided models for the study of erythropoiesis (Friend 1957, Rauscher 1962). The rodent sarcoma viruses of Kirsten, Harvey, and Moloney (Harvey 1964, Kirsten and Mayer 1967, Moloney 1966), the feline sarcoma virus of McDonough (McDonough et al. 1971), and the avian leukemia virus MC29 (Ivanov et al. 1964), to name a few examples, yielded oncogenes that figured prominently in the journey from viral to human oncogenes. Studies in animals also produced evidence for the occurrence of endogenous retroviruses which initially revealed themselves by their oncogenicity for uninfected animals. Otto Mühlbock found in 1965 that MMTV-induced mammary carcinoma in mice could be inherited as well as virally transmitted (Mühlbock 1965), and transmissible leukemogenic agents were obtained from X-ray-induced murine leukemias (Latarjet and Duplan 1962, Lieberman and Kaplan 1959).

Within a short time, technical advances accelerated the study of tumor viruses. The elucidation of the structure of DNA by Watson and Crick in 1953, fueled developments in microbial genetics and led to the reinterpretation of the virus concept itself. It soon became clear that certain types of virus can introduce parts of their own genetic material into a cell without killing it or inhibiting its multiplication. The virus material thus introduced may become actually integrated with the gene material of the recipient cell and behave as a new hereditary factor. Virus infection could thus lead to a permanent change in some cellular characteristics. This re-evaluation of the virus concept made it possible to understand how a tumor virus might change the regulated behavior of normal cells to the malignant proliferation characteristic of cancer cells.

Work accelerated when it was discovered that certain viruses could rapidly transform normal cells *in vitro*. This opened the way for direct studies on cancerous transformation of human cells, an approach previously hidden behind the walls of the living organism. Even Rous's own virus, previously regarded as lacking any importance for mammals, induces cancer under certain conditions in many different mammalian species.

Animal retroviruses with oncogenes (*onc* genes) are unique among viruses because they are not transmitted either in the germ line or by infection (Duesberg 1983). Indeed, they are only known because of their isolation from rare, spontaneous tumors. No epidemic of Rous sarcoma virus (RSV), avian myelocytomatosis virus (MC29) or the Kirsten, Harvey or Moloney murine sarcoma viruses have ever been reported (Duesberg 1983). Therefore, it was proposed in 1969 that transforming genes of retroviruses pre-exist in normal human cells as latent cancer genes that may be co-opted by retroviruses without *onc* genes or activated by carcinogens (Huebner and Todaro 1969). This view has since become known as the oncogene hypothesis.

In light of the discovery of reverse transcriptase, Temin proposed the protovirus hypothesis which postulated that normal cells contain potential (proto) oncogenes which by somatic mutation and retroviral or cellular reverse transcription could become viral or cellular oncogenes (Temin 1971). The association of this enzyme with viruses that could cause cancer lent support to the idea that their transforming genes had derived from oncogenic, cellular precursors. The critical distinction of the protovirus hypothesis from the oncogene hypothesis was that a qualitative rather than a quantitative change was postulated to convert normal cellular genes into oncogenes. Both hypotheses remained essentially untestable until viral *onc* genes were defined.

Major support for Huebner and Todaro was provided by Peter Duesberg and Peter Vogt. In a series of beautifully conceived and difficult-to-perform experiments, reported during the period 1970–1975, they succeeded in precisely defining a viral transforming gene (Duesberg and Vogt 1970, Duesberg et al. 1974, Wang et al. 1975). This made it possible to test the oncogene hypothesis directly. By 1982–83 the evidence from many studies that included sequencing a number of cloned cellular and viral genes was quite conclusive: exactly as the oncogene hypothesis required, the *onc* genes of retroviruses came from the chromosomes of their hosts. But there was the problem.

Although it resolved one difficulty—of how a highly infrequent virus-promoted natural cancer could be relevant to an extremely common human disease—the oncogene hypothesis raised another thorny issue. It would appear paradoxical for normal cells to have evolved a battery of known oncogenes which, when carried by retroviruses, are the fastest acting and most inevitably carcinogenic agents known (Duesberg 1983).

Comparisons between the (real) *onc* genes of retroviruses and the (hypothetically oncogenic) proto-*onc* genes of cells had indeed proved the cellular origin of transforming viral sequences. But the comparisons proved neither the structural identity nor the functional equivalence of these genetic homologs. In fact, what these numerous studies showed was that the viral and cellular genes differed substantially from one another in their structures. And because retroviruses contain special signals (called promoters that dictate how much of a gene's product a cell can make) that are hundreds of times stronger than the cell's own, the amounts of retroviral and cellular oncogenic proteins produced were found to differ greatly. The differences led to two models of oncogenesis: the qualitative and quantitative.

Both models held that viral *onc* genes arose from transduction of cellular genes. But the qualitative model required that the transduction event create an essentially new genetic entity, while the other viewed the transduced cellular proto-*onc* gene and the viral *onc* gene as basically the same, differing only in the quantity with which they, or their particular protein products, were present. According to the quantitative model, in a normal cell the proto-*onc* gene was silent, or expressed to a very low level, and in the cancer cell this latent oncogene was expressed more actively, i.e., activated. Or so went the version at the time. By 1981, Bishop endorsed the re-designation of proto-*onc* genes of cells as "(c) oncogenes" (Coffin et al. 1981) and referred to them as enemies from within the cell (Bishop 1981). At this point, mutation and oncogene theories of cancer had not yet fully merged.

According to Bishop, retrovirologists reached an uneasy peace by adopting a nomenclature in which viral oncogenes were known as v-*onc*'s, the cellular progenitors of v-*onc*'s as cellular oncogenes (c-*onc*'s), and each of the viral and cellular genes by terms derived from the names of the viruses in question: e.g., v-*src*, c-*src*, v-*ras*, c-*ras* (Bishop 1983). Oncogenes identified by transfection were known variously as "tumor genes," "cellular oncogenes," "transforming genes," or merely "oncogenes." The term "proto-oncogene" was used to denote either the cellular progenitors of retrovirus v-*onc*'s, or the cellular genes whose damage gives rise to the active oncogenes in tumor DNA.

Duesberg thought Bishop's re-designation was premature (Duesberg 1983). There was now ample sequence information to suggest differences between viral *onc* genes and proto-*onc* genes but still insufficient genetic and functional knowledge to determine whether cellular proto-*onc* genes were indeed cellular oncogenes. Therefore, Duesberg continued to use the term proto-*onc* rather than c-*onc* to emphasize this uncertainty (Bialy 2004).

When biologists use the word "function," it is almost always accompanied by the word "assay." In cancer biology, the best assay for whether a cell has been transformed is to inject its clonal descendants (in as small numbers as possible) into immunologically defective animals (so the injected material will not be eliminated by the immune system) and see if the animals develop malignant tumors. This was not so easily accomplished; therefore cell culture assays were used that were faster, less expensive, and served as indicators of whether putting the cells into animals were likely to give the expected result. Morphological transformation is one such important assay, where the ability of a cell to escape from its normal growth controls and form a visible colony on a laboratory dish is the experimental endpoint.

To expedite research—and in part to free modern cancer molecular biology of its complicated origins in oncogenic retroviruses—a new type of transformation assay emerged in the early 1980s. Instead of viruses, purified DNA was used to deliver genes, whose oncogenic activity was being investigated, into cells. The first results of these transfection assays, as they are known, heavily favored the qualitative model because no cloned proto-*onc* gene was able to transform primary cells that each of their viral *onc* counterparts could do so reproducibly and efficiently. To get around this problem, cancer molecular biologists introduced a variation of the naked DNA assay in which they replaced cultured primary cells by a special mouse cell line called NIH 3T3 developed by Green and Todaro in Boston (Todaro et al. 1963). The harm to cancer research caused by the widespread use of the 3T3 cell functional assay is difficult to over estimate (see Section 4.3).

As Gerald Dermer warned in 1983: "A fundamental disadvantage of using long-term cultures of embryonic rodent fibroblasts, such as 3T3 or 10T½, as normal cells may be that the cells do not age and are very different from the cells in our body. These immortal cell lines are used in assays to test the transforming activity of carcinogens and oncogenes because the cells lack variability and are considered to have moved through all but the last step in the progression from normal to malignant. Progression, however, suggests that the cells have acquired features of the malignant phenotype. If this is true, why are the cells used as normal controls?"

Primary cell cultures are those obtained from freshly dissected adult or embryonic animal tissue. Such cells are presumably as genetically normal as it is possible to obtain and are the first choice for testing the effects of an introduced gene. Very importantly they have the normal complement of chromosomes, and it is unfortunate that they cannot be maintained in this balanced genetic state for long. After a relatively few divisions, most of the explanted cells become disorganized, stop dividing and eventually die (Hayflick 1965). However, once in a while in rodent cells, a mutant cell appears which has the useful property of being able to continue to divide indefinitely. Significantly, spontaneous transformation of normal human cells in culture rarely—probably never—happens (Mamaeva 1998).

Unlike primary cultures, these immortalized cells, known as cell lines, are not genomically normal. Cell lines show diverse numbers of chromosomes as well as aberrant or marker chromosomes. The NIH 3T3 cell line has been described as "preneoplastic" (Duesberg 1983, Schafer et al. 1984) because it is genomically unstable (with a modal number of 68 chromosomes instead of the euploid number of 40 for the mouse) (American Type Culture Collection 1992) and can spontaneously transform into cancer cells within days (Rubin and Xu 1989). The unrecognized (or disregarded) significance of the preneoplastic nature of the highly aneuploid 3T3 cells has misdirected decades of cancer research.

In one set of highly regarded experiments using the 3T3 cells, Ed Scolnick working at the NIH fused the cellular prototype of the Harvey sarcoma virus transforming gene, *ras*, to a promoter from the virus, and thus created in the test tube something close to an artificial transforming retrovirus (Chang et al. 1982). When this was transfected into 3T3 (and only into 3T3) cells, colonies of cells tumorigenic for mice could be obtained—although this did not occur nearly as frequently as with the virus, where even when primary cells are used, every single infection leads to a cancerous transformation. Whether Scolnick's result supported the quantitative or qualitative model was a matter of scholarly debate at the time. What is important for us is that it and a few similar "landmark" publications involving proto-*ras* and 3T3 cells were the only experimental bases upon which the concept of functional oncogene was being built (Duesberg and Schwartz 1992). Importantly, all other proto-*onc* genes failed to transform even 3T3 cells (Bialy 2004) (Section 4.3).

Duesberg's 1983 characterization of proto-*onc* genes still holds:

Clearly, the identification of onc genes has moved viral carcinogenesis from the romantic into an academic age. The same cannot be said for the suspected role of proto-onc genes in cancer. As yet there is no functional and no consistent circumstantial evidence that proto-onc genes directly initiate and maintain cancer like the onc genes of retroviruses. There is also no such evidence that proto-onc genes encode one of the multiple initiation and promotion events that create virus-negative cancers (Cairns 1981, Logan and Cairns 1982)... . So far retroviral onc genes are the only proof that altered proto-onc genes can be cancer genes. There is as yet no conclusive example that an unaltered proto-onc gene can function as a cancer gene simply through enhanced transcription or gene amplification. Complete genetic definition and assays for the biological function of normal and mutated proto-onc genes are now necessary to understand their possible role in carcinogenesis.

(Duesberg 1983)

While Duesberg's article was being prepared for publication, two papers appeared, one in *Nature* (Waterfield et al. 1983) and the other in *Science* (Doolittle et al. 1983), which seemed to forcefully address the need for a functional assay for oncogenes. These were the first examples of an enterprise that is now denoted by the expression "functional genomics," which permitted molecular biologists to feel intellectually comfortable with the newly emerging oncogene, even though proto-*onc* and *onc* genes are clearly not the same genetic units (Bialy 2004). The papers showed that the Simian sarcoma virus *onc* gene, *sis*, derives from a gene that had been very recently cloned and shown to function as a "growth factor." Since cancer is a disturbance of a cell's normal growth controls, it seemed a small step to attribute true oncogene status to the c-*sis* proto-*onc* gene (platelet-derived growth factor or PDGF). All the pieces were now in place to guarantee that the once-clear differences between quantitative and qualitative interpretations of what an oncogene represented would be forever obscured. It only remained to change the word "activate" to mean mutation, without giving up its old quantitative meaning (Bialy 2004).

3.3 DOMINATE ONCOGENES

The structural similarity between viral *onc* genes and certain cellular genes led to countless problems because it implied they were functional equivalents. Since some viral oncogenes derive from cellular genes, whose normal functions appear to concern proliferation, offered a possible solution to this problem. It was proposed that proliferation genes could have their normal functions modified by mutation, functionally turning them into the *onc* genes of the extremely rare, transforming retroviruses to which they were related. Such mutated genes were said to be "activated" and appeared to be the answer to cancer.

The question, of course, was how would a mutated or "activated" gene cause cancer? In 1987, Bishop offered "three hypothetical explanations for how genetic damage [mutation of dominate proto-oncogenes] might cause the malfunction of a proto-oncogene or its product" leading to cancer (Bishop 1987):

1. The damage might cause the oncogene or its product to be continuously produced but the level of expression is no greater than the usual maximum.

2. The abnormality may be an overabundance of an otherwise normal gene product, the consequence, for example, of gene amplification, translocation into the vicinity of a strong transcriptional enhancer, insertion of retroviral DNA, or transduction into retroviral genome. This is known as the quantitative model of transformation.

3. Mutations might change the manner in which a protein acts. Examples include alterations in the substrate specificity of a protein kinase or in the specificity of a transcription factor. This is the so-called qualitative change of function.

"How," Bishop asked, "can we fit these themes into the context of cellular replication?" He offered the metaphor of a complex electrical circuit to convey his hypothesis. "The proliferation of cells is governed by an elaborate circuitry that reaches from the surface of the cell to the nucleus. The products of proto-oncogenes may represent some of the junction boxes in that circuitry: polypeptide hormones that act on the surface of the cell, receptors, and nuclear functions that may orchestrate the genetic response to afferent commands. What we now know of oncogenes allows us to view their actions as 'short circuits' at the corresponding junction boxes. This imagery is at best only a first approximation. For example, some proto-oncogenes may have roles in regulating differentiation or in the maintenance of fully differentiated cells rather than in cellular proliferation, a possibility that is not addressed by the circuitry envisioned here".

In 1982, Robert Weinberg and his colleagues at MIT and the NIH published a paper in *Nature* entitled "Mechanism of activation of a human oncogene" (Tabin et al. 1982)—the first use of "activation" to mean a qualitative change in the hypothetical proto-oncogene, rather than a quantitative change in the amount of its product. The paper so impressed the editors that they invited a special editorial from John Cairns to comment on how the "secrets of cancer" were now beginning to be revealed (Logan and Cairns 1982).

Weinberg showed that the genetic sequence of the proto-*ras* DNA from a bladder cancer cell line differs from the *ras* DNA of normal human cells by only one point mutation, which changes the normal amino acid glycine to valine in the proto-*ras* protein p21. Weinberg said this mutation was specifically responsible for transforming the 3T3 cells and concluded that it was also the cause of the bladder cancer from which the cell line derived. This astonishingly bold claim for the additional three carbon and six hydrogen atoms of valine became the basis for the hypothesis that point-mutations of proto-*ras* genes cause cancer (Logan and Cairns 1982, Reddy et al. 1982, Tabin et al. 1982).

The hypothesis assumes that point-mutations confer dominant transforming function to proto-*ras* genes that is equivalent to that of sarcoma-producing retroviral *ras* genes (Tabin and Weinberg 1985). It assumes further that the 3T3 cell-transformation assay measures a preexisting function of mutated cellular proto-*ras* genes. Consequently, point-mutated proto-*ras* genes were termed "dominantly acting oncogens" (Bishop 1983, Bishop 1991, Krontiris and Cooper 1981, Shih et al. 1981, Stanbridge 1990a,

Tabin and Weinberg 1985, Varmus 1984), dominate because they were supposed to change the phenotype even of the cells carrying the gene in just one of the two chromosomes (heterozygous). Subsequently, other proto-*onc* genes, such as proto-*myc* (Rabbitts et al. 1983, Westaway et al. 1984) and proto-*src* and the *src* genes of Rous sarcoma virus (Hunter 1987), and even genes that are not structurally related to retroviral oncogenes, such as certain anti-oncogenes (now called tumor suppressor genes), were also proposed to derive transforming function from point-mutations (Bishop 1991, Bookstein et al. 1990, Horowitz et al. 1989, Marx 1991, Stanbridge 1990b, Varmus 1984, Watson et al. 1987, Weiss et al. 1985).

However, it was soon shown the point mutation does not satisfy any of Bishop's three hypothetical mechanisms for causing cancer. The mutation does not lead to overproduction of the *ras* gene product (p21) in the 3T3 cells and does not change known biochemical properties of p21 (Finkel et al. 1984). Finkel et al. concluded: "A comparison of proteins encoded by normal human *ras* genes and by mutant *ras* H or *ras* K genes activated in human carcinomas revealed no changes in subcellular localization, posttranslational modification, or guanine nucleotide binding associated with activation. Subcellular fractionation indicated that both normal and activated *ras* proteins were associated exclusively with the membrane fraction. Furthermore, both normal and activated *ras* proteins exhibited similar degrees of posttranslational acylation. The K_D (dissociation constant) for dGTP binding was $1.0–2.2\times10^{-8}$ M, with no consistent differences between normal and activated *ras* proteins. In addition, a survey of 13 possible competing nucleotides revealed no differences in the specificity of nucleotide binding associated with *ras* gene activation. These results indicate that structural mutations which activate *ras* gene transforming activity do not alter the protein's known biochemical parameters and in particular do not affect the protein's intrinsic ability to bind guanine nucleotides".

Furthermore, the *ras* mutation has not been found in a survey of more than 60 primary human carcinomas (Duesberg 1985). The mutated human proto-*ras*, which transforms 3T3 cells, does not transform primary rat embryo cells (Land et al. 1983b, Ruley 1983) and more significantly, does not transform human embryo cells (Sager et al. 1983). The conclusion was clear: mutated proto-*ras* was neither necessary nor sufficient to transform normal cells (Duesberg and Schwartz 1992). Instead, cellular transformation was dependent on very high over-expression of either normal or mutant *ras* as a consequence of heterologous (i.e., viral) promoters that are not found in normal cells (Chakraborty et al. 1991). It followed that the transfection assay used by Weinberg and many others did not measure a genuine function of point-mutated proto-*ras* genes as existed in tumors, but instead was an expression artifact created during the transfection assay.

Such artifacts could be generated during transfection by substituting via illegitimate recombination the native proto-*ras* regulatory elements by artificial promoters derived from carrier and helper gene DNA (Chakraborty et al. 1991). Indeed, transformation of primary cells by cellular proto-*ras* genes depended on the presence of added viral helper genes or on other cellular genes linked to viral promoters (Lee et al. 1985, Ruley 1990, Schwab et al. 1985, Stone et al. 1987), or on the presence of retroviral promoters alone (Chakraborty et al. 1991). This recombination process was entirely analogous to the generation of retroviral *ras* genes, in which coding regions of normal proto-*ras* genes are recombined by transduction with heterologous retroviral promoters that enhance the transcription over 100-fold compared to proto-*ras* (Chakraborty et al. 1991, Duesberg 1987, Duesberg et al. 1989, Zhou and Duesberg 1990). In addition, transfection generated concatenated DNA multimers producing an artificial gene amplification that would also enhance the dosage of *ras* transcripts (Goff et al. 1982, Goldfarb and Weinberg 1981, Perucho et al. 1980, Robins et al. 1981).

The probable reason that proto-*ras* genes from tumors transform 3T3 cells, but not primary cells, is that mouse NIH 3T3 cells are much more readily transformed by exogenous genes, as well as spontaneously (Rubin and Xu 1989), than are embryo cells (Land et al. 1983a). Thus, the weak promoters acquired from random sources during transfection were sufficient to convert proto-*ras* genes with point-mutations into 3T3-cell transforming genes, but not into genes capable of transforming primary cells (Section 4.3).

The reason that point-mutated, but rarely normal, proto-*ras* genes (Taparowsky et al. 1982) are detected by transfection assays is that point-mutations enhance about 10- to 50-fold the transforming function imparted by heterologous promoters on proto-*ras* genes (Chakraborty et al. 1991, Cichutek and Duesberg 1986, Spandidos and Wilkie 1984, Velu et al. 1989). Thus, proto-*ras* genes derive their transforming function from heterologous promoters, and certain point-mutations merely enhance this transforming function (Section 4.3).

The ability of retroviral oncogenes, including those of Rous sarcoma, Harvey sarcoma, and MC29 and MH2 carcinoma viruses, to transform normal diploid animal cells were eventually shown to depend absolutely on transcriptional activity, rather than on mutations in the coding region (Chakraborty et al. 1991, Cichutek and Duesberg 1986, Cichutek and Duesberg 1989, Zhou and Duesberg 1988, Zhou and Duesberg 1989, Zhou and Duesberg 1990). This high transcriptional activity of retroviral oncogenes results from retroviral promoters (Duesberg and Schwartz 1992).

3.4   TUMOR SUPPRESSOR GENES

There are heritable and spontaneous retinoblastomas (Knudson 1985). Cytogenetic analyses of both have revealed chromosome 13 is either missing or altered in 20 to 25% (Benedict et al. 1983, Gardner et al. 1982). Less-well emphasized were the numerous other chromosome abnormalities present in all retinoblastoma tumors studied, probably because no aberration common to all tumors was found (Gardner et al. 1982). In 1986, Weinberg et al. (Friend et al. 1986) cloned a human DNA sequence on chromosome 13 that was missing or altered in about a third of 40 retinoblastomas and in 8 osteosarcomas. Consequently, the gene encoded in this sequence was termed the *rb* gene. Reportedly, the *rb* gene was unexpressed in all retinoblastomas and osteosarcomas

(Friend et al. 1986). On the basis of this, it was proposed that retinoblastoma arises from the loss of the *rb* gene, that somehow prevents or suppresses cancer (Knudson 1985). In the familial cases, the loss of one *rb* allele would be inherited and the second one would be lost due to spontaneous mutation. In the spontaneous cases, somatic mutations would have inactivated both loci. In the retinoblastomas with microscopically intact chromosome 13, submicroscopic mutations were postulated.

This anti-oncogene (tumor suppressor gene) hypothesis predicts that normal cells would constitutively express oncogenes that render the cell tumorigenic if both alleles of the corresponding suppressor gene were inactivated. This of course necessitates that the suppressor genes must be active at all times in normal cells. Another analysis of the primary retinoblastomas undertaken to test the hypothesis found deletions of the *rb* gene in only 4 of 34 tumors analyzed and transcripts of the *rb* gene were found in 12 out of 17 retinoblastomas and in 2 out of 2 osteosarcomas, casting doubt on the deletion hypothesis (Goddard et al. 1988). The remaining tumors had apparently normal *rb* genes. These results started the seesawing of the importance of the *rb* gene.

Under closer scrutiny, subsequent studies of retinoblastomas observed point-mutations and small submicroscopic deletions in *rb* genes that did not have macro-lesions (Bookstein et al. 1990, Dunn et al. 1989, Horowitz et al. 1989, Kaye et al. 1990). For example, both Weinberg et al. (Horowitz et al. 1989) and Lee et al. (Bookstein et al. 1990) reported a point-mutation in a splice sequence of the *rb* gene. In view of this, some believed that point-mutations or other minor mutations of the *rb* genes were sufficient for tumorigenesis (Dunn et al. 1989, Horowitz et al. 1989). However, Gallie et al. reported point-mutations and deletions of *rb* genes in only 13 out of 21 tumors (Dunn et al. 1989). In an effort to develop a functional assay, a DNA copy of the mRNA of the *rb* gene was cloned into a retrovirus and infection by this virus inhibited the growth of a retinoblastoma cell line *in vitro* (Bookstein et al. 1990, Hollingsworth and Lee 1991). However, two later studies reported that an intact, synthetic *rb* gene fails to inhibit tumorigenicity of human retinoblastoma and breast cancer cells in nude mice (Muncaster et al. 1992, Xu et al. 1991).

Clearly, the point-mutation hypothesis of the *rb* gene would never have emerged if the original chromosome deletion hypothesis had been confirmed. The point mutation explanation advanced the anti-oncogene hypothesis into a virtually inexhaustible reservoir of hypothetical cancer genes: any gene with any mutation in each of both alleles in a cancer cell could be a tumor suppressor or anti-oncogene. According to Weinberg, "one can cast a broad net for tumor suppressor loci by using a large repertoire of polymorphic DNA markers to survey...for repeated instances of LOH (loss of heterozygosity). Indeed, this genetic strategy has revolutionized the research field" (Weinberg 1991).

Over a dozen deleted or point-mutated anti-oncogenes are now considered to cause osteosarcomas, breast cancer, bladder cancer, lung cancer, colon cancer, Wilms' tumor, and neuroblastoma, in addition to retinoblastoma (Bishop 1991, Cooper 1990, Hollingsworth and Lee 1991, Stanbridge 1990b, Weinberg 1991). For example, a point-mutation in one of three genes of a colon cancer cell would signal an inactivated hypothetical colon cancer suppressor gene (Marx 1991, Nishisho et al. 1991). Further, the range of the *rb* suppressor gene has since been extended to other cancers, including small cell lung, bladder, prostate, breast carcinomas, and osteosarcoma (Cooper 1990, Hollingsworth and Lee 1991).

Nevertheless, the anti-oncogene hypothesis has been difficult to prove because: 1) the oncogenes that are said to be suppressed have not been named or identified (Stanbridge 1990a) and will be difficult to assay because all normal cells or animals should suppress them with the corresponding anti-oncogenes, and 2) transfection of an intact *rb* gene (Bookstein et al. 1990, Xu et al. 1991) has failed to revert transformed cells to normal and to suppress their tumorigenicity *in vivo* (Bookstein et al. 1990, Muncaster et al. 1992, Xu et al. 1991).

Likewise the hypothetical colon cancer suppressor gene p53 (once claimed to be a dominate oncogene (Finlay et al. 1989, Jenkins et al. 1985, Parada et al. 1984)) has failed to revert transformed cells to normal (Baker et al. 1990) and its complete absence has not affected the normal development of p53 knockout mice (Donehower et al. 1992). Nevertheless, 74% of these p53-free mice developed lymphomas and sarcomas at six months that probably derived from single cells, rather than through a systemic transformation as the anti-oncogene hypothesis would have predicted (Donehower et al. 1992).

A fanciful subdivision of the hypothetical tumor suppressors into caretaker, gatekeeper, and landscaper genes has been proposed (Kinzler and Vogelstein 1997, Michor et al. 2004). It has even been suggested (perhaps tongue-in-cheek) that the "distinction between different types of cancer genes may need to be expanded to include 'gatetakers' or 'carekeepers' " (Tomlinson and Bodmer 1999). When mutated, caretaker genes are said to cause genomic instability and aneuploidy. In contrast, gatekeeper genes encode proteins thought to restrain cell growth, while disrupting their function allows enhanced cell proliferation. Defective landscaper genes, as some imagine, foster a microenvironment conducive to tumor cell survival.

But, do aneuploidy preventing, caretaker genes (which includes mitotic checkpoint genes) really exist? This question is important because the chromosomal imbalance theory—going back to Boveri—states that any specific influence causing aneuploidy is carcinogenic by definition. Paulsen et al. claim to have "identified hundreds of [human] genes…that mediate genome stability" (Paulsen et al. 2009). Stirling et al. recently designated 692 yeast genes (~12% of the genome) as chromosome instability (CIN) genes. The authors said, "The breadth of the CIN gene list suggests that many biological processes protect genome integrity," and remarked in passing that mutation in one can lead to CIN (Stirling et al. 2011). But if there really were hundreds of genes protecting against aneuploidy, and mutation in any one could lead to chromosome instability, then a large fraction of an organism's cells would be chromosomally unstable all of the time, which is clearly wrong. Furthermore, a gene that prevents aneuploidy would be a true tumor suppressor gene and defective mutants would qualify as true cancer-causing genes, which should have already been discovered in the global search for such genes. However, as detailed in Sections 4.3 & 4.4.13, no mutant gene nor combination of genes has transformed normal cells into cancer cells.

Weaver and Cleveland published a select list of "Genes preventing aneuploidy that are mutated and/or misregulated in human cancers" (Weaver and Cleveland 2006). The list included 52 mutations in checkpoint genes (BUB1, BUBR1, BUB3, MAD1, MAD2) in 356 cancers, for an incidence of 15%, and 276 mutations in tumor suppressor genes (APC, BRCA1, BRCA2, Msh2) in 1828 cancers, also for an incidence of 15%. The 15% incidence of mutation is well below the background rate of 26% for tumors (Forbes et al. 2011). Tellingly, however, 75% of the checkpoint genes were over or under expressed compared to normal, and 45% of the tumor suppressor genes were likewise differentially expressed compared to normal. The substantial differential expression observed in cancer cells fits precisely within the framework of the chromosomal imbalance theory of cancer detailed in Chapters 5 & 6.

3.5 DRIVER GENES

In the age of whole cancer genome sequencing, it is now possible to describe the genome-wide somatic mutation content of a tumor sample, including structural rearrangements and non-coding variants. COSMIC (Catalogue of Somatic Mutations in Cancer) was designed to gather, curate, organize and present the world's information on somatic mutations in cancer and make it freely available (Forbes et al. 2011). COSMIC combines cancer mutation data from the scientific literature with the output from the Cancer Genome Project at the Sanger Institute UK (CGP is sponsored by the NIH (Bonetta 2005)). Genes are selected for full literature curation using the Cancer Gene Census (http://www.sanger. ac.uk/genetics/CGP/Census/), with a focus on those mutated by small point mutations in the coding domains, and more recently including those mutated by gene fusion.

As of July 2010, over 136,000 coding mutations had been found in almost 542,000 tumor samples. Of the 18,490 genes documented, 4,803 (26%) had one or more mutations. Full scientific literature curations are available on 83 major cancer genes and 49 fusion gene pairs (19 new cancer genes and 30 new fusion pairs added in 2010 alone) and this number is continually increasing. Key amongst these is p53, now available through a collaboration with the IARC p53 database. As of November 2010, there were 27,580 somatic and 597 germline mutations (IARC TP53 Database http://www-p53.iarc.fr/) among the 20,303 base pairs of the human p53 gene (NCBI database retrieval server http://www.ncbi.nlm.nih.gov/nucleotide/35213?) (Lamb and Crawford 1986). This bewildering complexity of virtually every aspect of p53 biology has left outsiders—and many within the field—bemused and confused (Horn and Vousden 2004).

Studies to date have revealed a complex genome, with approximately 40–80 amino acid-changing mutations present in a typical solid tumor (Bozic et al. 2010). One is compelled to ask, does the rapidly growing number of mutations signal a problem for the gene mutation theory or a potential bonanza? The answer seems to be both. On the bonanza side, hundreds—perhaps thousands—of mutations in "driver genes" were added to the compendium of cancer genes. Driver mutations are loosely defined as those that confer a selective growth advantage to the cancer cell. On the problematic side, it was simply untenable that all or even most of the vast number of mutations actually "drive" cancer.

The well-recognized concept of passenger mutation was dusted off and modified to deal with this problem. A "passenger mutation" is now defined as one which does not alter fitness but occurs in a cancer cell that coincidentally or subsequently acquired a driver mutation (Bozic et al. 2010). A passenger mutation, therefore, is found in any cell with a driver mutation. Youn and Simon recently offered an operational or statistical definition of a driver gene (Youn and Simon 2010). A driver gene contains mutations at a significantly higher rate than the background rate. The background mutation rate is estimated based on silent mutations which do not change amino acid encoding and which are therefore considered to be passenger mutations.

Driver mutations are strictly hypothetical because there is as yet no functional test for their contribution—if any—to carcinogenesis. Vogelstein acknowledged this problem at the 2010 Annual Meeting of the American Association for Cancer Research (AACR) in his talk titled: "The sequence of all 185,000 coding exons in each of 100 human tumors: What has it taught us?" He answered: "It is actually very difficult not only to tell whether a gene or specific mutation is a driver, but also whether that driver functions as a tumor suppressor or an oncogene. Functional studies—at least those available now— are not up to the task because too many genes, when expressed at high levels in normal cells, induce either cell growth or cell death in an unphysiologic fashion" (Tuma 2010).

Different studies of the same tumor type often report genetic "regions of interest" that are highly discordant. For example, two lung cancer studies in 2005, with similar sample sizes and analytic methods, reported 48 and 93 regions of interest, respectively (Tonon et al. 2005, Zhao et al. 2005). The overlap between the lists was <5%. Two possible explanations have been offered for such a disconcerting high level of discordance (Beroukhim et al. 2007). The first is that the true number of cancer-related genetic regions is extremely large, with each tumor containing only a small and variable subset of the alterations and each study detecting only a small subset of the regions. The second possibility is that many of the regions of interest reported in current studies are random events of no biologic significance, such as random passenger mutations.

The potential futility of cataloging hundreds of thousands of mutations was expressed by Beroukhim and colleagues: "Ultimately, the utility of systematic efforts to characterize the cancer genome is an empirical question. There are at least two potential concerns: on one hand, that the vast majority of cancer-related genes are already known with little left to learn; on the other hand, that cancer is hopelessly complicated, with a large number of cancer genes, each altered in a small fraction of tumors" (Beroukhim et al. 2007).

The lack of a functional test allows for subjective and wholly arbitrary criteria for distinguishing between driver and passenger mutations. In the absence of biological evidence, Futreal et al. explained, "The underlying rationale for interpreting a mutated gene as causal in cancer development is that the number and pattern of mutations in the gene are highly unlikely to be attributable to chance. So, in the absence of alternative plausible explanations," the argument goes, "the mutations are likely to have been selected because they confer a growth advantage on the cell population from which the cancer has developed" (Futreal et al. 2004). Thus, mutations in driver genes are difficult to distinguish from passenger mutations without resort to statistics.

For example, a recent study by Bozic et al. "considered a gene to be a tumor suppressor if the ratio of inactivating mutations (stop codons due to nonsense mutations, splice site alterations, or frameshifts due to deletions or insertions) to other mutations (missense and in-frame insertions or deletions) was >0.2. This criterion identified all well-studied tumor suppressor genes and classified 286 genes as tumor suppressors. We considered a gene to be an oncogene if it was not classified as a tumor suppressor gene and either (i) the same amino acid was mutated in at least two independent tumors or (ii) >4 different mutations were identified. This criterion classified 91 genes as oncogenes; the remaining 560 genes were considered to be passengers" (Bozic et al. 2010).

The authors went further quantifying "the selective growth advantage afforded by the mutations that drive tumor progression.... . The selective advantage is surprisingly small (0.4%) and has major implications for experimental cancer research... . For example, it shows how difficult it will be to create valid *in vitro* models to test such mutations on tumor growth; such small selective growth advantages are nearly impossible to discern in cell culture over short time periods." The miniscule advantage of individual driver mutations was interpreted (perhaps to the delight of biotech companies) to mean hundreds, perhaps thousands, of driver mutations were needed to produce an advanced malignancy within the lifetime of an individual (Bozic et al. 2010).

Gene Mutation Theory of Cancer

Perhaps the most damning evidence against the oncogene theory is the fact that the supposed human oncogenes do not transform true normal cells, which have a normal set of chromosomes. Furthermore, there is absolutely no evidence from observations of human tumors to indicate that the mutation of any proto-oncogene is essential for cancer initiation. In fact, in many tumors, all the supposed proto-oncogenes are normal; there are no oncogenes present.

Gerald Dermer, 1994 (Dermer 1994)

Originally, somatic mutation contributing to cancer pathogenesis included point mutations in individual genes, deletions, translocations, gene amplification, and changes in chromosome number and configuration, i.e., aneuploidy. With the advent of molecular biology, somatic mutation is nowadays generally limited to changes in the DNA sequence of individual normal genes. The gene mutation hypothesis, in contrast to the competing chromosomal imbalance theory, derived instant support from its conventional mechanism of phenotype alteration. Moreover, the gene mutation hypothesis attracted steady attention by adopting and adapting results from the rapidly evolving fields of sexual and later molecular genetics, which offered plenty of "doable" experiments (Fujimura 1996).

Ever since Morgan's first papers on Drosophila genetics appeared in 1910, gene mutation, rather than aneuploidy, was on everybody's mind as the mechanism of generating abnormal phenotypes (Morgan 1910). Moreover, Morgan and Bridges directly attacked Boveri's aneuploidy hypothesis, stating: "At present, however, reference to such possible sources, ... [i.e.] imperfect or irregular division of the chromosomal complex, ... is too uncertain to be of great value, for there are no instances where irregularities of this kind are known to give rise to prolific growth processes. The cancer-like or tumor-like growth shown by a mutant of Drosophila...is caused by a sex-linked Mendelian gene..." (Morgan and Bridges 1919).

The mutation hypothesis derived further support in 1927 when Muller, a former student of Morgan, had discovered that X-rays mutate genes (Muller 1927). Since X-rays were a previously known carcinogen, this discovery was interpreted as experimental support for the mutation hypothesis. It set off similar searches for mutagenicity of all carcinogens and for the corresponding cancer-causing mutations that still monopolize cancer research today (Alberts et al. 1994, Ames et al. 1973, Bishop 1995, Cairns 1978, Harris 1995, Knudson 2001, Lodish et al. 1999, Miller and Miller 1971, Muller 1927, Peltomaki et al. 1993, Pierce 2005, Pitot 1986, Varmus 1989b, Vessey et al. 2000, Vogelstein and Kinzler 1998, Vogelstein and Kinzler 2004).

According to Varmus in his Nobel Prize-wining year, there are six unifying concepts contributing to the mutant gene theory of cancer (Varmus 1989a):

1. Eukaryotic genomes are endowed with a substantial list of genes (presently numbering between 50 and 100) that may participate in neoplasia as a consequence of mutations.

2. Most of these genes, generally referred to as proto-oncogenes, have fundamental roles in the governance of cell growth or differentiation, have been highly conserved during evolution, and can be partitioned into gene families on the basis of their sequence, their function, or both.

3. The mutations that convert proto-oncogenes to active oncogenes range from subtle changes in sequence (e.g., single nucleotide substitutions) to gross rearrangements (insertions, deletions, gene amplifications, or chromosomal translocations); they may be induced by a variety of chemical, physical, or viral reagents; they may affect the

level of gene expression or the nature of the gene product, or both; and they appear to act in a dominant fashion.

4. Other cancer-inducing mutations appear to behave in a recessive fashion by inactivating genes now known as "tumor suppressor genes"; only a few such genes have been isolated so far.

5. Many viruses (most DNA viruses and some retroviruses) contain one or a few genes that can induce neoplastic change in animals or in cells grown in culture, but retroviral oncogenes have the particularly useful property of being derived directly from cellular proto-oncogenes.

6. Combinations of oncogenes are generally required to convert a normal cell to a tumor cell. Viral oncogenes or cellular oncogenes may act at different points in a growth regulatory network in which the products of several proto-oncogenes and proteins normally interact.

4.1 "CARCINOGENS ARE MUTAGENS"

A favorite explanation has been that oncogens [carcinogens] cause alterations in the genes of the ordinary cells of the body, ... somatic mutations as these are termed. But numerous facts, when taken together, decisively exclude this supposition.

(Rous 1967)

After his discovery that X rays—a previously known carcinogen—can mutate genes, Muller was the first to point out in 1927 that the "effect of X-rays, in occasionally producing cancer, may also be associated with their action in producing mutations" (Muller 1927). With more sensitive techniques, a growing number of carcinogens were shown to have mutagenic function (Bauer 1928, Braun 1969, Miller and Miller 1971). Even the chemically inert polycyclic aromatic hydrocarbons were found to react with DNA, although only after enzymatic oxidation (Brookes and Lawley 1964, Cairns 1978). The quest for mutagenic carcinogens reached a high point with Ames' slogan, "Carcinogens are mutagens" (Ames et al. 1973).

But in the excitement over matching carcinogens with mutagenic function it was simply disregarded that many, including the most effective, carcinogens were not mutagenic in established test systems, as for example the polycyclic hydrocarbons (Ashby and Purchase 1988, Berenblum and Shubik 1949, Burdette 1955). Even Rous's misgivings about the role of mutations was ignored. He warned: "The evidence as a whole makes plain though that some carcinogens induce somatic mutations whereas others do not, that some mutagenic agents fail to be carcinogenic, and that many substances closely related chemically to agents of both sorts do neither" (Rous 1959). Lijinsky also acknowledged that many carcinogens are mutagenic, but warned that carcinogens, which are mutagenic in test systems, "does not mean that a mutagenic process is involved" and that "the mutagenic reaction of carcinogens might be coincidental rather than causal: alternative mechanisms of carcinogenesis should be considered" (Lijinsky 1989).

4.2 RETROVIRAL ONCOGENES

Retroviruses with oncogenes (*onc* genes) transform susceptible cells in culture or in animals with the same kinetics as they infect them (Duesberg 1983, Duesberg 1985). Therefore, these viruses are by far the most direct and efficient natural carcinogens but they are never associated with healthy animals with functioning immune systems. The mechanisms by which the artificially activated, transgenic oncogenes and the naturally active oncogenes of DNA and RNA tumor viruses cause cancer have two characteristics in common: hyperplasias and long latency periods.

Early hyperplasias and dysplasias

The viral oncogenes induce systemic, mostly diploid hyperplasias or dysplasias within days to weeks after infection of cells in culture or in animals (Ahuja et al. 2005, Bell et al. 1990, Ewald et al.

1996, Hauschka 1961, Kato 1968, Mitelman 1974, Palmieri et al. 1983, Ponten 1976, Temin and Rubin 1958, Tooze 1973, Vogt and Dulbecco 1960, Webster et al. 1995, Zhou and Duesberg 1988, Zhou and Duesberg 1990). In contrast to autonomous cancers, these early hyperplasias are nonclonal with regard to integrated tumor viruses and reversible up to a certain point, either serologically (Tooze 1973) or if they are under the control of conditional oncogenes (Duesberg 2003, Ewald et al. 1996, Fried 1965, Martin 1970, Pelengaris et al. 1999, Shachaf et al. 2004). Making this point on physiological grounds, Bryan called Rous virus tumors, "viral hyperplasia of an extreme type" (Bryan 1960), and Ponten called "the oncogenic property of Rous sarcoma virus... more like that of an agent inducing benign overgrowth than one which gives autonomous malignant neoplasms" (Ponten 1976).

In contrast to autonomous cancers, these hyperplasias are not immortal *in vitro* or in transplantations in animals (Hellstroem et al. 1963, Palmieri et al. 1983). But during these early hyper- or dysplasias, the DNA and RNA tumor viruses induce nonclonal aneuploidies in a fraction of infected cells (Ewald et al. 1996, Kato 1967, Koprowski et al. 1962, Nichols 1963, Nichols et al. 1967, Ray et al. 1992, Stewart and Bacchetti 1991, Temin and Rubin 1958, Vogt and Dulbecco 1963, Yerganian et al. 1962). It is rarely emphasized that conventional carcinogens share with viral and artificially activated oncogenes the capacity to induce early hyperplasias or dysplasias together with nonclonal aneuploidy (Aldaz et al. 1987, Aldaz et al. 1988a, Aldaz et al. 1988b, Becker et al. 1971, Binder et al. 1998, Conti et al. 1986, Marquardt and Glaess 1957, Sanchez et al. 1986).

Late cancers

After latencies of about 6 months to 2 years, the viral oncogenes induce irreversible cancers with individual clonal karyotypes (Hellstroem et al. 1963, Kato 1968, Li et al. 2009, Mark 1967, Mark 1969, Mitelman 1974, Ponten 1976). These cancers have "infinite" life spans and are transplantable indefinitely, and hence called immortal. If tested, these clonal viral cancers often lack intact viral oncogenes or even fragments of viral oncogenes, or are independent of conditional oncogenes under non-permissive conditions (Ewald et al. 1996, Fried 1965), indicating that transformation is oncogene independent (de Lapeyriere et al. 1984, Kuznetsova et al. 1970, Lania et al. 1980, Marczynska and Massey 1986, Mora et al. 1977, Mora et al. 1986, Santarelli et al. 1996, Seif et al. 1983). Thus, the clonal cancers induced by transgenic oncogenes and by viral oncogenes are both oncogene independent. Nevertheless, it is the spectacular capacity of tumor viruses to induce hyperplasias and dysplasias without delay that has attracted much more attention than their capacity to induce clonal cancers with individual karyotypes after long latencies (Crawford 1980, Duesberg 1980, Temin 1980, Tooze 1973, Zhou and Duesberg 1988).

4.3 ARE "CELLULAR ONCOGENES" LIKE RETROVIRAL ONCOGENES?

Although the roster of proto-oncogenes is large and can be categorized in instructive ways, it is still not possible to be specific either about the number and kinds of oncogenes required to convert a particular cell into a cancer cell or about biochemical events that are crucial for transformation.

(Varmus 1989a)

There is no direct functional proof for the hypothesis that mutation of *ras* and other cellular genes, related to retroviral oncogenes, cause cancer. Cellular *ras* RNA in human cancer cells transcribed from either normal or mutated *ras* genes with native cellular promoters is expressed so poorly that it is practically undetectable (Duesberg and Schwartz 1992, Duesberg 1995, Hua et al. 1997), as for example in colon cancer cells with mutant *ras* genes (Rasnick and Duesberg 1999, Zhang et al. 1997). Mutant proto-*abl*, which has been associated with chronic myeloid leukemia (CML), stands out as another example of an inactive oncogene and for the various functions of its artificial derivatives.

Human CML proceeds in two distinct phases. The first is a chronic phase lasting on average 3–4 years in which undifferentiated and differentiated, functional myelocytes, granulocytes, and neutrophils are overproduced. Since the overproduced cells differentiate to functional blood cells, this phase of the disease is a clonal hyperplasia. In about 85% of CML cases these hyperplastic cells carry a clonal variant of chromosome 22, termed Philadelphia chromosome. The remaining CML cases have no Philadelphia chromosome (Nowell 1982, Sandberg 1990). The second phase of CML is a terminal leukemia of several months, termed blast crisis, in which new, autonomous clones of non-differentiating aneuploid myeloblasts take over, which also typically carry the Philadelphia chromosome. These cells are no longer functionally normal (Koeffler and Golde 1981a, Koeffler and Golde 1981b, Sandberg 1990).

In about 80% of CMLs with Philadelphia chromosomes, the variant chromosomes are generated by a reciprocal translocation in which a small piece of chromosome 9 is translocated to chromosome 22, and a smaller piece of 22 (Koeffler and Golde 1981a) goes to 9 (Rowley 1973). Since this translocation moves the coding region of the proto-*abl* gene to a promoter region from a gene termed *bcr* on chromosome 22, and since proto-*abl* is related to the oncogene of the murine Abelson leukemia virus, the hybrid *bcr-abl* gene is now thought to be the cause of CML (Heisterkamp et al. 1985).

However, there is a conceptual problem with this hypothesis. The Abelson virus carries a dominant oncogene, termed *abl*, which causes a polyclonal leukemia in mice that is fatal within a few weeks (Duesberg and Schwartz 1992, Weiss et al. 1985). But the chronic phase of CML is a hyperplasia, not a terminal leukemia. Logically, the *bcr-abl*-CML hypothesis postulates that a cellular mutant gene causes hyperplasia because this gene is related to a dominant retroviral oncogene. Experimental evidence confirms and extends the discrepancy between the cellular *bcr-abl* and viral oncogene theories of CML.

The transcripts of *abl* genes are barely detectable or even undetectable in CML patients by conventional hybridization with radioactive DNA probes (Gale and Canaani 1984). Therefore, transcripts are now typically detected by artificial amplification with the polymerase chain reaction (Bose et al. 1998). By contrast, transcription of the oncogene of Abelson virus in leukemic mice is 100- to 1,000-fold higher than that of the mouse or human *abl* genes (Duesberg and Schwartz 1992). Thus, the fast, viral leukemia with highly active *abl* genes is not a model for the slow, chronic phase of human CML with inactive *bcr-abl* genes.

The functional discrepancy between the Abelson virus oncogene and the cellular *bcr-abl* gene has been confirmed unintentionally by all efforts to prove the *bcr-abl*-CML hypothesis. For example, to generate a leukemia in mice with the *bcr-abl* of human CML, Baltimore et al. had to make the gene part of an artificial Abelson virus (Daley et al. 1990), which enhanced its activity 100- to 1,000-fold above its activity in CML (Duesberg and Schwartz 1992). Likewise Era and Witte had to rely on heterologous promoters derived from cytomegalovirus and a chicken actin gene in order to find "Bcr-Abl being the sole genetic change needed for the establishment of the chronic phase of CML" (Era and Witte 2000). However, these studies, as with their mutant *ras* gene antecedents, failed to consider that the cellular and pathogenic effects of these artificial *bcr-abl* constructs depended on 100- to 1,000-fold transcriptional activation compared to the inactive *bcr-abl* genes of human CML (Duesberg and Schwartz 1992). Thus, these studies confirm the lesson of the mouse Abelson virus—that a highly over-expressed *abl* gene is leukemogenic—but they say little about the function of the poorly expressed *abl* genes in the chronic phase of CML.

Moreover, since the discovery of the reciprocal translocation between chromosomes 22 and 9 in human CML (Rowley 1973), about 20% of Philadelphia chromosomes were shown to be translocations of chromosome 22 with chromosomes that do not carry *abl* genes, i.e., with chromosomes 2, 6, 7, 11, 13, 16, 17, 19, and 21 (Harris 1995, Nowell 1982, Sandberg 1990). According to Nowell, the discoverer of the Philadelphia chromosome (Nowell and Hungerford 1960), "These variants appear to have no significance with respect to the clinical characteristics of the disease, and so it appears that it is the displacement of the sequence of chromosome 22 that is of major importance, rather than the site to which it goes" (Nowell 1982). In other words, the mutation of proto-*abl* is not necessary for the generation of a Philadelphia chromosome nor for CML.

This leaves open the question whether mutation of proto-*abl* happens to be sufficient to initiate the chronic, hyperplastic phase of CML by some unknown mechanism that does not rely on high transcriptional activity. However, two facts suggest this is not the case:

1. Transgenic mice carrying a *bcr-abl* gene in every cell of their body, even with promoters that are much stronger than those of native *bcr-abl* genes, are not born with CML. Instead, many develop a non-CML type of leukemia after "long latency," because "BCR/ABL expression is not the sole cause of leukemia but rather predisposes for the cancer" (Voncken et al. 1995).

2. CML-specific, poorly expressed *bcr-abl* transcripts have recently also been detected in up to 75% of normal humans with the polymerase chain reaction (Biernaux et al. 1995, Bose et al. 1998). It follows that the *bcr-abl* gene is not sufficient to initiate even the chronic phase of CML.

Thus, the hypothesis that mutation of cellular genes related to retroviral oncogenes causes cancer, is unconfirmed. Nevertheless, the apparent functional proof for cellular oncogenes has fueled the cataloging of hundreds of mutated genes in cancer cells that are all now assumed to cause cancer either directly, as hypothetical oncogenes, or indirectly, as hypothetical tumor suppressor genes (Alberts et al. 1994, Boland and Ricciardiello 1999, Haber and Fearon 1998, Hanahan and Weinberg 2000, Lodish et al. 1999). Yet most of these mutant genes do not even transform 3T3 cells, but are still called "oncogenes" because they were first identified in cancer cells (Watson et al. 1987). Indeed, it has not been possible to isolate cellular genes from any cancer that transform normal human cells into cancer cells (Li et al. 2000), "after more than 15 years of trying" (Weitzman and Yaniv 1999).

However, the evidence that these mutations are neither necessary nor sufficient for cancer does not exclude the possibility that they, if present, play indirect roles in carcinogenesis as, for example, in clonal expansion (Cha et al. 1994) or in increasing the risk of aneuploidy. Indeed the transition from the chronic, preneoplastic phase of CML to the neoplastic phase, termed blast crisis, is preceded by and coincides with aneuploidy (Harris 1995, Sadamori et al. 1983, Sadamori et al. 1985), suggesting (but unproved) that the Philadelphia chromosome and/or its reciprocal counterpart may increase the risk of aneuploidization.

4.4 UPDATED GENE MUTATION THEORY IS POPULAR BUT UNCONFIRMED

The role of cellular oncogenes in carcinogenesis remains largely an inference.

(Bishop 1983)

Although hundreds of research papers have reported on the transforming powers of oncogenes, the biochemical mechanisms used to achieve these changes are still, with rare exceptions, quite obscure.

(Weinberg 1989)

...the genetic principles that underlie the evolution of cancer genomes and how combinations of mutations contribute to cancer phenotypes remain poorly defined.

(Anderson et al. 2011)

Berenblum and Shubik were some of the first to raise questions about gene mutation as the cause of cancer stating: "the theory has rested largely on the assumption that, given an irreversible change as the basis of carcinogenesis, the only known biological phenomenon to explain this would be a gene mutation. However, a closer examination of other common biological phenomena instantly reveals that this is not so" (Berenblum and Shubik 1949). And despite a potential conflict of interest with regard to the cancer gene of his sarcoma virus, in 1959, Rous concluded that, "the somatic mutation hypothesis, after more than half a century, remains an analogy: 'it is presumptive reasoning based on the assumption that if things have similar attributes they will have other similar attributes' " (Rous 1959). Rous's reservations about the hypothesis included non-genotoxic carcinogens, the slow action of carcinogens, and the inadequacy of known mutations to explain the many differences between cancer and normal cells.

As recently as 2009, Brash and Cairns felt the need to remind researchers that cataloging the "enemies within"—the hundreds of oncogenes and tumor suppressor genes—have yet to reveal "the mysterious steps in carcinogenesis" (Brash and Cairns 2009). They reminded researchers the picture that emerges from the classical studies of the epidemiology of human cancers and of experimental carcinogenesis in animals is hard to reconcile with what has been learned about mutagenesis in simple systems such as the bacteria. Initiation seems to be far too efficient to be simply mutagenesis of certain oncogenes and suppressor genes, and the subsequent time-dependent steps are even more obscure. "The prime mystery in carcinogenesis remains the very first step, because it is hard to imagine how the numerous genetic changes found in cancer cells could have been produced in any cell as the result of a single exposure to a DNA-damaging agent, or why months or years should have to elapse before the effect of these changes is observed".

Despite its current popularity, the gene mutation hypothesis has failed to meet many of its own predictions:

4.4.1 Gene mutation theory cannot explain non-mutagenic carcinogens and tumor promoters

Carcinogens are either chemical or physical agents that initiate carcinogenesis (Cairns 1978, Pitot 2002). Both chemical and physical carcinogens can be either mutagenic or non-mutagenic. Examples of non-mutagenic carcinogens are asbestos, tar, mineral oils, naphthalene, polycyclic aromatic hydrocarbons, butter yellow, urethane, dioxin, hormones, metal ions such as Ni, Cd, Cr, As, spindle blockers such as vincristine and colcemid, extra-nuclear radiation, carbon nanotubes and solid plastic or metal implants (Ashby and Purchase 1988, Berenblum and Shubik 1949, Burdette 1955, Duesberg et al. 2004b, Lijinsky 1989, Little 2000, Oshimura and Barrett 1986, Pitot 2002, Poland et al. 2008, Preussman 1990, Rous 1959, Scribner and Suess 1978, Zaridze et al. 1993). Moreover, the many agents that accelerate carcinogenesis, termed tumor promoters, are all by definition non-mutagenic (Pitot 1986), or not directly mutagenic, as for example croton oil and phorbol acetate (Iversen 1991b, Pitot 2002).

4.4.2 No cancer-specific gene mutations

No cancer-specific mutations have yet been found (Cooper 1990, Duesberg and Schwartz 1992, Haber and Fearon 1998, Hollstein et al. 1994, Little 2000, Strauss 1992, Vogelstein et al. 1988). "Although certain genes are frequently mutated in the cancers of particular tissues, no tumor has the same spectrum of mutations" (Frank 2004). According to a commentary titled "How many mutations does it take to make a tumor?", "There are no oncogenes or tumor suppressor genes that are activated or deleted from all cancers. Even tumors of a single organ rarely have uniform genetic alterations, although tumor types from one specific organ have a tendency to share mutations" (Boland and Ricciardiello 1999). When no specific mutations are found, other, as yet unknown, mutations are suggested to "phenocopy" the known mutations (Hanahan and Weinberg 2000), even though there is no functional evidence.

4.4.3 "Causative" mutations are not clonal and not shared by all cells of a tumor

Recent evidence shows that even known, hypothetically causative mutations, are not shared by all cells of the same tumor, e.g., mutant *ras* and the hypothetical mutant tumor suppressor gene p53 (Al-Mulla et al. 1998, Albino et al. 1984, Giaretti et al. 1996, Heppner and Miller 1998, Konishi et al. 1995, Kuwabara et al. 1998, Offner et al. 1999, Roy-Burman et al. 1997, Shibata et al. 1993, Yachida et al. 2010). Likewise, the spontaneous loss of the presumed oncogene, mutant *ras*, does not revert the phenotype of a cancer cell back to normal (Plattner et al. 1996). Thus, known oncogene and tumor suppressor gene mutations are not necessary for the maintenance and probably not even for the initiation of cancer, although they are present in some of its cells. Their non-clonality is predicted by the chromosomal imbalance theory (Sections 4.4.9 & 6.1.1).

4.4.4   Mutant genes do not transform normal cells into cancer cells

No mutant gene nor combination of mutant genes from cancer cells has been found that converts diploid human or animal cells into cancer cells, despite enormous efforts in the last 25 years (Akagi et al. 2003, Augenlicht et al. 1987, Duesberg and Schwartz 1992, Duesberg 1995, Duesberg et al. 2004b, Harris 1995, Harris 2005, Hua et al. 1997, Li et al. 2000, Li et al. 2002, Lijinsky 1989, Radford 2004, Schneider and Kulesz-Martin 2004, Stanbridge 1990a, Thraves et al. 1991, Weitzman and Yaniv 1999). On the contrary, several hypothetical mutant cancer genes, including *myc*, *ras*, and p53, have even been introduced into the germline of mice but these transgenic mice are initially healthy and are breedable, although some appear to have a slightly higher cancer risk than other laboratory mice (Donehower et al. 1992, Duesberg and Schwartz 1992, Hariharan et al. 1989, Li et al. 2000, Purdie et al. 1994, Sinn et al. 1987). For example, one study of the genes said to cause colon cancer reports that, "Transgenic pedigrees that produce K-*ras*Val12 alone, p53Ala143 alone, or K-*ras*Val12 and p53Ala143 have no detectable phenotypic abnormalities" (Kim et al. 1993).

There are other mouse strains with hypothetical cancer genes artificially implanted into their germline, and others with hypothetical tumor suppressor genes artificially deleted from the germline, which have survived many generations in laboratories with either the same or slightly higher cancer-risks than other lab mice (Donehower et al. 1992, Duesberg 2003, Duesberg et al. 2004b, Harris 2005). For instance, one group observed that, "Surprisingly, homozygosity for the *Apc*1638T mutation", an artificial null mutation of the hypothetical tumor suppressor gene *Apc*, "is compatible with postnatal life" and "animals that survive to adulthood are tumor-free" (Smits et al. 1999). Even more surprisingly, some mice with hypothetical cancer genes and others without hypothetical tumor suppressor genes fare even better than un-mutated controls. For example, the authors of one study state that, "Surprisingly [the] germline expression of an oncogenic *erbB2* allele (breast cancer gene, alias *Her2* and *Neu*)... conferred resistance to mammary tumorigenesis" (Andrechek et al. 2004). Yet another group reported that "unexpectedly" mice with null mutations of the retinoblastoma gene, *rb*, "developed fewer and smaller papillomas" than un-mutated controls (Renan 1993).

There are reports of tumors in mice that can be induced and even reversed experimentally via promoters that switch on and off hypothetical cancer genes, which have been artificially implanted into the germline (Shachaf et al. 2004, Weinstein 2002). But two questions have not been answered: 1) Why did only local, and thus possibly clonal, tumors appear in these "transgenic" mice, rather than systemic cancer? 2) Were these reversible tumors aneuploid or diploid hyperplasias? (Duesberg 2003, Shachaf et al. 2004). According to Harris, these "Experiments with transgenic animals are unanimous in their demonstration that oncogenes do not produce tumours directly, but merely establish a predisposition to tumour formation that ultimately requires other genetic changes which occur in a stochastic fashion" (Harris 1995). And even this predisposition may be an artifact of the ectopic position of the trans-gene in the chromosome rather than of its function.

As of 2004, at least 291 mutations in protein-coding genes (more than 1% of the human genome) have been proposed to cause cancer (Futreal et al. 2004). According to the Catalogue of Somatic Mutations in Cancer database and website, the tally of cancer genes as of 2009, was 412, of which 300 are said to be dominate (http://www.sanger.ac.uk/, (Bamford et al. 2004)). Three-quarters of these cancer genes were associated with leukemias, lymphomas and mesenchymal (connective tissue) tumors even though these account for less than 10% of human cancers. Since the most common cancers are epithelial in origin, there is considerable scope for the cataloging of many additional gene mutations associated with these cancers. Thus, the total number of human cancer genes remains a matter for speculation.

Most cancer genes have been identified and initially reported on the basis of genetic evidence (that is, the presence of somatic or germline mutations) and without biological information supporting the oncogenic effects of the mutations (Forbes et al. 2011, Futreal et al. 2004, Katsios and Roukos 2011, Park et al. 2010, Terwilliger and Hiekkalinna 2006, Wacholder et al. 2010). But, on September 16, 2005, J. Michael Bishop, who shared the 1989 Nobel Prize for the discovery of the cellular origin of retroviral oncogenes, confirmed that there was still no proven combination of mutant genes from cancer cells that is sufficient to cause cancer (from a seminar, "Mouse models of human cancer" at the Lawrence Berkeley National Laboratory, 2005).

Vogelstein and Kinzler closed an influential review of the mutation theory in 1993 (since cited in text books (Voet and Voet 1995)) as follows: "The genetics of cancer forces us to re-examine our simple notions of causality, such as those embodied in Koch's postulates: how does one come to grips with words like 'necessary' and 'sufficient' when more than one mutation is required to produce a phenotype and when that phenotype can be produced by different mutant genes in various combinations?" (Vogelstein and Kinzler 1993). The answer to Vogelstein and Kinzler's question is still open—two decades and many studies later.

4.4.5 Mutagenic carcinogens should cause instant transformation

Conventional mutation is immediate and just as stable as the parental genotype (Brookes and Lawley 1964, Fujimura 1996,

Griffiths et al. 2000, Lewin 1997, Muller 1927, Pierce 2005). In view of this it is surprising there are no fast carcinogens. Nevertheless, many carcinogens are very fast mutagens, as for example, X-rays, UV light and alkylating agents. But all carcinogens, mutagenic or not, are very slow—causing cancer only after exceedingly long "neoplastic latencies" (Foulds 1975, Pitot 2002) of many months to years in rodents, and of many years to decades in humans (Bauer 1949, Bauer 1963, Berenblum and Shubik 1949, Cairns 1978, Duesberg and Li 2003, Foulds 1975, Hueper 1952, Pitot 2002, Vogelstein and Kinzler 1993). As recently as 2009, Brash and Cairns reminded researchers "the time course of carcinogenesis is deeply mysterious" (Brash and Cairns 2009). Examples are: 1) the solid cancers, in survivors of atomic bombs in 1945, mainly developed 20 years after exposure to nuclear radiation (Cairns 1978), 2) the breast cancers, which developed in former tuberculosis patients only 15 years after treatments with X-rays in the 1950s (Boice and Monson 1977), and 3) the lung cancers, which developed in workers of a Japanese mustard gas factory only 30 years after it was closed in 1945 (Doi et al. 2002).

Similarly, the risk of lung cancer remains about 5–10 times higher for ex-smokers than it is for non-smokers, even decades after they stopped smoking (Cairns 1978, Cairns 2002, Doll 1970, Hittelman 2001). Thus, an initiated cell evolves only gradually to a visible cancer cell, long after exposure to carcinogen—much like a submarine volcano only gradually becomes a visible island (Bauer 1949, Bauer 1963, Foulds 1975). By contrast, the mutation theory would have predicted carcinogenesis as soon as the people in the above examples had received the doses of carcinogen that eventually caused their cancers.

Experimental carcinogenesis demonstrates even more directly that, once initiated, the evolution of cancer cells is an autonomous, if slow, process that is independent of further exogenous influences (Berenblum and Shubik 1949, Cairns 1978, Foulds 1975, Pitot 2002, Ruiz et al. 2005). Nevertheless, experimental carcinogenesis is accelerated by further carcinogens or tumor promoters (Cairns 1978, Iversen 1991a, Iversen 1991b, Pitot 2002, Rous 1967, Ruiz et al. 2005). This autonomous evolution continues in cancer cells and their descendents both *in vivo* and even *in vitro* (Berenblum and Shubik 1949, Foulds 1975, Hauschka 1961, Levan 1956, Rous 1967, Winge 1930). As a result, cancer cells progress independently within individual cancers, to form ever-more "polymorphic" (Caspersson 1964b) and phenotypically heterogeneous cancers with ever-more exotic karyotypes and phenotypes (Foulds 1975). Thus, "initiation" confers on cells a lifelong variability that can generate new phenotypes and karyotypes many cell generations or decades after it was established (Harris 2007).

Cairns summarized the problem with conventional gene mutation in *Cancer: Science and Society*: "The conspicuous feature of most forms of carcinogenesis is the long period that elapses between initial application of the carcinogen and the time the first cancers appear. Clearly, we cannot claim to know what turns a cell into a cancer cell until we understand why the time course of carcinogenesis is almost always so extraordinarily long" (Cairns 1978).

4.4.6 Gene mutation should have reproducible consequences

The effects of point mutations are immediate and reproducible. However, the progression of cancer is neither definite nor predictable but "follows one of alternative paths of development" (Braun 1969, Pitot 1986). According to Foulds: "different characters of a particular tumor undergo progression independently of one another. This rule leads to the more general proposition that the structure and behavior of tumors are determined by numerous unit characters that, within wide limits, are independently variable, capable of highly varied combinations, and apt to progress independently. This rule has proved valid and useful in diverse studies of neoplasia in animals and in man. The associations of characters such as growth rate, invasiveness, powers of metastasis, responsiveness to hormones, and morphologic characters in tumors of one general kind are highly varied and include many anomalous associations or 'dissociations' of characters, as exemplified by the 'locally malignant' and 'metastasizing benign' tumors of man. The rule still holds good when the analysis is pushed to the level of enzymes and histocompatibility genes, and investigations at this level confirm earlier suspicions about the individuality of tumors; probably no two tumors are exactly alike in every respect even when they are evoked by similar means from the same tissue and have the same general properties. Many wide generalizations about 'cancer' have broken down, as most of them eventually do, because they have not taken into account the independent variability of characters and the individuality of tumors. Moreover many characters often prominent in tumors are not essential to the neoplastic disease but are only incidental consequences or accompaniments of it..." (Foulds 1965). Foulds' characterization of the highly variable phenotypes of cancer is consistent with it being an aneuploidy syndrome (Section 6.1.8).

4.4.7 Non-selective phenotypes are not compatible with gene mutation

Cancer-specific phenotypes can be divided into two classes: those, which are selective, because they advance carcinogenesis by conferring growth advantages to cancer cells such as invasiveness, grossly altered metabolism and high adaptability via high genomic variability (Foulds 1975, Pitot 2002), and those which are not selective for growth (Bernards and Weinberg 2002, Duesberg et al. 2004b).

The non-selective phenotypes of cancer cells include metastasis, multi-drug resistance and *in vitro* immortality. Metastasis is the ability to grow at a site away from the primary tumor. Therefore, it is not selective at the site of its origin (Bernards and Weinberg 2002). Likewise, preexisting multi-drug resistance is not a selective advantage for natural carcinogenesis in the absence of chemotherapy. Yet, many cancers are intrinsically multi-drug-resistant (Doubre et al. 2005, Goldie 2001). Moreover, acquired multi-drug resistance protects against many more drugs than the cancer was ever exposed to (Duesberg et al. 2000a, Duesberg et al. 2001b, Schoenlein 1993). Even immortality is not a selective advantage for carcinogenesis, because many types of normal human cells can grow over 50 generations according to the Hayflick limit (Hayflick and Moorhead 1961), and thus many more generations than are necessary to generate a lethal cancer. Specifically, 50 cell generations beginning with a single cell would produce a cellular mass equivalent of 10 humans with 10^{14} cells each, assuming most survive (Duesberg and Li 2003).

4.4.8   Cancer causing genes are hard to reconcile with human survival

The spontaneous mutation rates of mammalian cells and the hundreds of hypothetical oncogenes and tumor suppressor genes postulated so far (Haber and Fearon 1998, Hanahan and Weinberg 2000, Lodish et al. 1999, Mitelman et al. 1997a) make it very difficult to reconcile with the survival of multicellular organisms. The spontaneous, net mutation rate (after proofreading) is about 1 in 10^9 nucleotides per mitosis (Lewin 1994, Li et al. 1997, Strauss 1992). Since the DNA of human and all other mammalian species is made up of about 10^9 nucleotides (O'Brien et al. 1999), then a mutation in every position of the human or mammalian genome can be expected in 10^9 cells. Considering that humans have 10^{14} cells (Cairns 1978, Strauss 1992), every human should contain 10^5 cancer cells if just one dominant oncogene existed that could be activated by just one point mutation. Since there are now hundreds of such genes and "activating" mutations are found in multiple positions of the same gene (Seeburg et al. 1984), cancer should be ubiquitous.

In response to this, the proponents of the mutation hypothesis now argue that it takes between 3 and 20 gene mutations to generate a human cancer cell (Lodish et al. 1999). Hahn and colleagues postulate that three mutant genes "suffice" to create a human tumor cell (Hahn et al. 1999), whereas Kinzler and Vogelstein postulate 7 mutations for colon cancer (Kinzler and Vogelstein 1996). However, their proposals create a new paradox: in view of the above mutation rates, cancer would be practically nonexistent (Jakubezak et al. 1996). For example, if 3 mutations were required, only 1 in $10^{9 \times 3}$ or 10^{27} human cells would ever turn into a cancer cell by spontaneous mutation, and if 7 were required, only one in 10^{63} would ever turn into a cancer cell. Thus only 1 in 10^{11} or in 10^{47} humans would ever develop cancer, since an average human life corresponds to about 10^{16} cells (Cairns 1978, Duesberg and Schwartz 1992). In other words, cancer would never occur. In view of this paradox, the proponents of the gene mutation hypothesis have postulated that malignant transformation depends on a "mutator phenotype" (Loeb 1991).

4.4.9 Mutator phenotype to the rescue

Several gene mutation theories are currently advanced to explain the genomic instability of cancer cells (Anderson et al. 2001, Balmain 2001, Hoeijmakers 2001, Loeb 1991, Loeb 2001, Loeb et al. 2003, Marx 2002, Pihan and Doxsey 2003, Rajagopalan et al. 2003, Sieber et al. 2003). According to a recent review: "In general, these theories assume that genomic instability is derived from mutations in genes that are involved in processes such as DNA repair and chromosomal segregation. The mutations of the 'mutator genes' have no direct selective advantage or disadvantage, only an effect on the mutation rates of other genes" (Sieber et al. 2003). However, there are four arguments against the necessity of such "mutator genes" (Breslow and Goldsby 1969, Loeb et al. 2003):

1) Mutator genes are only detected in a small minority of cancers (Anderson et al. 2001, Duesberg and Li 2003, Grosovsky et al. 1996, Hawkins et al. 2000, Hermsen et al. 2002, Lengauer et al. 1997, Lengauer et al. 1998, Marx 2002, Pihan and Doxsey 2003, Sieber et al. 2003, Strauss 1992, Tomlinson et al. 1996, Wang et al. 2002). Therefore, it is now claimed that the "mutator phenotype" is "transient", i.e., undetectable once a cancer cell is generated (Loeb 1997). Thus, the mutations that are thought to create the mutator phenotype are neither consistent nor clonal in cancers. In view of this, Rajagopalan et al. proposed, "epigenetic events that do not involve mutational changes in nucleotides could certainly have a significant role" (Rajagopalan et al. 2003).

Gene mutations, including mutator genes, would only be generated in those genomically unstable cells, in which teams of enzymes involved in the synthesis and maintenance of DNA are corrupted by aneuploidy (Section 6.1.6). Clonal mutations would derive from preneoplastic aneuploidy and non-clonal mutations would derive from neoplastic aneuploidy. This explains not only why gene mutations are common, but also why they are not consistently associated with genomically unstable cells and why they tend to be late and non-clonal.

2) Among the mutator genes that have been found, some, as for example the adenomatous polyposis coli (*Apc*) and the p53 tumor suppressor genes, are now considered consequences rather than causes of genomic instability (Haigis et al. 2002, Kobayashi et al. 2000, Offner et al. 1999, Tomlinson and Bodmer 1999). According to a review by Lengauer et al., "p53 mutations do not usually occur until much later (after initiation of carcinogenesis)" and thus "are unlikely to be its primary cause" (Lengauer et al. 1998).

3) There is as yet no functional proof for a direct role of mutator genes in carcinogenesis (Duesberg et al. 2000a, Duesberg and Li 2003, Harris 1995, Murnane 1996). For example, in experimental mice with a transgenic lambda phage as indicator of mutagenesis, "the frequencies of lambda cII-mutants were not significantly different in normal mammary epithelium, primary mammary adenocarcinomas, and pulmonary metastases" (Jakubezak et al. 1996). Moreover, mice with null mutations of p53 and with deletion mutants of *Apc* proved to be "surprisingly" procreative, and thus not genomically unstable (Donehower et al. 1992, Smits et al. 1999). In view of this, the study on p53 concluded, "an oncogenic mutant form of p53 is not obligatory for the genesis of many types of tumours" (Donehower et al. 1992), and the study on *Apc* concluded, "Most importantly, *Apc*1638T/1638T animals that survive to adulthood are tumor free" (Smits et al. 1999). "Surprisingly," mice with null mutations of the cyclin-dependent kinase 2, another hypothetical tumor suppressor and cell cycle control gene (Hahn and Weinberg 2002a), were recently

also found to be sufficiently stable to survive to adulthood—again cancer-free (Steinberg 2003).

Some studies of mice with mutated tumor suppressor genes point out that such mice have higher risks of cancer than untreated controls owing to their artificial mutator genes (Duesberg 2003), as for example the study on mice without p53 (Donehower et al. 1992). However, carcinogenesis in such animals is both age- and "strain-dependent" (Donehower et al. 1992), indicating that their artificial genes are not sufficient for carcinogenesis. A calculation of the cellular cancer risk of these mice makes this point even more obvious. Since cancers originate from single cells (Cairns 1978, Koller 1972, Ruddon 1981), and since mice consist of about 5×10^{10} cells and have renewed many of their cells by the time they develop cancers, their cellular cancer risk is less than 5×10^{-10}, in spite of the fact every cell lacks the tumor suppressor gene p53. This extremely low cellular cancer risk undermines the argument for a direct role of such genes in carcinogenesis.

4) The hypothesis, that an elevated rate of mutation is necessary for carcinogenesis, is also burdened by an inherent paradox, which was described in a 1999 review as follows: "an elevation of mutation rate must generally be regarded as a growth disadvantage to the cell. Models which disregard such mechanisms are particularly incompatible with the growing amount of data indicating that practically all neoplastic cells express some form of genomic instability" (Breivik and Gaudernack 1999). Tomlinson et al. expressed the same reservations about the mutator gene hypothesis saying, "The scenarios for a role of a raised mutation rate assume that there is no selective disadvantage to a cell in having an increased number of mutations. This may not be the case: for example, a deleterious or lethal mutation may be much more likely than an advantageous mutation. More subtly, an accumulated mutational load might induce apoptosis" (Tomlinson et al. 1996).

Thus, there is as yet no consistent correlative or functional proof for any of the mutation theories. Moreover, the suicidal consequences of persistent and autocatalytically escalating numbers of mutator genes are hard to reconcile with the long latency and many cell generations between initiation of preneoplastic genomic instability and cancer, and even harder with the immortality of cancer cells (Duesberg and Li 2003).

4.4.10 Gene mutation does not explain chromosome instability in cancer

The chromosomes of cancer cells are numerically and structurally unstable (Albertson et al. 2003, Gollin 2005, Koller 1972, Levan and Biesele 1958, Wolman 1986), so much so that no single tumor is composed of genetically identical cells (Fogh 1986, Foulds 1965, Lengauer et al. 1998, Levan and Biesele 1958, Rasnick and Duesberg 1999). This chromosomal instability of cancer cells is proportional to the degree of aneuploidy or chromosomal imbalance (Section 5.3.9), and is usually dominant in fusions with stable cells (Section 6.1.6).

Radiation-induced chromosomal and mutational instability can be found in 10–50% of clones that survive radiation exposure. In a review article on genomic instability, Pongsaensook et al., said: "These frequencies are too high to be explained by mutation of a single gene or even a large family of genes. The results indicate that genomic instability is unevenly transmitted to sibling subclones, that chromosomal rearrangements within unstable clones are non-randomly distributed throughout the karyotype, and that the majority of chromosomal rearrangements associated with instability affect trisomic chromosomal segments. Observations of instability in trisomic regions suggests that in addition to promoting further alterations in chromosomal number, aneuploidy can affect the recovery of structural rearrangements... . [T]hese findings cannot be fully explained by invoking a homogeneously distributed factor acting in *trans* ... [but instead] suggest that in addition to promoting further alterations in chromosomal number, aneuploidy can play a role in creating a de novo chromosomal rearrangement hot spot" (Pongsaensook et al. 2004).

4.4.11 Karyotypic–phenotypic cancer cell variation is orders of magnitude higher than gene mutation rates

The chromosomes of cancer cells are extremely unstable compared to those of normal cells. For example, 1 in 100 highly aneuploid human cancer cells loses or gains or rearranges a chromosome per cell generation (Li et al. 2005). Since humans contain 23 chromosomes, there are 23 possible specific aneusomies from a chance alteration of only one. Thus, the chance of finding any specific aneusomy is 1 in 2300 per cell generation, or approximately 10^{-3} per cell division. In agreement with this, up to 1 in 1000 aneuploid cancer cells spontaneously generates a specific new phenotype per cell generation—"at frequencies considerably greater than conventional mutation" (Shibata 2011, Wright 1999)—as for example drug-resistance (Duesberg et al. 2000a, Duesberg et al. 2001b, Harris 1995, Li et al. 2005, Tlsty 1997), or the ability to metastasize at "high rates" (Al-Hajj et al. 2003, Harris et al. 1982), or the loss of heterozygosity at rates of 10^{-5} per generation (Vogelstein and Kinzler 1998).

This inherent karyotypic–phenotypic variability of cancer cells is the reason why most cancers are "enormously" heterogeneous populations of non-clonal and partially clonal cells, which differ from each other in "bewildering" (Koller 1972) phenotypic and chromosomal variations (Heim and Mitelman 1995a, Schneider and Kulesz-Martin 2004)—even though most cancers are derived from a common, primary cancer cell and thus have clonal origins (Cairns 1978, Caspersson 1964b, Castro et al. 2005, Castro et al. 2006, Foulds 1975, Harris 1995, Hauschka 1961, Heim and Mitelman 1995b, Klein et al. 2002, Lengauer et al. 1998, Little 2000, Nowell 1976, Sandberg 1990).

By contrast, conventional mutation of specific genes is limited to 10^{-7} per cell generation for dominant genes and to 10^{-14} for pairs of recessive genes in all species (Holliday 1996, Lewin 1997, Marx 2002, Tlsty 1990, Tlsty 1997, Vogel and Motulsky 1986). Surprisingly, in view of the genetic theories of cancer, even the gene mutation rates of most cancer cells are not higher than those of normal cells (Duesberg et al. 2004b, Harris 1991, Holliday 1996, Lengauer et al. 1998, Marx 2002, Oshimura and Barrett 1986, Sieber et al. 2003, Strauss 1992, Tomlinson and Bodmer 1999, Tomlinson et al. 1996, Wang et al. 2002). In fact, "the difference between a cancer genome and its germ line sequence is approximately 100-fold less than the differences between genomes from different individuals" (Shibata 2011). While the rates of specific karyotypic (10^{-3}) and phenotypic (10^{-3}) variations of cancer cells are of the same order, they are 4 to 11 orders of magnitude higher than conventional mutation, and therefore not compatible with mutational theories.

4.4.12 Cancer phenotypes are too complex for conventional mutations

The complexity of most cancer-specific phenotypes far exceeds that of phenotypes generated by conventional mutation. Examples are the gross polymorphisms in size and shape of individual cells within individual cancers (Bauer 1949, Caspersson 1964b, Foulds 1975). Moreover, the kind of drug resistance that is acquired by most cancer cells exposed to a single cytotoxic drug is much more complex than just resistance against the drug used to induce it. Therefore, this phenotype has been termed "multi-drug resistance" (Duesberg et al. 2001b, Harris 1995, Schoenlein 1993). It protects not only against the toxicity of the challenging drug, but also against many other chemically unrelated drugs and is thus probably not due to prescient specific point mutations (Section 5.3.5).

Nevertheless, it has been argued that multi-drug resistance can be generated by singular genes (Kartner et al. 1983, Schoenlein 1993). However, it is biochemically implausible that a single protein could protect against many biochemically unrelated cytotoxic substances, such as DNA chain terminators, spindle blockers and inhibitors of protein synthesis all at once (Kartner et al. 1983, Schoenlein 1993). Moreover, it is improbable that only cancer cells would benefit from such genes, whereas normal cells of cancer patients remain vulnerable (Duesberg et al. 2000a, Duesberg et al. 2001b, Harris 1995).

Cancer-specific phenotypes such as grossly abnormal metabolism, metastasis, transplantability to heterologous species (Hauschka and Levan 1953) and "immortality" (Foulds 1975, Pitot 2002) are also likely to be polygenic, because all of these phenotypes correlate with the altered expression of thousands of genes (Aggarwal et al. 2005, Furge et al. 2004, Lodish et al. 2004, Pollack et al. 2002, Virtaneva et al. 2001) and with highly abnormal concentrations of thousands of normal proteins (Caspersson et al. 1963, Caspersson 1964b, Gabor Miklos 2005, Pitot 2002). In addition, the number of centrosomes is increased up to five-fold (from a normal of 2 to around 10) in highly aneuploid cancer cells, and their structures are often altered at the same time (Ghadimi et al. 2000, Lingle et al. 2002, Pihan et al. 1998, Pihan et al. 2003).

The complexities of these cancer-specific phenotypes, however, cannot be achieved by the low, conventional rates of gene mutations during the limited life spans of humans and animals. For example, it is virtually impossible that the up to five-fold increased numbers of centrosomes, which are observed in highly aneuploid cancer cells (Brinkley and Goepfert 1998, Lingle et al. 1998, Lingle et al. 2002, Pihan et al. 1998), would be the result of mutations that increase the numbers of the 350 different proteins that make up centrosomes (Doxsey et al. 2005). Thus, the mutation theory cannot explain the complex phenotypes of cancer, which better fit an aneuploidy syndrome.

4.4.13 Ubiquity of aneuploidy in cancer is not explained by the mutation theory

In 1990, Atkin and Baker asked: "Are human cancers ever diploid—or often trisomic?", because "...we are unaware of any studies using chromosome-banding techniques in which diploid metaphases are in the majority or otherwise strongly suspected of being tumor cells. From the several hundred tumors, including our own unpublished cases, that have now been studied, we can postulate that diploid human cancers, if they exist, must constitute significantly less than 1% of all cancers... . Studies on uncultured tumor material support the view that human cancers are always aneuploid. The chromosome changes are frequently complex and, moreover, show considerable variability, even among histologically similar tumors from the same site... . The absence of diploid cancers would then suggest that aneuploidy per se has some particular significance for malignancy" (Atkin and Baker 1990).

Rare exceptions to the coincidence between aneuploidy and cancer have been reported, as for example "diploid" colon cancers with mismatch repair deficiency (Lengauer et al. 1998). However, "array-based comparative genomic hybridization" analysis of what at first appeared to be diploid colon cancers has since indicated that about "5% of their entire genome" was segmentally aneuploid versus 20% of a control group of colon cancers without mismatch repair deficiency (Nakao et al. 2004). Colon cancers with "normal karyotypes" have also been described by Bardi et al. (Bardi et al. 2004). Again, further scrutiny revealed that these normal karyotypes were either from "hyperplastic polyps" (Bardi et al. 1997) or from "non-neoplastic stromal cells" (Bardi et al. 1993) or were considered to be misidentified tumor cells, showing "how dependent findings in solid tumor cytogenetics are on method" (Bomme et al. 1998) and (Bardi G., personal communication, 2004). Thus there is still no unambiguous evidence for diploid solid cancer.

Leukemias are regularly cited as examples of diploid cancers. However, as pointed out by Sandberg et al., the fact that aneuploidy has not been demonstrable in all cases of acute leukemia studied thus far does not detract from the intimate relationship between neoplasia and gross chromosomal abnormalities (Sandberg et al. 1962). Failure to demonstrate an aberrant stem line may be the result of relying too heavily on the determination of a single modal number (the most frequent number of chromosomes) which may be diploid even in the presence of absolutely increased numbers of aneuploid cells. This could be true if conditions *in vivo* were to favor mitosis of residual normal cells, since every cell population in acute leukemia must represent a mosaic.

The fallacy of relying on the modal number alone under *in vitro* conditions has been abundantly shown (Atkin and Baker 1990, Testa et al. 1985, Yunis 1984). Some of the leukemias studied by Sandberg et al. using primary bone marrow specimens showed diploid modes yet a significant increase of frank aneuploid metaphases. Another more obvious source of error is the lack of a sufficiently large sample of cells. Thus, two of the cases of myeloblastic leukemia in adults reported by Sandberg et al. had the diploid mode of 46, but the total number of countable metaphases was five and three, respectively. Finally, as stated by Hauschka (Hauschka 1963), pseudodiploidy may simulate a normal chromosomal constitution, a fact which requires painstaking analysis of apparently diploid cell populations with respect to individual chromosomes and has posed great difficulties in the evaluation of murine karyotypes because of the resemblance of the chromosomes to one another in that species.

There appears, however, to be agreement that aneuploid cells are dominant in the active stages of leukemia and represent true stem lines (Costa et al. 2003, Hauschka 1963, Reisman et al. 1964b, Sandberg et al. 1962, van den Berghe 1989, Yunis et al. 1981, Yunis 1984). As such, they meet the criteria required by the chromosomal imbalance theory that a change in the chromosomal constitution of the malignant cell is one of the basic alterations in acute leukemia. These criteria are: 1) modal distribution of the abnormal karyotype, 2) presence of the abnormal karyotype in the earliest phase of the disease that can be investigated, and at any time thereafter in which the marrow shows leukemic infiltration, 3) stability over long periods of time, 4) suppression during remission, and 5) restoration of the identical (or a closely related mode) in subsequent relapse (Reisman et al. 1964a).

In support of this hypothesis, there is evidence that aneuploidy arises prenatally through nondisjunction leading to acute lymphoblastic leukemia (ALL) in young children (Maia et al.

2003, Maia et al. 2004, Panzer-Grumayer et al. 2002). Modern techniques have shown that extra copies of chromosomes in ALL patients lead to increased expression of the associated gene sequences (Gruszka-Westwood et al. 2004). The authors concluded, "This is compatible with a functional interpretation of hyperdiploidy as a gene dosage effect." As with colon cancer, "there is no evidence for malignant plasma cells or precursors to have a normal karyotype. Chromosome changes found may be early events in the development of the disease, but karyotypic evolution starts early and evolves rapidly" (van den Berghe 1989).

Gained segments of chromosomes are typically rearranged either with the same chromosomes from which they were derived or with other chromosomes. The resulting hybrid chromosomes are termed marker chromosomes. Owing to their unique structure, marker chromosomes can serve as tracers for the origin of possibly metastatic cancer cells from primary cancers and for the origin of primary cancer cells from possibly aneuploid preneoplastic precursors (Koller 1972, Sandberg 1990). A typical example of a highly aneuploid cancer karyotype with numerous marker chromosomes, that of a breast cancer cell from the cell line MDA-231, is shown in Figure 4.1B. For comparison, Figure 4.1A shows the karyotype of a normal cell from a human male.

In addition, cancer cells often include extra-chromosomal forms of aneuploid segments of chromosomes, termed "amplicons", that are either microscopically detectable as "double minute" chromosomes (Heim and Mitelman 1995b, Sandberg 1990, Schimke 1984) or possibly "submicroscopic" (depending on the microscopic technique used) (Pauletti et al. 1990), with sizes as low as 1 megabase (Mb) (Nakao et al. 2004, Singer et al. 2000). But even extra- and intra-chromosomal amplicons (Liu et al. 1998) or deletions of only 1 Mb are still nearly as large as an entire *E. coli* chromosome of about 3 Mb. Aneuploidy is thus a much more massive genetic abnormality than the gene mutations that have also been found in cancer cells.

Finally, the functional consequences to the cell of chromosomal imbalance are readily measured. The abnormal expression of

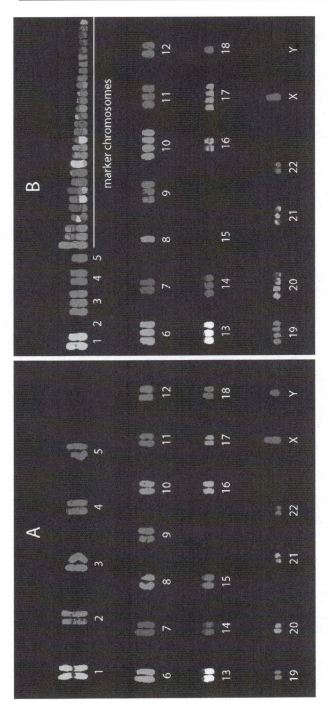

Fig. 4.1 Karyotypes of a normal and a highly aneuploid human cells.

Comparison of the two karyotypes shows that the normal cell (A) differs from the aneuploid cancer cell with numerous numerical and structural chromosomal alterations or aneusomies (B) (Duesberg et al. 2005, Rasnick 2002). Marker chromosomes (B) are structurally abnormal chromosomes, which are either rearranged intra-chromosomally or inter-chromosomally to form various hybrid chromosomes. Owing to their unique structure, marker chromosomes can serve as tracers for the origin of possibly metastatic cancer cells from primary cancers and for the origin of primary cancer cells from possibly aneuploid preneoplastic precursors (Koller 1972, Sandberg 1990).

Color image of this figure appears in the color plate section at the end of the book.

thousands of genes have been found to be proportional to the abnormal ploidy of the corresponding chromosomes in all cancer cells that have been tested by hybridizations of cellular RNAs with arrays of cellular genes (Aggarwal et al. 2005, Birkenkamp-Demtroder et al. 2002, Furge et al. 2004, Gao et al. 2007, Hertzberg et al. 2007, Masayesva et al. 2004, Pollack et al. 2002, Tsafrir et al. 2006, Upender et al. 2004, Virtaneva et al. 2001).

The Chromosomal Imbalance
Theory of Cancer

What can be the character of the alterations which render a cell and its progeny permanently neoplastic, and so self-assertive in behavior as to kill?

(Rous 1967)

These outcomes provide a direct test of the 100-year-old hypothesis that aneuploidy, a salient characteristic of solid tumors, drives tumorigenesis. The unambiguous answer is that not only can it do so, but it can also inhibit tumorigenesis and the cellular context is crucial. Both answers have important implications for human tumors.

(Weaver et al. 2007)

Cancer is a progressive somatic aneuploidy syndrome

The chromosomal imbalance theory (Table 5.1) states that initiation of aneuploidy coupled with the autocatalyzed progression of aneuploidy during cell division is necessary and sufficient to generate cancer on the rare occasions the cells survive. If the progression of aneuploidy continues unabated, cancer cells result, which behave much like new unicellular "species" with unique, ever-changing karyotypes and phenotypes. The very low frequencies of carcinogenesis reflect the very low probability of aneuploid cells acquiring reproductive autonomy—i.e. a new cellular species—by random karyotypic variations (Li et al. 2009). This mechanism explains the "conspicuously" long neoplastic latencies of carcinogenesis (Cairns 1978, Pitot 2002, Rous 1967, Yamagiwa and Ichikawa 1915).

Table 5.1 The chromosomal imbalance theory of cancer*

1. Cancer is a progressive somatic aneuploidy syndrome (Section 5.3).

2. Carcinogens and spontaneous mitotic errors produce nonspecific chromosomal alterations and aneuploidies (Section 5.3.2).

3. Aneuploidy is the steady source of karyotypic-phenotypic instability (Section 5.2).

4. The rate of chromosomal variations (chromosomal instability) is proportional to the degree of chromosomal imbalance (Section 5.3.9).

5. Since chromosomal alterations unbalance thousands of genes, they corrupt teams of proteins, including those that segregate, synthesize and repair chromosomes, simultaneously producing a heterogeneous mix of unique cellular and metabolic phenotypes (Section 5.3).

6. A gain in gene dosage over a substantial fraction of the genome is better tolerated than a loss (Hodgkin 2005, Lindsley et al. 1972). Therefore, hyperploidy is favored over hypoploidy (Section 5.3.11).

7. The survival advantage of the hyperploid cells (Point 6), coupled with the inherent chromosomal instability of aneuploid cells (Point 3), leads to the autocatalyzed progression of aneuploidy with each cell division (Sections 5.3.11 & 6.2).

8. In classical Darwinian terms, selection of viable chromosomal alterations encourages the evolution and spontaneous progression of neoplastic cells (Section 6.6). Thus, cancer cells evolve through a self-perpetuating chromosomal disorganization, which increases karyotypic entropy up to a maximum compatible with viability (Section 6.2.3).

*Chromosomal is defined here primarily by what is seen microscopically by classical cytogenetics (Heim and Mitelman 1995b, Sandberg 1990). It also includes amplicons, or deletions of chromosomes down to about 1 Mb, which are "submicroscopic" according to some (Pauletti et al. 1990) but microscopic according to other more recent techniques such as comparative genomic hybridization and gene array-based hybridization (Nakao et al. 2004, Pollack et al. 2002, Singer et al. 2000).

5.1 HEURISTIC EXPLANATION OF HOW CHROMOSOMAL IMBALANCE (AND NOT GENE MUTATION) GENERATES CANCER PHENOTYPES

It is perhaps best to begin with a global discussion of the conceptual divide separating the aneuploidy theory from gene mutation. The human genome project was driven substantially by the belief that there should be a fairly unique mapping (to use the mathematical concept) of the genotype onto the phenotype. This is conveyed by the idea of genetic programs determining phenotypes. If this were true, it should be possible to deduce or derive a cancer phenotype—or even whole organism—from a particular genome? Conversely, given the particular cancer (or organism), it should be possible to infer its genome? The various genome projects and countless micorarray experiments have proved this wrong.

According to the cancer researcher Vogelstein, there is no "normal [animal] cell with an abnormal karyotype" (Marx 2002). Thus, the complex aneuploidies can be expected to generate numerous abnormal phenotypes, which may include cancer. An analogy shows how unbalancing components changes characteristics. By unbalancing chromosomes, aneuploidy has the same effect on the phenotypes of cells as disrupting the assembly lines of a car factory on the characteristics and functioning of an automobile. Changes of assembly lines that essentially maintain the balance of existing components, alias genes, generate new, competitive car models. For example, the engine could be moved from the back to the front via adjustments in assembly lines without changing the balance of genes. Similarly, phylogenesis generates new species by regrouping old genes of existing species, without unbalancing the genome, into new numbers and structures of chromosomes (O'Brien et al. 1999).

However, if changes of assembly lines are made that alter the long-established balance and thus the stoichiometry of many components (e.g., genes), abnormal and defective products must be expected, as for example cars with five wheels or humans with trisomy of chromosome 21, which causes Down syndrome (Epstein 1986, Shapiro 1983). Although trisomy 21 is only a tiny aneuploidy compared to that of cancers (Sandberg 1990), it generates a spectrum of 80(!) Down syndrome phenotypes (Mao et al. 2003, Reeves 2000). Likewise, experimentally induced, congenital aneuploidies generate numerous abnormal phenotypes in drosophila, plants and mice, independent of gene mutation (Hernandez and Fisher 1999, Kahlem et al. 2004, Laffaire et al. 2009, Lindsley et al. 1972, Liu et al. 1998, Lyle et al. 2004, Matzke et al. 1999, Pavelka et al. 2010a).

By contrast, the effects of changing the phenotypes of the cell by mutation, without altering the karyotype, are much more limited than those resulting from changing the karyotype. Mutation without altering the karyotype is analogous to changing specific components of an existing car model: there could either be positive mutations, such as an improved carburetor, or negative mutations such as an unreliable ignition, or neutral mutations such as a new color. None of such mutations would generate an exotic new car model with unrecognizable properties. Indeed, the 1.42 million (Sachidanandam et al. 2001) or 2.1 million (Venter et al. 2001) point mutations that distinguish any two humans have not produced a new human species, nor have they even been sufficient to cause cancer in newborns. In view of this, such mutations are euphemistically called "polymorphisms".

Moreover, the functions of genes in biological assembly lines are strongly buffered against mutations: exceedingly rare activating mutations (Kacser and Burns 1981) are buffered down by normal supplies, and inactivating mutations are kinetically activated by increased supplies from un-mutated components of the assembly line (Cornish-Bowden 1999, Hartman et al. 2001, Kacser and Burns 1979, Rasnick and Duesberg 1999). But, there is no such buffering against aneuploidy. Thus, aneuploidy is inevitably dominant (Rancati et al. 2008), whereas mutation is nearly always recessive (Vogel and Motulsky 1986). It is for this reason that gene mutations could never generate new phylogenetic species or even new cancer cell-species in the absence of karyotypic alterations.

The remainder of this chapter and the next show how the chromosomal imbalance theory provides a coherent explanation of carcinogenesis that is independent of mutation, and explains each of the many distinctive features of carcinogenesis that are paradoxical from the standpoint of the gene mutation theory.

5.2 ANEUPLOIDY CAUSES CHROMOSOMAL INSTABILITY—THE HALLMARK OF CANCER

...increased ploidy per se, without extra centrosomes, can result in genomic instability.

(Storchova et al. 2006)

Multiple mutational and epigenetic mechanisms have been identified in which tumour cells lose mitotic fidelity; it could be argued that gene dosage effects caused by aneuploidy might be included among them so that, if it is obvious that CIN generates aneuploidy, it is also possible that aneuploidies generate CIN.

(Pacchierotti and Eichenlaub-Ritter 2011)

Shen recently declared: "[T]he accumulation of genomic alterations is not only a hallmark but also a driving force for tumorigenesis" (Shen 2011). "Accumulation" implies a rate of change of karyotype during cell division, which can be measured (Camps et al. 2005, Duesberg et al. 2000a, Fabarius et al. 2008, Klein et al. 2010, Lengauer et al. 1997, Li et al. 2009, Nicholson and Duesberg 2009). Most genomically unstable cells, above all cancer cells, differ from normal cells not only in abnormal chromosome numbers, but also in abnormal chromosome structures and gene mutations (Breivik and Gaudernack 1999, Grosovsky et al. 1996, Heim and Mitelman 1995b, Lengauer et al. 1998, Marx 2002, Matzke et al. 2003, Murnane 1996, Pihan and Doxsey 2003, Schar 2001). Thus, there is a coincidence of multiple mechanisms of genomic instability, particularly in cancer cells—one that alters chromosome numbers—one that rearranges chromosome structures and simultaneously some genes at the respective breakpoints—and one that mutates individual genes (Duesberg et al. 2004b).

Arbitrary categories are frequently considered conceptually and functionally unrelated types of genomic instability: chromosomal (CIN), microsatellite (MIN), and loss of heterozygosity (LOH). Among these, chromosomal instability (CIN) is the most prevalent form (Mitelman 1994, Mitelman 2011). Microsatellites are repeated sequences of DNA. Although the length of these microsatellites is highly variable from person to person, each individual has a set length. The appearance of abnormally long or short microsatellites in an individual's DNA is referred to as microsatellite instability (MIN). MIN is a condition manifested by damaged DNA due to defects in the normal DNA repair process. It used to be thought that the presence of CIN and MIN were mutually excluded in the same cancer but this has been shown to be wrong (Camps et al. 2006, Muleris et al. 2008). So-called LOH is due to chromosomal loss, with its putative significance being to increase the likelihood of deleting hypothetical tumor suppressor genes. It is telling that current gene mutation models of tumorigenesis only superficially consider the consequences of chromosome gain (Muleris et al. 2008).

Despite the importance of CIN for tumor initiation and progression, it is surprising that CIN is poorly defined and that the use of CIN is frequently inconsistent and imprecise. For example, CIN is used to describe cancers that are shown, by cytogenetics or flow cytometry, to have an aneuploid or polyploid karyotype. It has also been used to describe cells that harbor multiple structural chromosomal rearrangements. Others describe CIN as frequent alterations in chromosome number (Geigl et al. 2008). Many theories have been proposed to account for the timing and consequences of genomic instability in cancer, such as the mutator phenotype (Loeb 2001), telomere dysfunction (Artandi et al. 2000) and chromosomal imbalance (Duesberg et al. 2000a).

Genomic instability can be triggered by a change in chromosome number arising from either whole genome duplications (polyploidy) (Andalis et al. 2004, Mayer and Aguilera 1990, Song et al. 1995,

Storchova and Pellman 2004) and loss or gain of individual chromosomes (aneuploidy) (Duesberg et al. 1998, Duesberg et al. 2004b, Fabarius et al. 2003, Matzke et al. 1999, Storchova and Pellman 2004). This genome instability is manifested as rapid structural and epigenetic alterations that can occur somatically or meiotically within a few generations after heteroploid formation. The intrinsic instability of newly formed polyploid and aneuploid genomes has relevance for genome evolution and human carcinogenesis, and points toward recombinational and epigenetic mechanisms that sense and respond to chromosome numerical changes (Matzke et al. 1999).

A research group in Tel Aviv recently published a series of papers showing various constitutional autosomal trisomies (10 patients with trisomy 21, 2 with trisomy 18, and 2 with trisomy 13, 8 women with monosomy X) were associated with an increased frequency of non-chromosome-specific aneuploidy during cell division (Reish et al. 2006, Reish et al. 2011). The authors concluded: "our findings support the hypothesis that aneuploidy itself catalyses chromosomal instability. Constitutional aneuploidy, either X chromosome monosomy or autosomal trisomy arising from a single event in the early life of an organism, may give rise to new sporadic non-chromosome-specific losses of whole chromosomes…. It is possible that sporadic aneuploidy presented here in cells derived from constitutional trisomies is accountable for driving tumorigenesis, leading to initiation/progression of malignancy in the patients" (Reish et al. 2011).

Huettel et al. observed that experimentally induced trisomy of chromosome 5 in *Arabidopsis thaliana* led to truncated derivatives of the triplicated chromosome and disrupted the genome in a number of ways. The authors reasoned: "The trisomic chromosome may be vulnerable to breakage, particularly in vicinity of repetitive regions, and a truncated chromosome is more likely to be retained when two intact copies are present. The possibility of structural as well as numerical deviations in aneuploids underscores the need to perform array CGH for proper analysis and interpretation of the transcriptome data (Zanazzi et al. 2007). The formation and inheritance of chromosome structural variants in aneuploids might have evolutionary implications if restructured chromosomes are transmitted to progeny and eventually fixed in the population (Matzke et al. 1999). Enhanced structural instability of aneuploid genomes in somatic cells could have relevance for human cancer cells, which display progressive chromosome numerical and structural changes as the tumour evolves (Matzke et al. 2003, Nowell 1976)" (Huettel et al. 2008).

Huettel et al. further showed trisomy of one chromosome causes complex changes in gene expression. For example, most genes on trisomic chromosome 5 showed higher expression reflecting a dosage effect, but cases of apparent dosage compensation and even down-regulation were also observed. Trisomy of chromosome 5 perturbed expression to a lesser degree across the genome. "Genes involved in responses to stress and other stimuli were overrepresented among genes differentially regulated relative to the average chromosome trends, and transcription factors were overrepresented in the trans effects. The use of qRT-PCR to analyze expression of single genes demonstrated variable expression depending on the chromosome number and constitution, and on the features of individual genes... . The observed variations in gene expression probably depend on multiple factors including, but not limited to, changes in the dosages of regulatory molecules and epigenetic factors, and sensitivity of repetitive regions to copy number changes and gene silencing mechanisms" (Huettel et al. 2008).

Cancers are initiated and maintained by individual cells with aneuploid karyotypes, much like new species (Section 6.6). Such cancer-causing karyotypes are in flexible or dynamic equilibrium—destabilized by random aneuploidy and stabilized (within narrow limits of variation) by selection for viability-permitting oncogenic function. Together, the two competing forces form quasi-stable cancer-causing karyotypes depicted as red zones in Figure 5.1. At the same time, destabilizing aneuploidy generates non-neoplastic and nonviable variants, yellow zones in Figure 5.1. Occasionally, karyotypic variants evolve that encode new transforming functions such as drug-resistance (Duesberg et al. 2000a, Duesberg et al. 2001b,

Klein et al. 2010, Li et al. 2005, Swanton et al. 2009) or metastasis—processes that are typically part of tumor progression (Foulds 1969, Thompson and Compton 2010, Warth et al. 2009). Examples such as acquired drug-resistance or metastasis are depicted as branching red zones in Figure 5.1 (Sections 5.3.5 & 6.1.9).

Recently, Li et al. have shown that despite the inherent instability of aneuploidy and despite the added instability imposed by SV40-

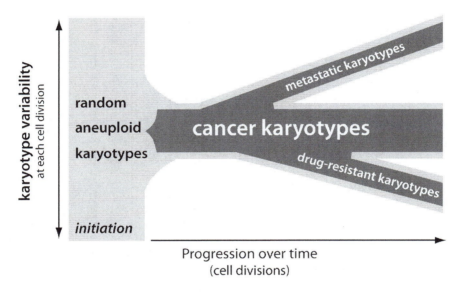

Progression over time
(cell divisions)

Fig. 5.1 Quasi-stable cancer and precancer karyotypes.
Cancers are initiated and maintained by individual cells with aneuploid karyotypes, much like new species. The tip at the left of the red zone signals the origin of the cancer-causing karyotype from an initiation pool of cells consisting of random aneuploid non-neoplastic karyotypes (yellow). The cancer-causing karyotypes are in flexible or dynamic equilibrium—destabilized by random aneuploidy and stabilized (within narrow limits of variation) by selection for viability and oncogenic function. Together, the two competing forces form quasi-stable average cancer-causing karyotypes (red zones) as the populations proliferate. The range of variability of the cancer-causing karyotypes is always accompanied by a range of non-neoplastic aneuploid variants (yellow zones). Occasionally, stochastic karyotypic evolutions generate new cancer-specific phenotypes, such as drug resistance and metastasis in a process termed tumor progression.
Color image of this figure appears in the color plate section at the end of the book.

activated genes used to induce transformation, the karyotype as a whole was selected for oncogenic function (Li et al. 2009). In the absence of virus-activated aneuploidogenic genes, the karyotypes of cell lines derived even from highly aneuploid cancers were considerably more stable (Li et al. 2005, Roschke et al. 2002) than those of the transformed clones (Li et al. 2009) and primary tumors. For example, Duesberg et al. found that between one third and two thirds of the cells of the established human colon and breast cancer lines HT29, SW480, and MDA-231 had identical whole-chromosome karyotypes over many generations, despite coexistence of some unstable marker chromosomes (Duesberg et al. 2006, Li et al. 2005).

It has been known since Boveri's discovery of the individuality of chromosomes that aneuploidy typically inhibits and impairs growth and development of non-cancerous cells and organisms (Harris 2007, Hauschka 1961, Lindsley et al. 1972). Recently, the adverse effects of aneuploidy on normal growth and development have been reinvestigated and extended to genetically altered animals (Cetin and Cleveland 2010, Holland and Cleveland 2009, Hughes et al. 2000, Pavelka et al. 2010b, Torres et al. 2010, Weaver and Cleveland 2007, Williams and Amon 2009). Because of the adverse consequences, it has been argued that aneuploidy is incompatible with cancer, even called "cancer's fatal flaw" (Williams and Amon 2009). The fact that aneuploidy impairs normal growth and development but is also ubiquitous in cancer has been called a "paradox" (Cetin and Cleveland 2010).

Trying to resolve the paradox, several researchers have recently concluded there must be aneuploidy-tolerating mutations in cancer cells (Baker and van Deursen 2010, Holland and Cleveland 2009, Torres et al. 2010, Williams and Amon 2009). Consistent with this line of reasoning, it was suggested that "identifying genetic alterations that permit cells to tolerate aneuploidy...will provide important insights into tumor evolution" (Williams and Amon 2009). However, this view does not take into account the perpetual karyotypic-phenotypic heterogeneity of the tumor cells that will vary under different selective conditions at much higher rates compared to conventional mutations in normal and cancer cells (Duesberg and Li 2003, Shibata 2011), the consequences of which are described in the next section.

5.3 CANCER IS A PROGRESSIVE SOMATIC ANEUPLOIDY SYNDROME

Interest has been growing in recent years in experimentally addressing what role aneuploidy plays in carcinogenesis. A flurry of experiments in yeast has led researchers to opposite conclusions. Recognizing the consequences to cancer research of which interpretation was correct, Judith Berman juxtaposed the results of two research groups in a short article in Nature, titled "When abnormality is beneficial" (Berman 2010). An extended quote from her commentary is worth repeating here:

> [T]he paper by Pavelka et al. (Pavelka et al. 2010b) adds fuel to a long-standing controversy, over whether aneuploid chromosomes are good or bad for cell proliferation, and highlights the point that 'good' and 'bad' are relative terms that are highly dependent on the conditions under which they are measured.
>
> Pavelka et al. exploited the facile genetics of the budding yeast Saccharomyces cerevisiae to produce a set of aneuploid strains in an unbiased manner. They first constructed strains that had three or five complete sets of whole chromosomes (triploids and pentaploids), instead of the usual two (diploids), and then induced them to undergo meiotic cell division. The odd numbers of starting chromosome sets ensured that a high frequency of spores would carry multiple aneuploidies. Other strengths of the study were the large number of genetically identical aneuploid strains generated (38), and the focus on strains that were stable and had undergone few cell divisions.
>
> The authors analysed the progeny soon after birth—before single-nucleotide mutations could accumulate, as assessed by whole-genome deep sequencing. They found that most strains had decreased growth rates in nutrient-rich media, as well as under several stress conditions. Notably, however, most of the aneuploid strains grew faster than their parent strain on transfer to at least one stress condition, such as exposure to a chemotherapeutic agent or an antifungal drug. So it seems that some combinations of aneuploid chromosomes proliferate better under stress conditions, despite having had no prior exposure to that condition.

Intriguingly, in a number of cases, different constellations of aneuploid chromosomes conferred a similar growth advantage. This indicates that there is more than one way to get the job done. Whole-genome deep sequencing confirmed that no single-nucleotide mutations had accumulated in the five isolates Pavelka et al. analysed for messenger-RNA and protein composition; thus aneuploidy alone was sufficient to confer the growth advantages.

Berman went on to say the results of Pavelka et al. "reaches different conclusions to those of Torres et al. (Torres et al. 2007, Torres et al. 2010), who reported that a single-nucleotide mutation in a deubiquitinating enzyme, which arose during the evolution of one aneuploid isolate, leads to improved proliferation of a few, but not all, strains. Pavelka et al. find that aneuploidies alone—without any mutations—can confer improved growth under some stress conditions. Both groups agree that, under conditions optimized by geneticists for growth in conventional laboratories, aneuploid cells usually divide less rapidly than cells with the normal chromosomal complement."

Berman then compared the experimental differences pointing to…

An important distinction in the methods used to generate the aneuploid strains [that] might explain the differences in the findings. Torres et al. engineered yeast cells with a haploid (single) set of chromosomes to carry one extra chromosome and then selected for faster growth using conventional lab conditions for 9–14 days—a time frame during which mutations are expected to accumulate. By contrast, Pavelka et al. analysed strains that often carried multiple aneuploid chromosomes and, importantly, minimized the number of generations before analysis.

Pavelka and co-workers also directly address a controversy concerning the role of excess proteins in aneuploid cells. Previously, Torres et al. (Torres et al. 2007) proposed that there is a specific set of genes and proteins that are regulated in response to aneuploidy in general. In their more recent study (Torres et al. 2010), they showed that some 20% of proteins exhibit levels that do not track with gene copy number, and that a large proportion of these proteins are members of macromolecular complexes. By contrast, other groups (Geiger et al. 2010, Springer et al. 2010) have found that the levels of most proteins generally reflect changes in chromosome copy number and that less than 5% of the proteins exhibit 'dosage compensation'—whereby the relative protein level is independent of gene-copy number. Pavelka et al. specifically test this hypothesis by quantitative mass spectrometry of about

2,000 proteins in each of five aneuploid strains and do not find compelling evidence for specific dosage compensation of protein-complex components.

Overall, these studies are consistent with the idea that aneuploidy is not a single, unique state and that all aneuploid strains do not share a single, common phenotype or protein profile. Rather, different aneuploid strains use different mechanisms for optimal growth under different conditions. This conclusion may be less satisfying than a single, simple answer, especially given the crucial implications for cancer cells: it remains unclear whether cancer cells divide uncontrollably because they are aneuploid and/or because they have accumulated mutations that allow them to tolerate aneuploidy. But it should be remembered that work on cancer cells themselves (Weaver and Cleveland 2007) suggests that not all aneuploidies are equal: aneuploidy can either promote or inhibit tumorigenesis, depending on the context. Pavelka and colleagues' work therefore supports the idea that, whereas mutations can facilitate the proliferation of aneuploid cells, aneuploidy itself can be sufficient to provide a growth advantage under a broad range of stress conditions.

5.3.1 Exact correlations between aneuploidy and cancer

Chromosomal alterations, alias aneuploidy, are ubiquitous in cancer (Atkin and Baker 1966, Atkin and Baker 1990, Hansemann 1890, Heim and Mitelman 1995b, Koller 1972, Sandberg 1990) (Section 4.4.13). However, Pagnigrahi and Pati warn, "experimental science does not recognize the notion of 'correlative proof of causation'. Two events may correlate, coexist and even co-progress from the initiation to the climax of a process; but no causal relationship can be inferred from such correlation" (Panigrahi and Pati 2009). But it is also obvious a causal relationship *cannot* be inferred in the absence of an extremely high correlation between a true cause and its effect. Thus, the objection of Pagnigrahi and Pati applies with greater force to specific gene mutations as causing cancer—indeed, to the multitude of genome wide association studies (Katsios and Roukos 2011, Manolio et al. 2009, Visscher and Montgomery 2009, Wacholder et al. 2010)—because of the much lower correlations that come nowhere near that of aneuploidy (Sections 4.4.2 & 4.4.3). The exact correlations with cancer is very powerful circumstantial evidence that aneuploidy is necessary for carcinogenesis, and particularly compelling since confirmed cases of diploid cancer, in which the tumor cells have balanced chromosomes, do not exist (Section 4.4.4). Even though exact correlations between cancer and aneuploidy have been reported since 1890, aneuploidy is currently not even mentioned in the cancer chapters of the leading textbooks of biology (Alberts et al. 1994, Cairns 1978, Lewin 1997, Lodish et al. 2004, Pierce 2005).

5.3.2 Origin of aneuploidy

As Ohno so confidently asserted, "malignant cells are either overtly or covertly aneuploid" (Ohno 1971). But what is the source of the aneuploidy seen in cancer? Aneuploidy may arise either spontaneously or by chemical induction of chromosome loss or gain during cell division in germ and somatic cells. Carcinogens such as the highly aneuploidogenic (aneugenic) virus-activated genes, or less efficient chemical or physical carcinogens and the chromosome instability syndromes (Section 5.3.6), all induce aneuploidy (Duesberg et al. 2000b, Duesberg et al. 2004b, Duesberg 2005, Duesberg et al. 2005, Duesberg 2007, Li et al. 2009, Nicholson and Duesberg 2009, Oshimura and Barrett 1986, Rajagopalan and Lengauer 2004, Saggioro et al. 1982).

All trisomies and most monosomies are thought to be generated by non-disjunction, which is the failure of sister chromatids in mitosis or of paired chromosomes in meiosis to migrate to opposite poles at cell division (Aardema et al. 1998). Cytokinesis failure has been suggested as a mechanism responsible for aneuploidy in cancer cells (Gisselsson 2005, Masuda and Takahashi 2002). Indeed, a number of studies have shown that polyploidy, induced by experimentally inhibiting cytokinesis, can lead to malignant transformation and tumorigenesis (Fujiwara et al. 2005, Nguyen et al. 2009). However, as Silkworth et al. (Silkworth et al. 2009) and Ganem et al. (Ganem et al. 2009) recently demonstrated, cytokinesis failure per se would not be sufficient to explain the rates of chromosomal instability in cancer cells. These authors proposed instead that multipolar spindle assembly followed by spindle pole coalescence represents a major mechanism of chromosome instability. The two groups showed that merotelic kinetochore (one kinetochore of a mitotic chromosome attached to both poles of the mitotic spindle) attachments can easily be established in multipolar prometaphases. Most of these multipolar prometaphase cells would then bi-polarize before anaphase onset, and the residual merotelic attachments would produce chromosome mis-segregation due to anaphase lagging chromosomes (Ganem et al. 2009, Silkworth et al. 2009).

5.3.3 Carcinogens induce aneuploidy

Studies which investigated the function of carcinogens have shown they cause aneuploidy (Duesberg et al. 2000b, Duesberg and Rasnick 2000, Duesberg and Li 2003, Duesberg et al. 2004b, Fabarius et al. 2002). When cultured human lung cells are exposed to asbestos, individual fibers are engulfed into the cytoplasm where they induce significant mitotic aberrations leading to chromosomal instability and aneuploidy. MacCorkle et al. demonstrated that intracellular asbestos fibers induce aneuploidy and chromosome instability by binding to a subset of proteins that include regulators of the cell cycle, cytoskeleton, and mitotic process. Moreover, pre-coating of fibers with protein complexes efficiently blocked asbestos-induced aneuploidy in human lung cells without affecting their uptake by cells (MacCorkle et al. 2006). Their results strongly indicate that aneuploidy occurs largely via protein-binding/sequestration/anchoring that likely interfere with the dynamics and regulation of spindle assembly/disassembly and chromosome movement. In addition to protein binding, fiber size has been clearly shown to be an important factor that correlates with the genotoxicty of asbestos. Smaller fibers have been shown to enter lung cells, where they associate with the cytoskeleton, are transported along microtubules, and can physically interfere with mitotic apparatus assembly and chromosome segregation (Ault et al. 1995, Cole et al. 1991, Jensen et al. 1996, MacCorkle et al. 2006).

Oncogenic viruses induce extensive reversible diploid hyperplasias soon after infection *in vitro* and *in vivo* (Ponten 1976) (Section 4.2). If the hyperplasias are allowed to persist, transformation to malignancy usually occurs. The relatively high odds of 1 per 100,000 cells per month and short latent periods (the origin of the clonal stem cell preceded colony formation by 1–2 months) of viral transformation reflect the high and persistent levels of karyotypic fluidity achieved by the virus-activated genes, compared with the relatively low fluidity of cells rendered aneuploid by conventional carcinogens (Li et al. 2009). This is because the virus-activated genes are self-replicating and thus permanent, in contrast to transient non-biological carcinogens (Li et al. 2000). The high karyotypic fluidity of cells carrying aneuploidogenic viral genes also explains their high mortality from fatal karyotypes, (Hassold 1986, Hernandez and Fisher 1999, King 1993, Li et al. 2009, Saksela and Moorhead 1963, Torres et al. 2008, Weaver et al. 2007).

The diversity of the cancers induced by the non-viral aneugens holds true for SV40 and other polyomaviruses (Ahuja et al. 2005, Li et al. 2009, Tooze 1973). Indeed, the neoplastic diversity of SV40 tumors in animals has recently been called a "cellular uncertainty principle" (Ahuja et al. 2005). The same is found to a lesser degree for the diversity of tumors induced by 6 retrovirus-activated genes (Fabarius et al. 2008, Kendall et al. 2005). Viral and non-viral aneugens are necessary only for the initiation of aneuploidy and transformation but not its maintenance (Duesberg 2003, Ewald et al. 1996, Li et al. 2009, Rous 1967, Sotillo et al. 2007). As soon as aneuploidy is induced, its autocatalytic aspect persistently leads to ever-changing candidate karyotypes (Section 5.3.11).

5.3.4 Carcinogenesis from aneuploidy is much more probable than via mutation

The genetic targets of carcinogens are over 1000 times larger than a gene, and thus equivalent to the size of chromosomes (Duesberg et al. 2004b). Since mutagenic carcinogens, like radiation, mutate, at half-lethal doses, a specific gene in only 1 out of 10^5–10^6 cells (Grosovsky et al. 1996, Shapiro 1983), but induce genomic instability in 1 out of 2–30 surviving cells (Duesberg et al. 2004b), the target causing genomic instability is about 10^3–10^6 times bigger than a gene. It follows that a chromosome, rather than a gene, is the target of these carcinogens. Moreover, non-mutagenic carcinogens can neither generate mutations nor aneuploidy by attacking DNA, because they are not "genotoxic". But, non-mutagenic carcinogens, as for example the polycyclic hydrocarbons, cause aneuploidy by corrupting the spindle apparatus (Ashby and Purchase 1988, Berenblum and Shubik 1949, Burdette 1955, Duesberg et al. 2000a, Duesberg et al. 2004b, Lijinsky 1989, Little 2000, Oshimura and Barrett 1986, Pitot 2002, Preussman 1990, Rous 1959, Scribner and Suess 1978, Zaridze et al. 1993). Thus initiation of carcinogenesis is not dependent on gene mutation.

5.3.5 Multi-drug resistance and other complex phenotypes are more probable from aneuploidy than mutation

Chromosomal variation alters cancer-specific phenotypes at rates that are 4 to 11 orders magnitude faster than conventional gene mutation (Duesberg et al. 2005, Nicholson and Duesberg 2009). Indeed, cancer based on spontaneous, somatic mutation would practically not exist (Section 4.4.11). Thus, phenotype variation in cancer cells is independent of mutation.

Genomic plasticity intrinsic to aneuploidy is a major source of drug resistance seen in pathogenic yeast (Polakova et al. 2009, Selmecki et al. 2009) and the parasite *Leishmania* (Leprohon et al. 2009, Ubeda et al. 2008). In view of this and the massive aneuploidy driving cancer cells, chromosomal alterations have been proposed as the cause of multi-drug resistance (Duesberg et al. 2000a, Duesberg et al. 2001b, Klein et al. 2010, Lee et al. 2011, Li et al. 2005, Swanton et al. 2009, Thompson and Compton 2010). To test this hypothesis Duesberg et al. pursued two experimental questions: First, could aneuploid mouse cells from which multi-drug resistance genes had been deleted (Allen et al. 2000) still become drug-resistant? In accordance with their prediction they found that aneuploid mouse cells minus all known multi-drug resistance genes do indeed become multi-drug resistant (Duesberg et al. 2000a, Duesberg et al. 2001b). Second, does drug-resistance correlate with resistance-specific chromosomal alterations? Indeed, this too was confirmed recently (Li et al. 2005). They concluded that multi-drug resistance is chromosomal and thus multigenic in origin.

In further support for a genetic basis of the phenotype–karyotype correlations of cancers, several researchers have recently found that the gene expression profiles of thousands of normal genes are directly proportional to the copy numbers of the respective chromosomes (Birkenkamp-Demtroder et al. 2002, Furge et al. 2004, Gao et al. 2007, Hertzberg et al. 2007, Masayesva et al. 2004, Pollack et al. 2002, Tsafrir et al. 2006, Upender et al. 2004). In other words the individual karyotypes determine the individual phenotypes of cancers. This is consistent with the complex phenotype–karyotype relations of aneuploidy syndromes and normal species (O'Brien et al. 1999).

5.3.6   Preneoplastic aneuploidy

As recently as 2005, Harris declared, "Most cytogeneticists would take the view that the temporal relationship between the appearance of aneuploidy in a tissue and the emergence of a tumor had long ago been settled in the 1960s and 1970s" (Harris 2005). In 1959, Levan reviewed the experimental evidence and came to the conclusion that aneuploidy reflected continuous structural remodeling of the karyotype until eventually a variant was produced that could grow progressively *in vivo* (Levan 1959).

Intrigued by the aneuploidy in cancer and the long neoplastic latencies, many researchers have analyzed cancer-prone tissues for preneoplastic genetic and chromosomal alterations, particularly aneuploidy (Atkin 1997, Duesberg et al. 2004b, Gibbs 2003, Marx 2002, Sieber et al. 2003, Spriggs 1974). The first consistent evidence for preneoplastic aneuploidy was obtained by Caspersson in 1960s for cervical tissues (Caspersson 1964a), which has since been confirmed (Atkin 1997, Harris 2005, Spriggs 1974). Similar studies have also found aneuploidy prior to carcinogenesis in precancerous tissues and neoplasias of the head, throat, colon, lung, breast, skin, pancreas, prostate, gonads, esophagus and the cervix (Ai et al. 1999, Ai et al. 2001, Balaban et al. 1986, Böcking and Chatelain 1989, Bomme et al. 1998, Hermsen et al. 2002, Heselmeyer-Haddad et al. 2005, Hittelman 2001, Langenegger 2009, Looijenga et al. 1999, Luttges et al. 2001, Micale et al. 1994, Moskovitz et al. 2003, Nasiell et al. 1978, Pihan and Doxsey 2003, Pilch et al. 2000, Rabinovitch et al. 2001, Rubin et al. 1992, Saito et al. 1995, Shih et al. 2001, Umayahara et al. 2002, Willenbucher et al. 1999).

The meticulous karyological studies of Spriggs and colleagues (Spriggs et al. 1962, Spriggs et al. 1971, Spriggs 1974) and Stanley and Kirkland (Stanley and Kirkland 1968) on cancerous and pre-cancerous conditions of the human uterus established that: 1) aneuploidy was present in the epithelia long before any tumour appeared, and 2) when frank cancers did emerge, the mode of chromosome numbers narrowed and marker chromosomes appeared, supporting the view that the tumors were clonal outgrowths arising from a background of cells with disordered chromosome complements. Harris noted that, "The work of Spriggs et al. on carcinoma of the cervix uteri is of special interest in a clinical context. The variation in the size and shape of the cell nuclei seen in pre-cancerous epithelium is a reflection of the underlying aneuploidy. It is clear that karyological disorder may be present in the cervical epithelium whether a tumour is produced or not" (Harris 2005).

The presence of aneuploidy in histopathologically normal cells from the likely histologic origin of ovarian cancer suggests that this type of chromosomal instability is among the earliest of genetic events in the natural history of ovarian tumorigenesis (Pothuri et al. 2010). Even childhood lymphoblastic leukemia is preceded by the appearance of aneuploidy *in utero* (Section 4.4.13). Moreover, multinational epidemiological studies have found that the relative cancer risk of people can be predicted from the degree of chromosomal aberrations of peripheral lymphocytes (Bonassi et al. 2000, Hagmar et al. 1998).

On the experimental side, Rachko and Brand completed a detailed study in 1983, of chromosomal changes during neoplastic development in foreign body tumorigenesis in mice. Their experimental procedure allowed the isolation and characterization of definitive preneoplastic cells. Their studies indicated that preneoplastic cells at the earliest time they could be detected, perhaps at initiation, showed numerical changes in chromosomes (Rachko and Brand 1983). Experiments undertaken to study the origin of aneuploidy in animals treated with carcinogens have also found aneuploidy prior to cancer in the liver, skin and subcutaneous tissues of carcinogen-treated rodents (Bremner and Balmain 1990, Clawson et al. 1992, Conti et al. 1986, Danielsen et al. 1991, Fabarius et al. 2008, Marquardt and Glaess 1957, Rachko and Brand 1983, Van Goethem et al. 1995). Likewise, treatments of diploid human and animal cells *in vitro* with carcinogens were found to generate aneuploidy long before transformation. Unexpectedly, this preneoplastic aneuploidy proved to be variable in subsequent cell generations—creating "delayed" genomic instability or even "delayed reproductive death" (Barrett 1980, Benedict 1972, Connell and Ockey 1977, Connell 1984, Cooper et al. 1982, Cowell 1981, Duesberg et al. 2000b, Duesberg et al. 2004b, Fabarius et al. 2002, Freeman et al. 1977, Holmberg et al. 1993, Little 2000, Trott et al. 1995, Vanderlaan et al. 1983, Walen and Stampfer 1989, Wright 1999).

Aneuploidy also precedes transformation of human and animal cells infected by SV40 and other DNA tumor viruses (Li et al. 2000, Ray et al. 1990, Wolman et al. 1980). Even spontaneous transformation of cells *in vitro* is preceded by aneuploidy (Cram et al. 1983, Hayflick and Moorhead 1961, Levan and Biesele 1958). When Duesberg et al. tested preneoplastic aneuploidy with regard to its role in cancer, they found that experimental preneoplastic aneuploidy always segregated with subsequent morphological transformation and tumorigenicity (Duesberg et al. 2000b, Fabarius et al. 2002). Based on these results, they concluded that aneuploidy initiates carcinogenesis. This conclusion is directly supported by the high cancer risks of heritable chromosome instability syndromes and of congenital aneuploidies.

Chromosome instability syndromes

Heritable diseases that predispose to abnormally high rates of systemic aneuploidy, termed "chromosome instability syndromes", include Fanconi's anemia, Bloom's syndrome, Ataxia Telangiectasia, Xeroderma, Werner's and other syndromes. These chromosome instability syndromes also predispose to high rates of cancer and generate cancers at younger age than in normal controls (Section 6.1.3). In these syndromes, heritable mutations function as genetic aneugens and carcinogens (Duesberg et al. 2005).

Congenital preneoplastic aneuploidy

Minor congenital aneuploidies are viable, while major congenital aneuploidies are lethal (Dellarco et al. 1985, Hassold 1986). The best known examples are Down syndrome, Retinoblastoma, Wilms tumor, Klinefelter's syndrome and others, summarized by Sandberg (Sandberg 1990). Just like the chromosomal instability syndromes, the congenital aneuploidy syndromes carry high cancer risks and generate cancers at younger age than in diploid controls (Hasle et al. 2000, Koller 1972, Sandberg 1990). The 20–30 times higher-than-normal incidence of leukemia in Down syndrome is one of the best-studied examples (Hasle et al. 2000, Koller 1972, Sandberg 1990, Shen et al. 1995, Zipursky et al. 1994). The same is true for congenital aneuploidy in mice, in which an artificial duplication of only one megabase of chromosome 11 was found to induce lymphomas and other tumors after latencies of several months (Liu et al. 1998).

The gene mutation theory, on the other hand, neither predicts nor explains the presence of preneoplastic aneuploidy—except, perhaps indirectly, by postulating the generation of cancer genes via chromosomal rearrangements (Duesberg et al. 2005). Here again, the evidence for cancer-specific mutations is missing (Duesberg et al. 2004b). According to a recent review by Little, "While radiation-induced cancers show multiple unbalanced chromosomal rearrangements, few show specific translocations or deletions as would be associated with the activation of known oncogenes or tumor suppressor genes" (Little 2000).

5.3.7   Cancer-"specific" (non-random) aneusomies

Albert Levan summarized malignancy as the genetic adaptation of cells to a new mode of life (Levan 1969). He went on to say: "Although apparently haphazard, the chromosome variation in cancer has been demonstrated in all carefully analyzed instances to be governed by strict rules... . The chromosome variation in malignancy is of a specific kind: it generally oscillates around average karyotypes, and each cancer cell population is characterized by one predominant karyotype, the stemline karyotype, and in addition often one or more sideline karyotypes. It is true that during certain periods, the stemline may become less predominant, for instance after drastic environmental changes, as after explantation in tissue culture of a tumor cell population firmly adapted to the conditions of *in vivo* environment, but if the population survives long enough, a definitive stemline will again form. The fact that the chromosome variation in tumors is never haphazard but gathers around stemlines and sidelines is compatible with the idea that the development of each tumor takes place according to an evolutionary pattern: the most viable karyotype prevails at all times".

In a short article in 2007, Weaver and Cleveland concluded that, "aneuploidy resulting from chromosomal instability drives an increase in both benign and cancerous tumors, indicating that it is clearly not inconsequential. The long latency and incomplete penetrance of these tumors suggests that only a small subset of the large number of possible abnormal combinations of chromosomes is capable of inducing transformation. It also suggests that the chromosomal complements capable of transformation are more complex than gain or loss of one or a few chromosomes and require multiple generations of segregational errors to evolve" (Weaver and Cleveland 2007).

Despite the karyotypic instability of cancer cells and heterogeneity of cancers, partially specific or "nonrandom" chromosomal alterations, also termed aneusomies for individual chromosomes, have been found in cancers over the last fifty years

(Atkin and Baker 1966, Atkin 1986, Atkin 1991, Balaban et al. 1986, Bardi et al. 1993, Bardi et al. 1997, Fabarius et al. 2002, Heim and Mitelman 1995a, Johansson et al. 1994, Koller 1972, Oshimura et al. 1986, Pejonic et al. 1990, Pejovic et al. 1990, Virtaneva et al. 2001, Yamamoto et al. 1973, Zang and Singer 1967, Zang 1982). Specific aneusomies have even been linked to distinct events of carcinogenesis (Table 5.2). The majority of these nonrandom chromosomal alterations have been detected in cancers since the 1990s by the use of array-based genomic hybridization, rather than by identifying specific aneusomies cytogenetically (Dellas et al. 1999, Gebhart and Liehr 2000, Heim and Mitelman 1995b, Heselmeyer-Haddad et al. 1997, Jiang et al. 1998, Nupponen et al. 1998b, Patel et al. 2000, Petersen et al. 1997, Pollack et al. 2002, Reith 2004, Richter et al. 1998, Ried et al. 1999, Weber et al. 1998).

Table 5.2 Specific aneusomies have been linked with the following distinct events of carcinogenesis

Stages of human cancer	(Dellas et al. 1999, Fujimaki et al. 1996, Heselmeyer-Haddad et al. 2005, Hoglund et al. 2001b, Katsura et al. 1996, Koller 1972, Patel et al. 2000, Ried et al. 1999, Wilkens et al. 2004)
Invasiveness	(Hoglund et al. 2001b, Meijer et al. 1998, Wilkens et al. 2004)
Metastasis	(Al-Mulla et al. 1999, Aragane et al. 2001, Bockmuhl et al. 2002, Hermsen et al. 2002, Knosel et al. 2004, Nakao et al. 2001, Nishizaki et al. 1997, Petersen and Petersen 2001)
Drug-resistance	(Li et al. 2005, Schimke 1984, Tlsty 1990)
Transplantability to foreign hosts	(Hauschka and Levan 1953)
Cellular morphologies	(Vogt 1959)
Abnormal metabolism	(Hauschka and Levan 1958, Koller 1972)
Cancer-specific receptors for viruses	(Koller 1972, Vogt 1959)

Moreover, in cases where it has been tested, cancer-specific gene expression profiles are directly proportional to the dosages of the corresponding chromosomes (Aggarwal et al. 2005, Furge et al. 2004, Gao et al. 2007, Gruszka-Westwood et al. 2004, Lodish et al. 2004, Pollack et al. 2002, Upender et al. 2004, Virtaneva et al. 2001). Cancer-"specific" or nonrandom chromosomal alterations, however, are neither predicted nor explained by mutational theories of cancer. In fact, they are a direct challenge to the gene mutation theory, because specific chromosomal alterations generate specific phenotypes, independent of mutation. Down syndrome, with over 80 specific phenotypes caused by trisomy 21 without any gene mutation, is a classic example (Epstein 1986, Kahlem et al. 2004, Laffaire et al. 2009, Lyle et al. 2004, Mao et al. 2003, Reeves 2000, Shapiro 1983).

5.3.8 Clonal aneuploid karyotypes: stability within instability

Because the karyotypes of cancers are aneuploid and thus unstable (Section 5.2), their stability would depend on constant karyotypic selection for viability and oncogenic function. Owing to the inherent chromosomal instability of aneuploidy, the progeny of clonal cancers typically evolve sub-clonal and non-clonal aneusomies over time, generating karyotypic diversity within tumors (Duesberg et al. 2004b, Fabarius et al. 2003, Fabarius et al. 2008, Koller 1972, Nowell 1976, Reshmi et al. 2004, Wilkens et al. 2004, Winge 1930, Wolman 1986). Even so, the karyotypes of highly aneuploid cancers are clonally identifiable (Foulds 1969), hence stable over time, despite chromosomal instability (Balaban et al. 1986, Koller 1972, Loeper et al. 2001, Mitelman 2011, Reeves et al. 1990, Tsao et al. 2000, Wang et al. 2006, Wolman 1983). Wolman acknowledged this in an influential review in 1983, saying that the karyotypes of cancers are "surprisingly stable" despite "karyotypic progression" (Wolman 1983).

The HeLa cell line, which was derived from a human cervical cancer in 1951, is a primary example. The line has apparently maintained an average karyotype with a per cell chromosome number of around 78 and with line-specific chromosome copy numbers and marker chromosomes for over 50 years in cell culture (American Type Culture Collection 1992, Kraemer et al. 1971, Macville et al. 1999, Nelson-Rees et al. 1980). Reshmi et al. found that, even though the modal number of HeLa cells is conserved in cancer cell lines, the chromosomes within are "not necessarily the same" (Reshmi et al. 2004). Gusev et al. termed this paradox "stability within instability" (Gusev et al. 2001) and Albertson et al. commented in a review that "these cells do show substantial cell-to-cell variability but the average genotype is stable" (Albertson et al. 2003).

Researchers continue to report that the karyotypes of cell lines derived from human cancers are "unexpectedly" (Eshleman et al. 1998), "relatively" (Roschke et al. 2002), and "remarkably" (Albertson et al. 2003, Reshmi et al. 2004, Roschke et al. 2002) more stable than predicted from the chromosomal instability of cancer cells (Camps et al. 2005, Gorringe et al. 2005, Grigorova et al. 2004, Gusev et al. 2001, Heng et al. 2006b, Li et al. 2009, Nicholson and Duesberg 2009, Roschke et al. 2002). It was noted with surprise that human clear cell sarcomas maintained the same karyotypes in serial transplantations in nude mice (Crnalic et al. 2002), and that rat tumors maintained the same karyotypes over years of serial transplantations in rats (Nichols 1963, Wolman et al. 1977). Moreover, comparative genomic hybridizations and conventional cytogenetic analyses, tracing the karyotypes of individual cancers over multiple stages of carcinogenesis up to 17 years (Al-Mefty et al. 2004, Balaban et al. 1986, Kuukasjarvi et al. 1997, Loeper et al. 2001, Nupponen et al. 1998a, Reeves et al. 1990, Richter et al. 1998, Walch et al. 2000, Wang et al. 2006, Weber et al. 1998), have identified in primary lesions "karyotypes that were as complex as their paired relapses" (Jin et al. 2005).

The "infectious cancers" are fascinating natural examples of the immortality of "fully speciated cancers" (Section 6.6) with individual clonal karyotypes (Vincent 2010). For example, the "canine venereal tumor" (Makino 1974, Murgia et al. 2006) and the facial cancer of the Tasmanian devil (Pearse and Swift 2006) are naturally passed among histocompatible animals. The karyotypes of the canine and Tasmanian tumors are basically the same in all cases that have been tested. These cancer cells have thus been stable in countless natural transmissions—just like microbial parasites. According to Vincent, "The acquisition of germ line properties by cancer cells clearly indicates they have transcended the host and become something different" (Vincent 2010).

The karyotypic stability of cancers is especially evident from the clonality of chromosome-specific copy numbers revealed by comparative genomic hybridization (Baudis 2007, Gebhart and Liehr 2000, Kallioniemi et al. 1994b), and by clonal marker chromosomes (Koller 1972, Mitelman et al. 1997b). Figure 5.2 shows 20 primary bladder cancer cells had unique karyotypes (figure kindly provided by Peter Duesberg (Duesberg et al. 2011)). Nevertheless, the karyograph revealed regions of similarity. The most notable feature shared by every cell was the complete absence of intact chromosomes 1 and 13. The insert in Figure 5.2 shows the karyogram of one of the cancer cells lacking chromosomes 1 and 13. Cells cannot live without at least one copy of each chromosome (Mamaeva 1998). Therefore, chromosomes 1 and 13 appear as part of several marker chromosomes labeled in the red region on the abscissa in Figure 5.2.

Recently, Navin et al. developed a method called Sector-Ploidy-Profiling (SPP) to study the clonal composition of breast tumors (Navin et al. 2010). SPP involves macro-dissecting tumors, flow-sorting genomic subpopulations by DNA content, and profiling genomes using comparative genomic hybridization (CGH). Breast carcinomas display two classes of genomic structural variation: monogenomic and polygenomic. Monogenomic tumors appear to contain a single major clonal subpopulation with a highly stable chromosome structure. Polygenomic tumors contain multiple clonal tumor subpopulations, which may occupy the same sectors, or separate anatomic locations. In polygenomic tumors, heterogeneity can be ascribed to a few clonal subpopulations, rather than a series of gradual intermediates.

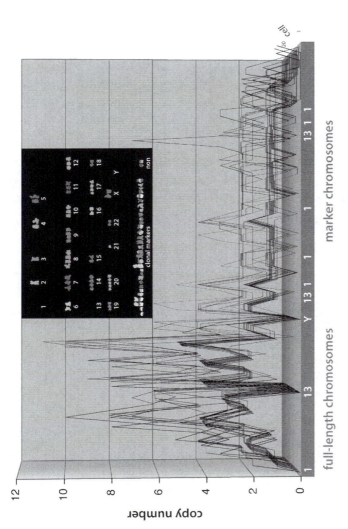

Fig. 5.2 Clonal aneuploid karyotypes indicate stability within instability.

Twenty cells from a primary bladder cancer had unique karyotypes. Nevertheless, the karyograph reveals regions of similarity. The abscissa is the full-length chromosome number (blue region) and the marker chromosome designation (red region). The ordinate is the number of copies of each chromosome. The z axis represents the metaphases of 20 different cells. The most notable feature shared by every cell was the complete absence of intact chromosomes 1 and 13. The insert shows the karyotype of one of the cancer cells lacking chromosomes 1 and 13 (blue region). Cells cannot live without at least one copy of each chromosome. Therefore, the missing chromosomes 1 and 13 appear as part of several marker chromosomes (red region).

Color image of this figure appears in the color plate section at the end of the book.

The clonal evolution models for tumor progression are consistent with the polygenomic tumor subpopulations. The primary assumption of the clonal evolution models (monoclonal and polyclonal) is that the majority of cancer cells are capable of unlimited proliferation. This assumption contrasts with the fundamental assumption of the cancer stem cell hypothesis (Tan et al. 2006), which states that only a rare subpopulation of tumor cells is capable of unlimited proliferation, while the vast majority are only capable of normal cell division potential. In the polygenomic tumors, Navin et al. observed that the majority of chromosome breakpoints are persistent throughout the tumor in all subpopulations, suggesting that the majority of cells are capable of unlimited proliferation.

5.3.9 Chromosomal instability is proportional to the degree of aneuploidy

The rates of specific chromosomal variations can reach 4–11 orders of magnitude higher than those of conventional gene mutations (Section 4.4.11). It follows that the karyotypic heterogeneity of cancers is a consequence of the inherent chromosomal instability of aneuploidy. In preliminary tests of the chromosomal imbalance theory, Duesberg et al. observed that aneuploidy catalyzes chromosomal variations in proportion to the degree of imbalance (Duesberg et al. 1998, Duesberg et al. 2000a, Fabarius et al. 2003, Li et al. 2005, Rasnick and Duesberg 1999). Indeed, numerous correlations have confirmed the principle that the degree of malignancy of cancer cells is proportional to their degree of aneuploidy (Balaban et al. 1986, Bardi et al. 1993, Böcking and Chatelain 1989, Camps et al. 2005, Castro et al. 2005, Castro et al. 2006, Choma et al. 2001, Duesberg and Li 2003, Duesberg et al. 2004b, Foulds 1975, Frankfurt et al. 1985, Fujimaki et al. 1996, Grimwade et al. 2001, Hauschka 1961, Hoglund et al. 2001b, Johansson et al. 1994, Katsura et al. 1996, Koller 1972, Kost-Alimova et al. 2004, Mitelman et al. 1997b, Nasiell et al. 1978, Nowell 1976, Pejovic et al. 1990, Roschke et al. 2003, Schoch et al.

2001, Spriggs 1974, Umayahara et al. 2002, Wilkens et al. 2004, Winge 1930).

5.3.10 Chromosomal instability drives cancer progression without gene mutation

It has been suggested that aneuploidy affects the rate of DNA replication in cells (Hand and German 1975), and increases genetic instability resulting in further chromosomal alterations (Goh et al. 1978, Kaneko et al. 1981) (Section 5.2). Aldaz et al. postulated that most papillomas progress to the carcinoma stage but at different speeds, which may be regulated by the degree of gross chromosomal abnormalities arising in these lesions (Aldaz et al. 1987, Conti et al. 1986). "The speed of such neoplastic progression may be related to specific chromosomal alterations. It is possible that random chromosomal changes occur constantly in papillomas and carcinomas, but the involvement by chance of specific chromosomes is what can confer a selective advantage to a particular clone, which becomes the stem line during a transient period of time in the life of the tumor. The possibility of finding this same event occurring simultaneously in different foci and at different rates of progression in the same tumor may be the reason the interpretation of karyotypic changes in solid tumors is complicated" (Aldaz et al. 1987).

Nevertheless, aneuploidy and other forms of chromosomal abnormality of cancer cells are generally considered as being "secondary" events (Harris 1995, Harris 2005, Heim and Mitelman 1995b, Johansson et al. 1996, Koller 1972)—secondary to hypothetical primary mutations (Harris 2005, Hede 2005, Knudson 2001, Lengauer and Wang 2004, Levan et al. 1977, Michor et al. 2005, Nowell 1976, Pennisi 1999, Rajagopalan et al. 2004, Rajagopalan and Lengauer 2004, Tomlinson and Bodmer 1999, Zimonjic et al. 2002). However, recent studies of the experimental induction of chromosomal instability have led some commentators to say, "these results emphasize the potential tumorigenic potential of aneuploidy, although the underlying mechanisms remain to be defined" (Chandhok and Pellman 2009). However, Boveri in 1914, had already offered an explanation of the origin of chromosome instability and its role in the progression of cancer: "[T]he creation of certain abnormal chromosome combinations so perturbs the equilibrium within the nucleus that particular chromosomes go on changing under the influence of changes in other chromosomes. One group of chromosomes might eventually preponderate and perhaps even suppress the activity of others. It is therefore understandable that a malignant tumor that is at first closely similar to its tissue of origin progressively becomes less so and eventually becomes completely unrecognizable" (Harris 2007).

Gao et al. observed that malignant progression is equivalent to the shifting of phenotypes. They demonstrated how changes in karyotype and subsequently the transcriptome can mediate the gene expression changes necessary for phenotypic determination (Gao et al. 2007). "[T]he fold increase or decrease in the chromosome content ratio is virtually the same as the transcriptome ratio for all comparisons, and the chromosome content transcriptome ratios over all chromosomes average approximately one. Even the sub-chromosomal regions of derivative chromosomes, when considered as part of the copy number of a specific chromosome, markedly influence the transcriptome ratio of that chromosome, and, dramatically, the numerical ratios of the chromosome content are virtually the same as the transcriptome of the specific chromosome region". The authors concluded "there is a direct quantitative correlation between chromosome content and a proportional change in the transcriptome.... . The chromosome content, therefore, in delivering the transcriptome in direct proportions, delivers a specific level of gene expression for each gene."

Grade et al. concluded very much the same thing in 2007: "We and others have...conducted analyses that allowed for the simultaneous mapping of genomic copy number changes and average gene expression levels in colorectal cancers and model systems thereof. These studies now suggest that chromosomal aneuploidies, and thus genomic imbalances, result in an alteration of transcriptional activity that is correlated to the variation in genomic copy number" (Grade et al. 2007).

5.3.11 Autocatalyzed progression of aneuploidy

Aneuploidy in general leads to less viable cells compared to euploid precursors (Boveri 1914, Hodgkin 2005, Lindsley et al. 1972, Rancati et al. 2008, Torres et al. 2007, Williams et al. 2008), and a gain in genetic material is better tolerated than a loss (Atkin and Baker 1990, Brewer et al. 1999, German 1974, Lindsley et al. 1972, Rancati et al. 2008, Rasnick and Duesberg 1999, Sandler and Hecht 1973, Torres et al. 2007). Therefore, hyperploidy is favored over hypoploidy. The survival advantage of the hyperploid cells, coupled with the inherent genetic instability of aneuploid cells (Section 5.2), leads to the autocatalyzed progression of aneuploidy with each cell division (Rasnick and Duesberg 1999, Rasnick 2000).

Recently, Reish et al. demonstrated that various constitutional autosomal trisomies are associated with an increased production of random aneuploidy (Reish et al. 2011). "[T]he altered replication pattern of disomic genes and sporadic non-chromosome-specific aneuploidy shown in monosomies or trisomies accords with the notion that an unbalanced chromosome complement alters genetic balance, including the proper control of chromosomal segregation (Antonarakis et al. 2004), and leads to a cascade of chromosomal alterations resulting in new abnormal chromosome complements... . These independent observations support previous evidence showing that aneuploidy is an autocatalytic process".

Experimentally induced lagging chromosomes produces chromosomal instability in various cancer cell lines. Thompson and Compton recently demonstrated that the rate of chromosome mis-segregation thus induced is similar between the near diploid RPE-1 and pseudodiploid HCT116 cells on the one hand and the triploid HT29, MCF-7 and Caco2 cells on the other (Thompson and Compton 2008). "These findings indicate that aneuploidy can cause CIN and that CIN may be a self-propagating type of genome instability" (Thompson and Compton 2010). Interestingly, similar rates of chromosome mis-segregation were sufficient to induce significant deviation in the modal chromosome numbers in colonies of the triploid cell lines but failed to do so in the near diploid RPE-1 and HCT116 cells. The authors concluded that, "Collectively, these data show that elevating chromosome missegregation rates alone in human cultured cells is not sufficient to convert stable, near-diploid cells into highly aneuploid cells with karyotypes that resemble those of tumor cells" (Thompson and Compton 2008). So the question is: how are viable highly aneuploid (triploid) cancer cells produced?

The 60–90 chromosomes that are typical of late-stage cancer cells (Shackney et al. 1995a) probably do not arise by a gradual, stepwise increase in the level of aneuploidy as a consequence of the autocatalyzed chromosomal instability of aneuploid cells. The stepwise progression of aneuploidy would, in general, produce fewer and fewer viable cells, that inhibit the progression to cancer. There is an alternative to the stepwise model of carcinogenesis that is based on chromosome doubling and subsequent loss of chromosomes (Giaretti and Santi 1990, Shackney et al. 1995a).

Oksala and Therman have described numerous routes to the production of polyploid cells (Oksala and Therman 1974). While tetraploid cells are comparatively stable, both *in vitro* and *in vivo*, they are, nevertheless, associated with a degree of chromosomal instability leading to random loss of chromosomes (Cooper and Black 1963, Giaretti and Santi 1990, Moorhead and Saksela 1965, Ohno 1971, Shackney et al. 1989, Shackney et al. 1995a, Shackney et al. 1995b, Shankey et al. 1993). The process of chromosome doubling followed by chromosome loss has been shown in numerous sequential studies of normal rodent fibroblasts undergoing spontaneous neoplastic transformation *in vitro,* in SV40 malignant transformation of normal human fibroblasts in culture, in established human epithelial cancer cell lines grown in tissue culture, and in sequential studies of early human bladder cancer (Shackney et al. 1995a).

Interestingly, near tetraploid cancer cell lines may continuously produce an appreciable number of near-diploid daughter cells (Erenpreisa and Cragg 2010, Erenpreisa et al. 2011, Ohno 1971), which in future generations revert back to the near tetraploid state. The production of pseudodiploid cancer cells from aneuploid precursors may contribute to the occasional reports that some malignant tumors apparently maintain the normal diploid complement of the species.

The hypertriploidy–hypotetraploidy characteristic of later-stage cancer cells (Wolman 1983) can either result directly from the tetraploidization of hypodiploid cells with DNA indices greater than 0.75 but less than 1, or alternatively, by the tetraploidization of diploid and hyperdiploid cells followed by chromosome loss (Giaretti and Santi 1990, Spriggs 1974). Tetraploidization preserves the nuclear balance of the near-diploid aneuploid cells, which should promote the viability of the cells with double the previous number of chromosomes. A DNA index of 0.75 appears to be the lower limit of viability for hypodiploid cells (Atkin and Baker 1966, Atkin 1974, Giaretti and Santi 1990, Shackney et al. 1995a). The tetraploidization of these barely viable cells produces the DNA index of 1.5 that is characteristic of the lower limit aneuploidy peak seen in flow cytometry studies of malignant cancer cells.

These results indicate that the two-step model of carcinogenesis (Pitot 1986) corresponds to two levels of aneuploidization. The initiation step in carcinogenesis is the production of non-cancerous, aneuploid cells with near diploid karyotypes. In such cells, the level of aneuploidy is below the threshold for cancer. In the promotion step (dependent upon cell proliferation) the threshold of aneuploidy for cancer is reached or exceeded by the tetraploidization of near diploid cells followed by loss of chromosomes. Regardless of the mechanistic details of carcinogenesis, carcinogens are aneuploidogens (aneugens). The power of a carcinogen is predicted to be proportional to its ability to cause aneuploidy.

5.4 QUANTITATIVE ANALYSIS OF HOW ANEUPLOIDY GENERATES NEW CELLULAR PHENOTYPES

Only when we effectively study functional interactions in a kinetic way will we begin to understand what we are doing.

(Horrobin 2003)

In spite of more than a century of evidence conclusively demonstrating a very high association between chromosomal imbalance and cancer—even to the point of a one-to-one correspondence (Section 5.3.1)—there remains the persistent and nagging question: is aneuploidy a consequence or the cause of cancer? Many cancer researchers are waiting for definitive experiments to settle the issue. However, the answer is not to be found in molecular genetics but rather in new conceptual and mathematical approaches, i.e., looking at old data in new ways, as did Copernicus, Kepler, Galileo, Newton, and Einstein.

Hanahan and Weinberg "imagine that cancer biology and treatment—at present, a patchwork quilt of cell biology, genetics, histopathology, biochemistry, immunology, and pharmacology—will become a science with a conceptual and logical coherence that rivals that of chemistry or physics" (Hanahan and Weinberg 2000). Eleven years later, Hanahan and Weinberg "continue to foresee cancer research as an increasingly logical science, in which myriad phenotypic complexities are manifestations of a small set of underlying organizing principles" (Hanahan and Weinberg 2011). For that to happen, cancer research will have to become mathematical. Unfortunately, most researchers are disinclined and ill-prepared to consider the quantitative consequences of chromosomal imbalance. They are usually content with finding over- and under-expressed genes in tumor cells—though often without any mention of magnitude.

Horrobin recently bemoaned, "One of the distressing aspects of modern genomics and molecular biological studies is that they are almost entirely kinetic-free zones. Their practitioners are merely sketching out anatomically what pathways might be possible, and are not describing functionally what pathways actually do take place *in vivo*. Only when we get to that state of functional knowledge will the medical benefits begin to come through" (Horrobin 2003). This section introduces the mathematical tools needed to investigate how quantitative changes to the genome caused by chromosomal imbalance affect the overall metabolic activity (hence phenotype) of a cell.

The chromosomal imbalance theory asserts that the multiplicity of structural and functional cancer-specific phenotypes are the direct consequence of aneuploidy. Supporting this view, Lindsley et al. showed that aneuploidy readily produced numerous complex phenotypes of Drosophila, including reduced survival, small size, and a variety of morphological abnormalities such as rough eyes, abnormal wings and bristle patterns, and a misshapen abdomen. The authors concluded that, "The primary point of theoretical interest to emerge from these studies is that the deleterious effects of aneuploidy are, in the main, caused by the additive effects of genes that slightly reduce viability and not by the individual effects of a few aneuploid-lethal genes among a large array of dosage insensitive loci" (Hodgkin 2005, Lindsley et al. 1972, Sandler and Hecht 1973).

Recent "experimental data have pointed to the conclusion that aneuploidy is a main driving force for the observed adaptive evolution" in yeast (Rancati et al. 2008) and that fitness ranking between euploid and aneuploid cells is dependent on context and karyotype, providing the basis for the notion that aneuploidy can directly underlie phenotypic evolution and cellular adaptation (Pavelka et al. 2010a). "Changes in karyotype and the transcriptome can mediate the gene expression changes necessary for phenotypic determination" (Gao et al. 2007). Aneuploidy, alters the expression of many genes by a small extent (direct effect) and a few genes by a large extent (indirect effect) and can thus be viewed as a large-effect mutation (Rancati et al. 2008, Springer et al. 2010, Upender et al. 2004). The results from studies of aneuploidy in miaze "suggest that during the development of mature tissues, relatively mild quantitative initial effects of gene dosage imbalance lead to fixed qualitative changes in gene expression patterns" (Makarevitch and Harris 2010). Likewise, trisomy of the smallest human chromosome has the dramatic consequence of Down syndrome, which "cannot be explained as a direct result of one or some very few loci on chromosome 21" (Shapiro 1983). Since Down syndrome is phenotypically much less aberrant than cancer, the threshold of aneuploidy necessary for cancer is expected to be relatively high (Section 6.2.3).

A theory for the analysis of phenotypes generated by complex assembly lines of genes (metabolic control analysis) was developed by Kacser and Burns (Kacser and Burns 1973, Kacser and Burns 1981) and independently by Heinrich and Rapoport (Heinrich and Rapoport 1973, Heinrich and Rapoport 1974). According to this theory, the control of normal phenotypes is distributed to varying degrees among all the genetic components of complex systems (Fell and Thomas 1995, Fell 1997, Heinrich and Rapoport 1973, Heinrich and Rapoport 1974, Kacser and Burns 1973, Kacser and Burns 1979, Kacser and Burns 1981, Kacser 1995). The approach of Kacser and Burns was adapted to assess the role of aneuploidy in determining the phenotypes of cancer cells. The results show that transformation to a cancer cell requires quantitative alterations of massive numbers of genes, which is exactly what aneuploidy does.

The comprehensive biochemical phenotype of a cell (i.e., the metabolic flux) is determined by the action and interaction of all of its active components (Fell 1997, Kacser and Burns 1981). Since the production of gene products is, on average, proportional to gene dose (Aggarwal et al. 2005, Gao et al. 2007, Grade et al. 2007, Hieter and Griffiths 1999, Leitch and Bennett 1997, Matzke et al. 1999, Oshimura and Barrett 1986, Pavelka et al. 2010a, Rancati et al. 2008, Rasnick and Duesberg 1999, Springer et al. 2010, Upender et al. 2004), the metabolic flux of normal cells can be determined from the species-specific pool of genes. As originally proposed by Kacser and Burns, all active genes of a cell contribute an approximately equal share of the biochemical flux of the cell, and they are all kinetically connected within and even among the distinct biochemical assembly lines of a cell (Fell 1997, Kacser and Burns 1981).

In the spirit of Kacser and Burns, the mathematical formulations below are kept as simple and intuitive as possible. Since all gene products are enzymes, or substrates and products of enzymes, or modifiers of enzyme activity, variations in the levels of gene products of biological systems are kinetically equivalent to changes in effective enzyme concentrations (Heinrich and Rapoport 1973, Heinrich and Rapoport 1974, Kacser and Burns 1973, Kacser and Burns 1981). For algebraic convenience, the simple straight chain of enzymes will be used in this analysis. However, the results apply equally to systems of interlocking pathways, cycles, feedback loops (Kacser and Burns 1973, Kacser and Burns 1981), regulatory cascades (Kahn and Westerhoff 1991), and control of gene expression (Westerhoff et al. 1990), except that the formulations become more tedious (Kacser and Burns 1979).

At steady state, the biochemical phenotype of a cell that is generated by n enzymatic steps can thus be described by Scheme 5.1 (Rasnick and Duesberg 1999).

$$X_1 \underset{\longleftarrow}{\overset{E_1}{\longrightarrow}} S_1 \underset{\longleftarrow}{\overset{E_2}{\longrightarrow}} S_2 \ldots S_{n-1} \underset{\longleftarrow}{\overset{E_n}{\longrightarrow}} X_2$$

Scheme 5.1 Coupled enzymatic reactions

In this scheme, X_1 is the "source" (of nutrients) and X_2 is the resulting comprehensive phenotype or "sink", and E_i is the enzyme concentration for the ith step in the cellular assembly line (Kacser and Burns 1981). Using the fact that at steady state each intermediate flux is equal to the overall flux of a connected system, Equation 5.1 was derived by Kacser and Burns (Kacser and Burns 1981) for the overall steady state flux, F, for the production of X_2 according to Scheme 5.1. The variable F is the metabolic phenotype of a cell. The K values are equilibrium constants, the K_m values are Michaelis constants, and the V values are maximum rates (Rasnick and Duesberg 1999).

$$F = \frac{X_1 - \dfrac{X_2}{K_1 K_2 \ldots K_n}}{\dfrac{K_{m_1}}{V_1} + \dfrac{K_{m_2}}{V_2 K_1} + \ldots + \dfrac{K_{m_n}}{V_n K_1 K_2 \ldots K_{n-1}}} \tag{5.1}$$

Equation 5.1 can be simplified. Since all terms in the numerator of Equation 5.1 are constants, they can be combined into a single constant term C_n, which represents the environmental and constitutive parameters for the specific system or phenotype being considered. Further, since $V_i = E_i k_{cat(i)}$, all the V_i terms are proportional to their respective enzyme concentrations. Each fraction in the denominator of Equation 5.1, then, can be replaced by the composite e_i terms, all of which are proportional to enzyme concentration. The e_i terms represent the functions of the n gene products contributing to the flux. These modifications result in the simple Equation 5.2 that gives the overall metabolic output or flux for a normal cell composed of n individual functions or genes.

$$F = \frac{C_n}{\dfrac{1}{e_1} + \dfrac{1}{e_2} + \ldots + \dfrac{1}{e_n}} \tag{5.2}$$

Equation 5.2 can be rearranged to Equation 5.3, which shows that the reciprocal of the cellular metabolic phenotype, F, multiplied by a constant is the linear combination of the reciprocals of all n elemental phenotypes e_i that comprise a cell.

$$\frac{C_n}{F} = \frac{1}{e_1} + \frac{1}{e_2} + \ldots + \frac{1}{e_n} \tag{5.3}$$

For a system as complex as a diploid cell, the number of gene products, n, necessary to determine its phenotype is on the order of tens of thousands. For systems this complex, the $1/e_i$ terms make only small individual contributions and can be approximated by replacing them with $1/\bar{e}$, the mean of all the $1/e_i$ terms. Making this substitution in Equation 5.3 gives Equation 5.4, which can be used to describe the metabolic phenotype, F_d, of a normal, diploid cell for a given environment.

$$\frac{C_n}{F_d} = \left(\frac{1}{e} + \frac{1}{e} + \ldots + \frac{1}{e}\right)_n = \frac{n}{e} \qquad (5.4)$$

Aneuploid cells

The effects of aneuploidy on the collective biochemical phenotype of a cell can be quantified if we determine how the flux of a normal cell is altered in proportion to the dosages of the aneuploid genes (Aggarwal et al. 2005, Gao et al. 2007, Matzke et al. 1999, Rasnick and Duesberg 1999, Upender et al. 2004). Therefore, Equation 5.4 was modified to calculate the effects of aneuploidy on the phenotypes of eukaryotic cells, which increases or decreases substantial fractions of the genes, but not necessarily all genes, of a cell (Rasnick and Duesberg 1999). If only a subset of the n cellular genes is involved, the fluxes in Equation 5.4 can be partitioned into those that are affected by aneuploidy (m) and those that are not (n−m) to give Equation 5.5.

$$\frac{C_n}{F_a} = \frac{n-m}{e} + \frac{m}{\pi e} \qquad (5.5)$$

F_a is the metabolic phenotype of a eukaryotic cell resulting from aneuploidy. The number of genes experiencing a change in dosage due to aneuploidy is m. The variable π is the ploidy factor, reflecting the change in the number of copies of m genes. For example, $\pi=1.5$ for trisomy of m genes. The difference n−m is the number of genes not experiencing aneuploidy. The relative effect of aneuploidy compared to normal diploid cells can be obtained by dividing Equation 5.5 by Equation 5.4 to give Equation 5.6.

$$\frac{F_d}{F_a} = \frac{n-m+\dfrac{m}{\pi}}{n} = 1 - \frac{m}{n} + \frac{m}{n\pi} \qquad (5.6)$$

To further simplify Equation 5.6, the normal, diploid metabolic phenotype was set to $F_d=1$, and the quotient m/n was replaced with ϕ, which is just the fraction of the cell's gene products experiencing changes in dosage due to aneuploidy relative to the normal cell. These modifications give the dimensionless Equation 5.7, where F_a is now the relative flux, equivalent to the metabolic phenotype of the aneuploid cell.

$$\frac{1}{F_a} = 1 - f + \frac{f}{p} \qquad (5.7)$$

The $1-\phi$ term represents the fraction of unaffected gene products. The composite term ϕ/π is the fraction, ϕ, of gene products undergoing a π-fold change in expression. The relation of the cellular activity or biochemical flux to the DNA index of aneuploid cells can be estimated since the production of gene products is proportional to gene dose (Aggarwal et al. 2005, Gao et al. 2007, Hieter and Griffiths 1999, Leitch and Bennett 1997, Matzke et al. 1999, Rasnick and Duesberg 1999, Upender et al. 2004). Thus, the DNA or RNA index is given by Equation 5.8 (Rasnick and Duesberg 1999).

$$DNA_{index} = 1 - f + f\,p \qquad (5.8)$$

Fig. 5.3   A graphical representation of Equation 5.7 showing how aneuploidy changes phenotypes.

The normal diploid metabolic phenotype F_d is perturbed by varying the ploidy factor π and the aneuploid fraction ϕ to produce an ensemble of aneuploid metabolic phenotypes F_a. The phenotypes (F) of polyploid cells with balanced karyotypes fall on a straight line ($\phi=1$, broken green line), with haploids at $\pi=0.5$, diploids at $\pi=1$, triploids at $\pi=1.5$ and tetraploids with $\pi=2$, differing by equal increments of 0.5 F units. An ensemble of aneuploid metabolic phenotypes, F_a, was produced by varying the ploidy factor, π, and the fraction of the normal chromosome set ($0<\phi<1$, black and blue lines) according to Equation 5.7. $F_a>1$ represents positive aneuploidy, corresponding to gain-of-flux relative to the diploid cell, and $F_a<1$ represents negative aneuploidy, corresponding to loss of biochemical flux. Specific examples of aneuploid phenotypes

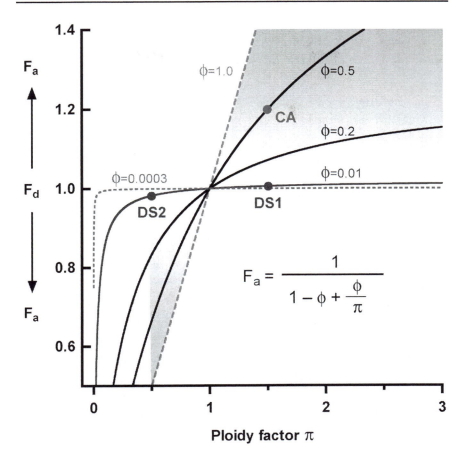

$$F_a = \cfrac{1}{1 - \phi + \cfrac{\phi}{\pi}}$$

Fig. 5.3 Contd. ...

are Down syndrome (ϕ=0.01, blue line) with trisomy (π=1.5) of chromosome 21, Fa=1.006 (DS1), and monosomy (π=0.5) of chromosome 21, F_a=0.98 (DS2). DS2 is more severe than DS1, consistent with the general principle that a lost of gene dose is more deleterious than a gain. Another example is a typical, near triploid colon cancer (CA, red dot) with an average of 69 chromosomes, corresponding to ϕ=0.5, π=1.5, and F_a=1.2. The effect on the phenotype of increasing or decreasing the functional dosage from π=0 to 3 of the seven genes (ϕ=0.0003) thought to cause colon cancer is indicated by the dotted red line. For π>0.05, the metabolic phenotype described by the dotted red line nearly coincides with that of the normal diploid cell, which is far from sufficient to generate cancer. The shaded area represents the cancer phenotypes.

Color image of this figure appears in the color plate section at the end of the book.

The product $\phi\pi$ is a measure of the increase or decrease in the gene products themselves. The general forms of equations 5.7 & 5.8 are given by Equations 5.9 & 5.10.

$$\frac{1}{F_a} = 1 - \sum\phi_i + \sum\frac{\phi_i}{\pi_i} \tag{5.9}$$

$$DNA_{index} \text{ or } RNA_{index} = 1 - \sum\phi_i + \sum\phi_i\pi_i \tag{5.10}$$

A graphical representation of Equation 5.7, where the normal diploid metabolic phenotype, F_d, is perturbed by varying the ploidy factor π and the aneuploid fraction ϕ to produce an ensemble of aneuploid metabolic phenotypes, F_a, is shown in Figure 5.3. The variable ϕ defines the shape of the curve as well as the limiting metabolic flux at the plateau for an aneuploid fraction $\phi<1$. The ploidy factor π determines the specific values of F_a within the limits set by ϕ. Since all ploidy increments are quantal (i.e., additions or deletions of whole chromosomes as well as segments) they generate steps rather than a continuous curve. However, since the number of genes is large and any subset of chromosomes may be aneuploid in a given cell the resulting π values are practically continuous. However, the discrete nature of the changes in gene dose due to aneuploidy has profound consequences that cannot be ignored, as discussed in Sections 6.2.1 & 6.2.3.

Figure 5.3 also shows that for $\pi<1$, there is a decline in F_a, indicating a loss of flux compared to the normal phenotype, and for $\pi>1$, there is a gain. The slopes are steeper for $\pi<1$ than for $\pi>1$, which is consistent with a loss of gene dose being more deleterious than a gain (Section 5.3.11). The shaded areas of Figure 5.3 indicate aneuploidies that exceed a threshold for cancer (Section 6.2.3). The positions of two specific examples of human aneuploidies (trisomy or monosomy of chromosome 21, i.e. Down syndrome) and a typical, pseudo-triploid colon cancer with 69 chromosomes (Sandberg 1990) are identified in Figure 5.3. Since chromosome 21 represents about 1.8% of the haploid human genome ($\phi=0.018$), trisomy ($\pi=1.5$) only changes the metabolic phenotype from $F_d=1$ to $F_a=1.006$ (DS1 in Figure 5.3) and monosomy changes it to $F_a=0.98$ (DS2 in Figure 5.3). Both F values lie outside the shaded region for cancer (Figure 5.3). But the pseudo-triploid colon cancer with 69 chromosomes ($\phi=0.5$, $\pi=1.5$) would generate a flux of about $F_a=1.2$, and would thus readily surpass the threshold (Section 6.2.3) for a cancer causing aneuploidy (shaded area of Figure 5.3).

Polyploid cells

Equation 5.4 can also be used to describe the phenotypes of polyploidization, i.e., all integral multimers of the complete haploid chromosome set of a cell. Since the production of gene products is proportional to gene dose, as described above, haploidization of a diploid cell will halve the dose of the ē gene products, producing a haploid flux, $F=0.5$, which corresponds to the biochemically rather inert gametes (Figure 5.3) (Hieter and Griffiths 1999). According to the same equation, the F values, and thus the biochemical activities, of polyploid cells are increased in proportion to their degrees of polyploidization (see broken green line in Figure 5.3). For example, the F value of tetraploid liver cells would be 2, that of 8-ploid and 16-ploid heart muscle cells would be 4 and 8, respectively, and that of 16-ploid and 64-ploid megakaryocytes would be 8 and 32, respectively (Hieter and Griffiths 1999).

Diploid cells with gene mutations

Equation 5.7 can also be used to investigate directly the effect of gene mutation on the biochemical flux—i.e., the probability of generating abnormal phenotypes by gene mutations, including those proposed to cause cancer. Because virtually all enzymes and functions of cells are integrated into kinetically linked biochemical assembly lines, and work *in vivo* at only a small fraction of their capacity (Kacser and Burns 1981), rare positive or activating mutations of enzymes or of hypothetical oncogenes are very effectively buffered *in vivo* via supplies and demands of un-mutated upstream and downstream enzymes. For example, transfecting 10 to 50 copies of any one of the five enzymes of the tryptophan pathway into yeast increases the yield of tryptophan no more than 2–30% (Cornish-Bowden 1995).

The dotted red line of Figure 5.3 graphically demonstrates the effects of mutating the dosage of 7 genes ($0<\pi\leq3$), as is postulated for colon carcinogenesis via oncogenes (Kinzler and Vogelstein 1996). As can be seen in Figure 5.3, this line almost coincides with the metabolic phenotype $F_d=1$ of a normal diploid cell. Based on Equation 5.7, the effect on the cellular phenotype of changing the dosage of any seven kinetically linked genes by mutation is negligible, because 7 gene mutations out of about 25,000 human genes (Collins et al. 2004) represents $\phi=0.0003$. Seven genes simply do not have the power to transform a diploid cell into an aneuploid cancer cell. Comparing the phenotypic power of mutations in a handful of genes to the metabolic consequences of aneuploidy is analogous to the difference between chemical and nuclear explosions.

It may be argued that the mutant genes that cause cancer are "dominant" (Alberts et al. 1994), i.e., independent of others, and highly pleiotropic, affecting the function of many others. Conceivably, some genes that govern differentiation could play such roles (Bailey 2000, Fell 1997). However, the currently known, hypothetical cancer genes are not dominant, because they do not transform normal diploid cells in culture nor in transgenic animals, which carry these genes in their germ line. The only known exceptions are the genes of viruses that transform or kill cells, without delay and with single hit kinetics, owing to truly dominant viral promoters that increase the functional π values of these genes up to about 1,000 (Section 4.3). This is exactly the reason why biotechnologists use viral promoters in synthetic vectors designed to maximize gene expression.

5.4.1 The effect of aneuploidy on genomic stability can be quantified

The metabolic output or flux of a normal cell is in balance with its genetic content. However, the black curves ($\phi=0.2$ and 0.5) in Figure 5.3 show there is an imbalance between the cellular activity and the DNA content of aneuploid cells. Since cancer-specific genetic instability has been shown experimentally to be proportional to the degree of genetic imbalance of aneuploid cells (Section 5.3.9), the stability index S (Equation 5.11) is defined as the total cellular metabolic activity F_a (Equation 5.7) divided by the DNA index (Equation 5.8). S ranges between 0–1, where 1 signifies a balanced genome (Rasnick and Duesberg 1999).

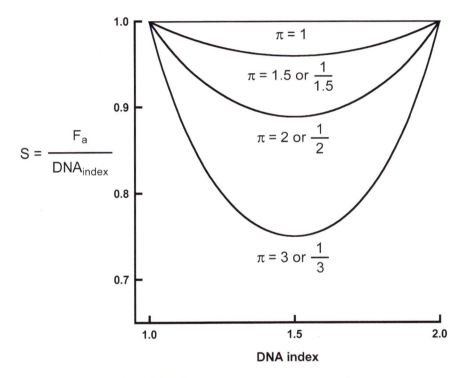

Fig. 5.4 Aneuploid DNA index = 1.5 catalyzes the greatest chromosomal instability.

In a normal cell the chromosomal content is balanced, guaranteeing the metabolic output or flux is in balance with its genetic content ($\pi=1$). Imbalance due to aneuploidy ($\pi\neq1$) leads to chromosomal instability. Plots of the stability index function S (equation 5.11) for various values of π versus the DNA index show that the cellular metabolic activity, F_a, of an aneuploid cell is more in balance with its genetic content only when the DNA index is 1 or 2. The greatest genetic imbalance is at DNA index=1.5, exactly halfway between the stable values.

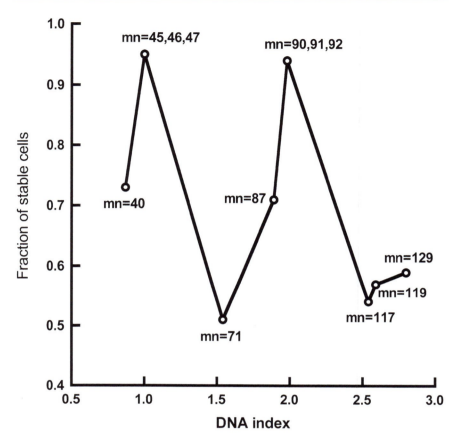

Fig. 5.5   **Karyotypic stability of colon cancer cell lines in agreement with stability index function S (Equation 5.11).**

The karyotypes of colon cancer cell lines are unstable in proportion to their degree of chromosomal imbalance, mirroring the graph of the stability index function S (Figure 5.4). The DNA index was calculated by dividing the modal number of chromosomes (mn) by 46, the normal complement of human chromosomes. The karyotypes close to a balanced number of chromosomes, i.e. $\pi=1$ (diploid) or $\pi=2$ (tetraploid), were the most stable. The modal numbers of the cell lines were: mn=40 (SW837), mn=45 (HCT116), mn=46 (DLD1, pseudodiploid), mn=47 (SW48), mn=71 (HT29), mn=87 (LoVo), mn=90 (2×HCT116), mn=91 (DLD1×HCT116), mn=92 (2×DLD1), mn=117 (DLD1×HT29), mn=119 (SW480), mn=129 (2×HT29, mn=142 but this hybrid lost chromosomes) (Lengauer et al. 1997).

$$S = \frac{F_a}{DNA_{index}} = \frac{1}{\left(1 - \phi + \phi\pi\right)\left(1 - \phi + \frac{\phi}{\pi}\right)} \qquad (5.11)$$

Figure 5.4 is a graph of Equation 5.11, which shows that the most stable aneuploid cells will have DNA indices near 1 (pseudo-diploid) and near 2 (pseudo-tetraploid). The least stable DNA index is 1.5, halfway between the stable values.

Lengauer et al. measured the genetic instability of a number of human colon cancer cell lines (Lengauer et al. 1997). The authors were unaware that genetic instability is a function of the level of aneuploidy. However, when their data were graphed (Figure 5.5), the results readily confirmed the predictions of the stability index function (Rasnick and Duesberg 1999). Pseudo-diploid (HCT116, DLD1, SW48) and pseudo-tetraploid (2 × HCT116, DLD1 × HCT116, 2 × DLD1, from fused cells) cells were the most stable and the HT29 colon cancer cell line with a DNA index=1.5 was the least stable. The observed 50% genetic instability of the triploid HT29 cells was due to the perpetual conflict between the most economical production of translation products on the one hand and the maintenance of chromosomal balance on the other (Rasnick and Duesberg 1999).

6

Theory of Chromosomal Imbalance Solves Mysteries and Paradoxes

One of the most puzzling aspects of the tumor problem is concerned with the multiplicity of diverse physical, chemical, and biological agencies that are capable of bringing about essentially the same end result.

(Braun 1969)

Jean Marx began a 2002 article: "Cancer cells are chock-full of mutations and chromosomal abnormalities, but there's no agreement among researchers on how incipient cancer cells accumulate so many changes" (Marx 2002). Of the three competing theories of cancer—1) Point mutations (Chapter 4), 2) self-perpetuating chromosomal variations (Chapter 5), and 3) the recent hybrid theory of point mutations causing aneuploidy and chromosomal instability—only the second explains the massive genomic abnormalities of cancer as a somatic aneuploidy syndrome. Table 6.1 compares the power of the chromosomal imbalance and gene mutation theories to explain a broad range of cancer phenomena.

6.1 CARCINOGENESIS IS DEPENDENT ON ANEUPLOIDY AND NOT MUTATION

In view of the karyotypic and phenotypic individuality of cancers, and the correlations of these phenotypes with the abnormal

Table 6.1 Explanatory Power of Competing Theories of Cancer

Phenomenon	*Explained*		*Section*
	Aneuploidy	*Mutation*	
Cancer is not heritable	Yes	No	6.1.2
Clonal origin of cancer	Yes	Yes	3.1
			5.3.8
No specific gene mutation	Yes	No	4.4.2
			6.2.4
No transforming gene mutation	Yes	No	4.4.4
			5.4
Non-clonal oncogenes and tumor suppressor genes	Yes	No	4.4.3
			4.4.9
			6.1
Non-mutagenic carcinogens	Yes	No	4.4.1
			5.3.4
Long latencies from carcinogen to cancer	Yes	No	4.4.5
			5.3.6
			6.1.3
			6.2.2
1,000-fold age bias of cancer	Yes	No	6.2.1
			6.2.2
Autonomous growth	Yes	No	5.4
			6.1.9
			6.5
Abnormal cellular and nuclear morphology	Yes	No	5.3.7
			6.1.9
Abnormal metabolism	Yes	No	4.4.7
			5.4
			6.1.9
			6.3
Warburg effect	Yes	No	6.3
Abnormal gene expression	Yes	No	5.4
			6.1.9
Complex phenotypes	Yes	No	4.4.11
			5.3.5
			6.1.7
			6.1.9

Table 6.1 Contd. ...

Phenomenon	Explained		Section
	Aneuploidy	*Mutation*	
Invasiveness	Yes	No	4.4.7
			5.3.7
			6.1.9
Metastasis	Yes	No	4.4.7
			5.3.7
			6.1.9
Spontaneous tumor disappearance	Yes	No	7.3.1
Hayflick limit	Yes	No	6.2.1
"Immortality"	Yes	No	4.4.9
			4.4.11
			6.1.4
Ubiquitous aneuploidy	Yes	No	4.4.13
Preneoplastic aneuploidy	Yes	No	5.3.6
			6.1.5
DNA indices of 0.5–>2	Yes	No	6.2.3
Karyotypic or "genetic" instability	Yes	No	4.4.13
			5.2
			5.4.1
Dominance of phenotypes of genomically unstable neoplastic and preneoplastic cells	Yes	No	6.1.5
Multi-drug resistance	Yes	No	4.4.7
			5.3.5
			6.1.8
			6.2.4
Too many and abnormal centrosomes	Yes	No	4.4.11
			6.1.7
Non-clonal karyotypes and phenotypes	Yes	No	6.1.4
Cancer-"specific" (non-random) aneusomies	Yes	No	5.3.7
			6.1.6
Carcinogens induce aneuploidy	Yes	No	5.3.3
Absence of cancer vaccines	Yes	No	6.5
Spontaneous progression of malignancy	Yes	No	5.3.10
			5.3.11
			6.2

expressions of thousands of normal genes, it is unclear what role mutations in 3–7 "specific" proto-oncogenes, which are shared by many but not all cancers of the same type (Bishop 1995, Vogelstein and Kinzler 1993, Weinberg 2007), could contribute to carcinogenesis (Section 5.4). The low levels of expression of hypothetical cellular oncogenes compound this problem. For instance, mRNAs of cellular oncogenes are typically undetectable in cancers without artificial amplification (Duesberg et al. 2007, Zhang et al. 1997) (Zhang and Vogelstein, personal communication) and are not even consistently mutated in cancer cells (Duesberg and Schwartz 1992, Duesberg and Li 2003, Duesberg 2003, Kallioniemi et al. 1992, Ledford 2010, Loeb et al. 2003). It is probably for this reason oncogenes these days are rarely even mentioned or specifically discussed in gene-expression studies of cancers (Birkenkamp-Demtroder et al. 2002, Furge et al. 2004, Gao et al. 2007, Hertzberg et al. 2007, Masayesva et al. 2004, Pollack et al. 2002, Tsafrir et al. 2006).

6.1.1 Mutations of cancer cells as a consequence of aneuploidy

In 1994, Prehn asked, if the different normal cellular phenotypes are so dependent upon varied patterns of expression in the genome, why is it that when something goes amiss and an abnormal phenotype—a neoplasm—appears, we presume that the cause lies in mutation rather than in an induced abnormality in the pattern of gene expression (Prehn 1994)? This led him to think that, "Despite the plethora of 'oncogenes' and 'tumor suppressor genes,' the hypothesis that cancer is usually the result of genomic mutations may be wrong" (Prehn 1994). To Prehn, the presence of point mutations in cancer cells was not due to their importance but to their limited biological significance, making them irrelevant. He concluded, "it may be more correct to say that cancers beget mutations than it is to say that mutations beget cancers." Thus, he concluded that mutations do not cause cancer but rather cancer phenotypes result from the confused or aberrant patterns of normal-gene expression abundantly present in aneuploid cells (Sections 5.3.7 and 6.1.6).

Cancer-specific aneuploidy can generate gene mutations by the same mechanism that varies the structures of chromosomes, e.g., by unbalancing teams of DNA repair enzymes (Fabarius et al. 2003). In addition, aneuploidy is mutagenic because it renders DNA synthesis error-prone by unbalancing nucleotide pools (Das et al. 1985). Thus, the simplest explanation of the many mutations of cancer cells would be that these mutations are consequences of aneuploidy and thus not necessary for carcinogenesis (Prehn 1994, Rasnick and Duesberg 1999, Rasnick 2000). This hypothesis explains why mutations of proto-oncogenes are frequently not detectable (Wang et al. 2002) or are non-clonal in cancers, why they do not transform normal cells into cancer cells, and do not undermine the lives of transgenic mice. Thus, mutation of cancer cells is a consequence of aneuploidy, rather than a cause.

Based on the roles of chromosomal variation and mutation in four distinct cancer-specific events—1) initiation (Section 5.3.2), 2) generation of complex phenotypes (Section 5.3.5), 3) high rates of karyotypic–phenotypic variations (Section 6.1.4), and 4) generation of mutations via aneuploidy (above)—it follows that chromosomal carcinogenesis does not depend on somatic gene mutation. In response to this conclusion, it may at least be argued the cancers associated with heritable cancer-disposition syndromes depend on mutation—although sporadic cancers do not. In the following, however, is shown that even the heritable mutations of cancer-disposition syndromes cause cancers only via aneuploidy.

6.1.2 Cancer is not heritable because aneuploidy is not

I am puzzled as to why most people other than professional geneticists seem uninterested in the high levels of non-concordance among identical twins for common diseases. And even the geneticists are more interested in the concordance than the non-concordance. For almost every common disease, whether it be inflammatory, malignant, degenerative, psychiatric or any other type, when one identical twin is affected the other twin is not affected from 40–90% of the time.

(Horrobin 2003)

The aneuploidy theory predicts the absence of cancer in newborns and non-identical cancer risks in twins (with the exception of the rare production of prenatal aneuploidy leading to childhood leukemia (Maia et al. 2003, Maia et al. 2004, Panzer-Grumayer et al. 2002)), because aneuploidy is the initiating cause of cancer and is not heritable, as was originally shown by Boveri (Boveri 1902/1964). Aneuploidies are not heritable, because they corrupt developmental programs (Epstein 1986, Shapiro 1983) and are usually fatal (Hassold 1986, Hernandez and Fisher 1999). Only some very minor congenital aneuploidies, such as Down-like syndromes and syndromes based on abnormal numbers of sex chromosomes, are sometimes viable, but only at the cost of severe physiological abnormalities and of no, or very low fertility (Bauer 1963, Griffiths et al. 2000, Sandberg 1990, Vogel and Motulsky 1986). Thus, ontogenesis is nature's checkpoint for normal karyotypes.

While cancer is not heritable there are heritable cancer-disposition syndromes, which predispose to high risks of non-systemic aneuploid cancers (Knudson 2000, Knudson 2001, Lodish et al. 2004). In other words, these heritable mutations are genetic equivalents of carcinogens, which increase the cancer risk by inducing random aneuploidy at high rates. This view is supported by the presence of systemic aneuploidy in patients prior to carcinogenesis (Sandberg 1990), as for example in mosaic variegated aneuploidy (Hanks et al. 2004, Kajii et al. 1998), retinoblastoma and other chromosomal eye syndromes (Chaum et al. 1984, Howard 1981, Squire et al. 1985), ataxia telangiectasia and Fanconi anaemia (Pathak et al. 2002, Wright 1999), Bloom syndrome (German 1974), Gorlin-syndrome (Shafei-Benaissa et al. 1998), and xeroderma (Chi et al. 1994, Lanza et al. 1997, Vessey et al. 2000). Even childhood leukemia appears to result from prenatal aneuploidy (Maia et al. 2003, Maia et al. 2004, Panzer-Grumayer et al. 2002).

The hypothesis that systemic aneuploidy defines cancer risks is also supported by the epidemiological studies described above, which have shown that this risk corresponds directly with the degrees of chromosomal aberrations in peripheral lymphocytes (Duesberg et al. 2005). Thus, the abnormally high rates of carcinogenesis in heritable cancer disposition syndromes are dependent on the abnormally high rates of systemic aneuploidizations that are generated by these heritable mutations. These heritable aneuploidy syndromes confirm and extend the chromosomal imbalance theory of carcinogenesis.

6.1.3 Long neoplastic latencies are due to slow progression of aneuploidy

No matter what carcinogen is used and how often it is applied, cancers only develop after "conspicuously" (Cairns 1978) long latent periods of many months to decades (Berenblum and Shubik 1949, Foulds 1969, Pitot 2002, Rous 1967). Classical clinical observations and animal experiments, beginning with Yamagiwa and Ishikawa in 1915 (Yamagiwa and Ichikawa 1915), had shown that carcinogens cause cancer only after long neoplastic latencies of many months to decades, but the reason for the inevitable latencies remained unsolved (Bauer 1963, Cairns 1978, Duesberg et al. 2005, Nettesheim and Marchok 1983, Pitot 2002, Preston et al. 2007, Rous 1967, Vogelstein and Kinzler 1993, Yamagiwa and Ichikawa 1915). More recent research has revealed that carcinogens induce random aneuploidy without delay, but cancers appear with clonal karyotypes only after long delays, as predicted by the chromosomal theory and speciation hypothesis (Section 6.6).

Such preneoplastic aneuploidy has been observed:

1) in humans after exposure to atomic radiations (Awa 1974) and in human cells in which "a surprisingly high proportion of T-cells with stable and often complex irradiation-induced chromosome aberrations are able to proliferate and form expanding cell clones *in vitro*" (Holmberg et al. 1993),

2) in the hyperplastic livers of mice fed butter yellow in 1957 (Marquardt and Glaess 1957), in "preneoplastic" lesions in the liver, spleen and thymus of mice treated with dimethylbenzanthracene

(Stich 1963) or on the skin in the form of precancerous papillomas (Conti et al. 1986),

3) in rats treated with nitrosamine and other chemicals that induce liver cancer, "to identify the importance of chromosome versus genome mutations" (Van Goethem et al. 1995) or in hyperplasias of rats treated with dimethylbenzanthracene to induce mammary cancer (Takahashi et al. 1977), and

4) in Syrian hamster cells treated with carcinogens *in vitro* (Gibson et al. 1995), or in untreated mouse and Chinese hamster cells growing *in vitro* prior to acquiring tumorigenicity spontaneously (Kraemer et al. 1983, Levan and Biesele 1958).

According to the chromosomal imbalance theory, the long neoplastic latencies from initiation to cancer reflect the time needed to evolve cancer-generating chromosome alterations by means of the autocatalyzed progression of aneuploidy (Sections 5.3.6 & 6.2 and Figures 5.1 & 6.1). The theory states that chromosomal evolutions are slow because preneoplastic aneuploidies are typically minor, i.e., are near-diploid, and thus only weak catalysts of chromosomal variation (Rancati et al. 2008, Rasnick and Duesberg 1999, Rasnick 2000, Rasnick 2002, Torres et al. 2007, Williams et al. 2008). Moreover, many non-neoplastic aneuploidies are likely to be fatal to a cell due to non-viable chromosome combinations (Boveri 1914, Duesberg et al. 2004b, Hassold 1986, Hernandez and Fisher 1999, Little 2000, Sandberg 1990, Wright 1999). Therefore, it is unlikely preneoplastic cells would form large clonal populations promoting further evolutions (Figure 6.1). The non-clonality of the preneoplastic aneuploidies also hides any abnormal phenotypes with low levels of aneuploidy, and because phenotypes of single cells are hard to recognize.

By contrast, the chromosomal imbalance theory predicts relatively short neoplastic latencies in patients with congenital aneuploidies and with chromosomal instability syndromes and thus cancer at young age (Section 6.1.2). This follows because the pool of aneuploid cells is much higher in these conditions than in normal counterparts. Neoplastic "progression" of established cancer cells, however, is predicted to be faster than the

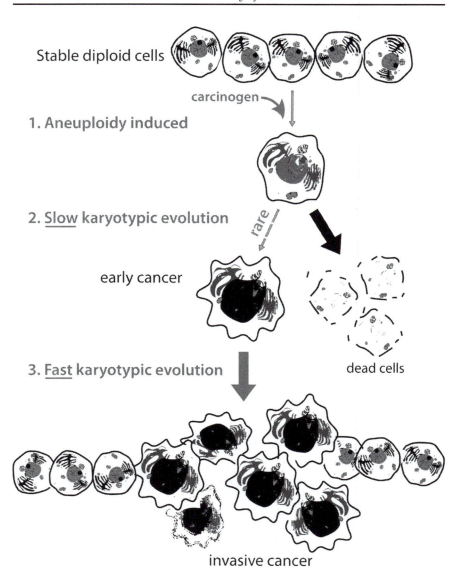

Stable diploid cells

carcinogen

1. Aneuploidy induced

2. <u>Slow</u> karyotypic evolution

rare

early cancer

dead cells

3. <u>Fast</u> karyotypic evolution

invasive cancer

Fig. 6.1   Three steps in the evolution to invasive cancer.

1) Initiation: A carcinogen or a spontaneous accident induces random aneuploidy by various mechanisms, e.g. nondisjunction, breaking and rearranging chromosomes, etc. 2) Slow pre-neoplastic chromosomal evolutions: By unbalancing thousands of genes, aneuploidy corrupts teams of proteins that segregate, synthesize and repair chromosomes. Aneuploidy is therefore a steady source of chromosomal variations, from

Fig. 6.1 Contd. ...

chromosomal evolutions during the preneoplastic phase for two reasons: 1) The highly aneuploid neoplastic cells, through their selective phenotypes, will generate large "clonal" populations with high probabilities of survival and further variations. 2) The high level of aneuploidies of most cancer cells catalyze much higher rates of chromosomal variations than those of preneoplastic cells (Section 6.2).

Karyotypic disorganization and variability are, of course, biologically limited by requirements for essential metabolic functions (Camps et al. 2005, Chiba et al. 2000, Duesberg et al. 2004b, Stock and Bialy 2003), also termed an "optimized genome" (Roschke et al. 2002). The chromosomal imbalance theory predicts a certain convergence of chromosomal evolutions to a point of dynamic equilibrium, at which maximal karyotypic disorganization or entropy coincides with maximal variability and adaptability that still leads to viable cells (Duesberg et al. 2005, Rasnick 2000, Rasnick 2002) (Section 6.2.3). According to this theory, maximal chromosomal variability would correspond to DNA indices near 1.7, approximately three times the haploid set (Section 6.2.3).

Fig. 6.1 Contd. ...

which, in classical Darwinian terms, neoplastic karyotypes eventually evolve. The initial low level of aneuploidy catalyzes a slow progression of pre-neoplastic chromosomal evolutions. While chromosomal imbalance is necessary for progression it also retards it because many aneuploid cells die from loss of both copies of a chromosome and non-viable chromosome combinations. 3) Fast neoplastic evolutions: Once a neoplastic chromosome combination evolves, subsequent karyotypic variations are accelerated, because neoplastic cells are generally more aneuploid and thus more adaptable than pre-neoplastic cells and can form locally large pools by outgrowing normal cells. Thus, neoplastic cells evolve independently within tumors forming ever-more heterogeneous and malignant phenotypes such as invasiveness, metastasis and drug-resistance at high rates. In sum: Malignancy can be seen as a consequence of autonomous chromosomal evolutions that increase karyotypic entropy to its biological limits, at or near a DNA index of 1.7 (Section 6.2.3).

Color image of this figure appears in the color plate section at the end of the book.

Near triploid aneuploidy offers an optimal average redundancy of one spare chromosome for each normal chromosome pair, and thus sufficient redundancy to compensate for any losses or genetic mutations of a given chromosome (Duesberg et al. 2004b). Accordingly, the karyotypes of most malignant cancer cells are or "converge" (Chiba et al. 2000, Hoglund et al. 2001a, Oikawa et al. 2004) near three times the haploid set of chromosomes (Atkin 1964, Duesberg et al. 2000a, Duesberg and Li 2003, Duesberg et al. 2004b, Giaretti 1994, Johansson et al. 1996, Koller 1972, Kraemer et al. 1971, Lengauer et al. 1997, Levan and Biesele 1958, Rasnick and Duesberg 1999, Rasnick 2000, Rasnick 2002, Roschke et al. 2002, Roschke et al. 2003, Sandberg 1990, Shackney et al. 1995a). The long-established, commercially available human cancer cell lines are models of such stably unstable karyotypes with karyotypic entropies close to their biological limits of aneuploidy (Camps et al. 2005, Castro et al. 2005, Castro et al. 2006, Kraemer et al. 1971, Roschke et al. 2002, Roschke et al. 2003, Schneider and Kulesz-Martin 2004).

A mystery unresolved by gene mutation is why the neoplastic latencies are very species-dependent, namely over 10-fold shorter in rodents than in humans (Fusenig and Boukamp 1998, Holliday 1996, Kuroki and Huh 1993, Soballe et al. 1996, Urano et al. 1995, Vogelstein and Kinzler 1993). Similarly, what makes the age bias of cancer compatible with the lifespan of an animal, i.e., grants cancer-free decades to humans (Figure 6.2) but only a few years to rodents (Cairns 1978, Holliday 1996). Differential mutation rates or growth rates are not the answer, because the rates of conventional mutations are highly conserved in all species (Lewin 1997, Vogel and Motulsky 1986) and the cells of humans and rodents grow at about the same rates.

Based on recent studies it appears the low chromosomal stability of aneuploid rodent cells compared to that of equally aneuploid human cells may hold a clue to this puzzle (Duesberg et al. 2000a, Duesberg et al. 2001b, Duesberg et al. 2006, Fabarius et al. 2003, Li et al. 2005). The evidence obtained so far, suggests that the chromosomal stabilities not only of normal but also of

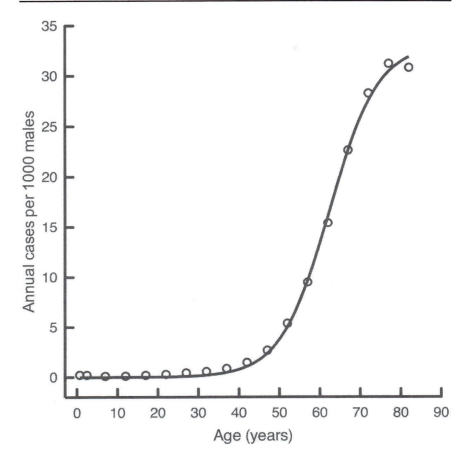

Fig. 6.2   Age specific incidence of invasive cancers of males in the United States in 2001.

The dominant contributors to the total number of invasive cancers are solid tumors. The growth is approximately exponential between ages 50–70 and then levels off (Section 6.2.2). Data for the figure are from the National Program of Cancer Registries (http://www.cdc.gov/cancer/npcr/index.htm). Because cancer is primarily a disease of old age it is compatible with an acquired, but not with an inherited disease.

cancer cells are species-specific. In view of these species-specific chromosomal stabilities, Holliday proposed that the genetic control of chromosomal stability is at least two times more robust in humans than in rodents (Holliday 1996). Section 6.4 presents an alternative proposal.

6.1.4 High rates of karyotypic–phenotypic variations and "immortality"

The inherent chromosomal instability of cancer cells is directly predicted by the chromosomal imbalance theory (Sections 5.2 & 5.4.1); it is confirmed by numerous correlations (Section 5.3.1), and is mechanistically linked to aneuploidy by the proportionality between the instability and the degree of aneuploidy detected experimentally (Section 5.3.9). Further, the inherent chromosomal instability of aneuploid cells is entirely consistent with the critical observation of Holmberg et al. that, "an increased frequency of sporadic chromosome aberrations was only observed in irradiated cells with aberrant karyotypes and not in irradiated cells with normal karyotypes, which suggests that the 'genomic instability' in these clones is associated with the abnormal karyotype rather than with the radiation exposure as such" (Holmberg et al. 1993).

Early on, researchers characterized cancers as immortal, because they could be transplanted indefinitely from animal to animal or cultivated indefinitely *in vitro* (Hayflick 1965, Levan 1956, Rous 1967, Van Valen and Maiorana 1991). According to Hauschka, "tumor karyotypes have competitive survival value and will be constant for thousands of cell generations" (Hauschka 1961). But a coherent theory of immortality was not offered.

Hayflick, on the other hand, proposed a mechanism of immortality in an influential study in 1965: "it could be argued that escape from the inevitability of aging by normal cells *in vivo* and diploid cell strains *in vitro* is only possible when such cells acquire, respectively, properties of transplantable tumors or heteroploid cell lines. One of the common denominators of these latter two systems is heteroploidy (usually modally distributed)" (Hayflick 1965). Heteroploidy is synonymous with aneuploidy.

Immortalization is an *in vitro* phenomenon resulting from the conditions of cell culture. Only after a period of adaptation and selection does an immortal clone sometimes arise from a primary cell culture" (Dermer 1983, Mamaeva 1998). An early description of "immortalization" by the cytogeneticist Koller explains the process, saying: "It seems that malignant growth is composed of competing clones of cells with different and continuously changing genotypes, conferring the tumor with an adaptable plasticity against the environment. The bewildering karyotypic patterns reveal the multi-potentiality of the neoplastic cell; while normal cells and tissues age and die, through their inherent variability, tumor cells proliferate and survive" (Koller 1972). Thus, owing to their cellular heterogeneity cancers survive negative mutations and cytotoxic drugs via resistant subspecies, analogous to bacteria and other single cell organisms (Levan 1969) (Sections 5.3.5 & 6.1.9).

Aneuploidy explains the "immortality" of cancer cells via the diversity of phenotypes that are constantly generated *de novo* by the inherent karyotypic instability of aneuploid cells. Owing to the inherent instability of aneuploidy, populations of cancer cells are in fact "polyphyletic" zoos (Hauschka and Levan 1958) of chromosomally distinct species (Section 6.6). Such populations of cancer cells are relatively "immortal" via subspecies that can survive mutations or conditions that are lethal to the majority of the cells of a cancer, as for example cytotoxic drugs. By contrast, homogeneous populations of diploid cells would either all survive or all die in a given challenging condition.

6.1.5 Karyotypic evolution of cancer

Heng et al. used multiple color spectral karyotyping to trace individual cells within representative populations that are stage-specific during the immortalization process (Heng et al. 2006b). The authors summarized five salient points relevant to carcinogenesis in a review article the same year (Heng et al. 2006a):

1. The karyotypic evolution during immortalization revealed the dynamic interplay between stochastic non-clonal chromosome aberrations (NCCAs) and different clonal chromosome aberrations (CCAs). The initial genomic changes (represented by NCCAs) occur at random when the genome is unstable. The degree of stochastic changes can reach extremely high levels right before the crisis stage

(karyotypic chaos) where none of the cells are the same in a given population. This surprising observation challenges current methodologies of studying cancer cells that have assumed that the majority of the cells are the same and that heterogeneity represents a minority of the cell population.

2. The evolutionary process of cancer can clearly be divided into two phases as judged by the karyotypic patterns: the discontinuous phase (marked by elevated non-clonal events, NCCAs and transitional clonal events, transitional CCAs) and the stepwise continuous phase (marked by stepwise clonal evolution and stable CCAs). The key event that separates these two phases is the cell crisis stage. Different from previous models of cancer evolution, the data demonstrate that stochastic karyotypic aberrations rather than sequential recurrent aberrations are the basis for early evolution. The unpredictable genotypes of cancer are caused by the stochastic nature of the initial phase of cancer progression.

3. The degree of genomic instability can be monitored by the degree of stochastic chromosomal changes. In particular, the NCCAs were found to be important indicators of chromosomal instability. Conversely, clonal aberrations do not correlate with genomic instability.

4. Karyotypic heterogeneity is caused by stochastic NCCAs and their interplay with CCAs. By comparing the frequency and types of NCCAs and CCAs during the immortalization process, the NCCA/CCA cycle corresponds well with cancer progression. When NCCAs dominate, a cell population is within an unstable "struggling to survive" phase coupled with high levels of genomic instability and increased genomic heterogeneity. When CCAs dominate, a cell population is within a relatively stable "growth" phase displaying greater stability and dominant pathways. Cancer progression occurs through multiple cycles of NCCAs/CCAs, with the cycle also recurring in response to drug treatments (Heng et al. 2006b).

5. Multiple cycles of NCCAs/CCAs are needed for a normal cell to first turn cancerous and then to further progress into advanced cancer cells. A particular CCA can be formed stochastically from NCCAs during the cancer evolutionary process. After a certain time period of growth, the CCA population will then be replaced by the NCCA population, until the next stage where new CCA populations form and became dominant. When both the genome and environment are stable, the various CCA populations often share some karyotypic signatures and the transition takes a much longer period of time. When the genome is unstable, regardless of whether it is due to internal factors or induced, the CCA populations are usually drastically different, demonstrating the stochastic nature of karyotypic evolution.

Heng et al. concluded that "NCCAs provide the material and the opportunity for cancer evolution to occur and CCAs are the end products of a given stage of evolution as defined by a specific selected NCCA and its environment. For a majority of solid tumors the importance of a given CCA is limited in terms of tracing a common path of progression, as late stage CCAs are often not shared by different tumors of the same cancer type. The complexity of CCAs, however, is of value as it reflects the clonal diversity and the selective results of NCCAs. NCCAs and their dynamic interplay among various combinations of NCCAs/ CCAs is what drives and shapes cancer progression. This agrees with the observations generated from a systematic analysis of the literature that the cancer karyotypes in late phases display more heterogeneity, as karyotypic evolution is a highly disorganized process during the late stages resulting in the disintegration of pathways (Hoglund et al. 2002). Such increased karyotypic heterogeneity contributes to the loss of long-range correlations that is also reflected by increased NCCAs and newly emerging CCAs" (Heng et al. 2006a).

Genomically unstable cells not only generate but can even lose new, abnormal phenotypes at very high rates compared to normal diploid cells. For example, while cancer cells may mutate to drug and multi-drug resistance at rates of up to 10^{-3} per mitosis, many can mutate back at similar high rates (Duesberg et al. 2000a, Duesberg et al. 2001b). By contrast, the phenotypes of normal diploid cells, which are controlled by haploid genes, mutate spontaneously only at rates of 10^{-6} to 10^{-7} and those controlled by diploid genes only at rates of 10^{-12} to 10^{-14} (Duesberg et al. 2000a, Duesberg et al. 2001b, Grosovsky et al. 1996).

Even malignancy is reversible at high rates in the non-selective condition of cell culture. For example, extraction of "less virulent clones" from highly malignant ascites tumors was achieved by Hauschka et al. (Hauschka 1961). Hsu derived even benign cultures from a virulent rat Novikoff hepatoma (Hsu 1960). Eagle et al. obtained "loss of neoplastic properties" in clonal derivatives of a human carcinoma cell line in parallel with "characteristic" karyotype alteration (Foley et al. 1965). Simi et al. and Sachs et al. also observed loss of the transformed phenotypes together with specific chromosomal alterations at high rates in transformed Chinese hamster cells, as well as subsequent reversion to malignancy (Bloch-Shtacher and Sachs 1977, Rabinowitz and Sachs 1972, Simi et al. 1992). And Fidler and Hart observed reversibility and back-reversibility of metastatic phenotypes of mouse melanoma cells at relatively high rates (Fidler and Hart 1982).

6.1.6 The phenotypes of genomically unstable cells are usually dominant

In contrast to the recessive phenotypes of hereditary instability syndromes (Section 6.1.2) such as xeroderma pigmentosum, Fanconi's anemia, hereditary non-polyposis colon cancer, hereditary hyperplastic polyposis and Bloom syndrome (Hawkins et al. 2000, Hoeijmakers 2001, Murnane 1996, Ruddon 1981, Sieber et al. 2003), the phenotypes of genomically unstable neoplastic and preneoplastic cells are usually dominant, either directly or after a delay of several cell generations (see Section 6.1.10 for exceptions.) The evidence for this dominance was gained from the following combinations of experimental fusions between cells with and without certain instability-specific phenotypes:

1. Instant or delayed tumorigenicity is dominant in hybrids consisting of normal and cancer cells (Harris 1995). The delayed tumorigenicity typically follows a spontaneous loss and re-assortments of various chromosomes (Duesberg and Rasnick 2000, Harris 1995, Stanbridge 1990a).

2. Delayed "reproductive death" of irradiated Chinese hamster cells is dominant in hybrids with un-irradiated counterparts (Chang and Little 1992).

3. Immortality is dominant in all "hybridomas" made from normal antibody producing immune cells and immortal aneuploid mouse myeloma cells (Harris 1995).

4. Chromosomal instability of highly aneuploid colon cancer cells is dominant in cell hybrids made with near-diploid, relatively stable chromosomal counterparts (Duesberg et al. 1998, Duesberg et al. 2004b, Fabarius et al. 2003, Lengauer et al. 1997).

6.1.7 Cancer-"specific" chromosomal alterations

The theory that cancer results from selections of random mutations among genomically unstable cells predicts a continuum of cancers with increasing degrees of genomic abnormalities. However, known cancers and even tumorigenic cell lines fall into a near diploid class of low instability and two highly aneuploid classes of high instability—a relatively rare (near 1.5 N, DNA index=0.75) class and a very common (near 3 N, DNA index=1.5) class (Atkin 1964, Duesberg et al. 1998, Flagiello et al. 1998, Giaretti 1994, Johansson et al. 1996, Lengauer et al. 1997, Remvikos et al. 1995, Sandberg 1990).

The presence of "specific" or nonrandom chromosomal alterations in cancer (Sections 5.3.7 & 6.1.5) is correlative proof for the aneuploidy theory in terms of Koch's first postulate (i.e., aneuploidy is abundantly present in all cancers but not in normal cells). Functional proof that cancer-specific aneuploidy generates malignancy in terms of Koch's third postulate (i.e., induction of aneuploidy in normal cells produces the spectrum of pre-malignant and malignant phenotypes) could be derived from evidence that the degree of malignancy is proportional to the degree of aneuploidy (Sections 5.3.6 & 5.3.9).

In addition, gene expression in cancer cells is directly proportional to the gene dosage generated by the respective chromosomal alterations (Gao et al. 2007), which indicates that specific aneusomies carry out specific functions (Aggarwal et al. 2005, Auer et al. 2004, Furge et al. 2004, Lodish et al. 2004, Miklos and Maleszka 2004a, Pollack et al. 2002, Virtaneva et al. 2001, Zhang et al. 1997). It is for this reason that thousands of metabolic and structural proteins are over- or under-expressed in cancer cells (Auer et al. 2004, Caspersson et al. 1963, Caspersson 1964b, Miklos and Maleszka 2004a, Pitot 2002, Ruddon 1987).

6.1.8 Cancer is a progressive somatic aneuploidy syndrome with complex phenotypes

Conventional genetic theories cannot explain the generation of the complex, polygenic phenotypes of cancer (Section 4.4.6). By contrast, viewing cancer as a progressive somatic aneuploidy syndrome naturally accounts for its heterogeneous phenotypes (Section 5.3), which are typical of aneuploidy syndromes in general. Thus, the complexity of cancer-specific phenotypes is due to the complexity of the genetic units that are varied (Hoglund et al. 2004), namely chromosomes with thousands of genes. Accordingly, the complex phenotypes of cancer cells have recently been shown to correlate with over- and under-expressions of thousands of genes (Aggarwal et al. 2005, Furge et al. 2004, Gruszka-Westwood et al. 2004, Lodish et al. 2004, Pollack et al. 2002, Virtaneva et al. 2001, Zhang et al. 1997). This in turn confirms the long-known over- and under-productions of thousands of normal proteins by cancer cells (Cairns 1978, Caspersson et al. 1963, Gabor Miklos 2005, Pitot 2002), and the observation that the overproduction of centrosomes by cancer cells is proportional to the degrees of aneuploidy (Ghadimi et al. 2000, Lingle et al. 2002).

6.1.9 Non-selective phenotypes such as multi-drug resistance

Conventional genetic theories explain the evolution of cancer cells by cancer-specific mutations and Darwinian selections (Section 4.4.7). But this mechanism cannot account for the non-selective acquired phenotypes of cancer cells, such as metastasis, intrinsic multi-drug resistance, and immortality, all of which are readily explained by the chromosomal imbalance theory (Section 5.3.5).

6.1.10 Paradox of karyotypic stability-within-instability of cancers

The chromosomal imbalance theory resolves the apparent paradox of the stability-within-instability of cancers (Section 5.3.8 & 6.1.5). It explains, for example, why the modal chromosome numbers of cancers remain stable even though the chromosomes making up that count may vary widely (Section 6.2.3). "The remarkable karyotypic stability of established tumor cell lines in culture over many generations in many different laboratories supports this idea. These cells do show substantial cell-to-cell variability but the average genotype is stable" (Albertson et al. 2003). Further, Castro et al. analyzed 79 solid tumor types with at least 30 karyotypes (Castro et al. 2006) from the Mitelman Database of Chromosome Aberrations (Mitelman 2011). Their results demonstrated that as cancer progresses, the number of recurrent karyotypes decreases with increasing karyotypic diversity correlating with malignancy.

Karyotypic stability within instability also explains many classical idiosyncrasies of cancer and carcinogenesis, as illustrated by the following six examples:

Similar cancers have similar karyotypes and transcriptomes

It has been demonstrated, especially by comparative genomic hybridizations, that similar cancers from the same tissues of origin have very similar karyotypes and transcriptomes (Baudis 2007,

Gebhart and Liehr 2000, Kuukasjarvi et al. 1997, Nupponen et al. 1998a, Richter et al. 1997, Richter et al. 1998, Weber et al. 1998). It follows that cancer karyotypes determine cancer phenotypes, analogous to aneuploidy syndromes, and similar species of the same taxonomic groups having similar karyotypes (e.g., rodents) (King 1993, O'Brien et al. 1999).

The proportionality between the degree of aneuploidy and malignancy

For examples, see reports and reviews by Wilkens et al. (Wilkens et al. 2004), Duesberg et al. (Duesberg et al. 2005, Nicholson and Duesberg 2009), Foulds (Foulds 1969), Wolman (Wolman 1983), Balaban et al. (Balaban et al. 1986), Mitelman et al. (Mitelman et al. 1997b), and Doak (Doak 2008).

Correspondence between high rates of phenotypic and karyotypic alterations

Take, for example, the high rates at which cancer cells acquire drug and multi-drug resistance correlating with high rates of karyotypic alterations (Sections 5.3.5 & 6.1.4). Gene mutations are excluded because their rates are orders of magnitude lower than phenotypic alterations (Section 4.4.11).

Immortality of cancer cell populations is an evolutionary phenomenon

Individual cancer cells are quite labile (Foulds 1954, Rous and Kidd 1941). It is the population of cells that is immortal, not the cells themselves (Section 6.1.4). The notion of the immortal cancer cell comes from the *in vitro* production of immortal cell lines. Immortal cell lines are new species that have assumed the properties of single-cell organisms (Levan 1969) such as bacteria and yeast. "In such populations the genetic adaptation to the environment is never concluded: periods of relative stability, in which the stemline of the population is at ease with the environment, alternate with periods of upheaval, when the utmost chromosomal variability of the population is mustered to overcome the threat" (Ising and Levan 1957) (Section 6.1.5). The high rates of karyotype variation coupled with constant selection for viability and proliferative advantages explain how aneuploid cancer cell populations attain immortality (Duesberg and Rasnick 2000, Duesberg et al. 2005, Nicholson and Duesberg 2009). These processes are truly analogous to events in natural selection, and the sequence of stemline karyotypes in a tumor may form identifiable evolutionary series (Section 6.6).

Evolution of cell populations is immensely accelerated as compared with evolution of complex organisms. This is understandable since changes in complex organisms must preserve an organized multicellular soma as well as meiosis and fertilization. Tumor progression, on the other hand, is not limited by metazoan constraints. As a model system of natural evolution (Section 6.6), cancer may help visualizing evolutionary events that are inaccessible in complex organisms (Ising and Levan 1957).

The loss of the transformed phenotype by alteration of the karyotype

Consider the obliteration of the oncogenic phenotype by fusion with normal cells or by experimental chromosome transfer (Duesberg et al. 2000a, Harris 1995, Ko et al. 2005, Saxon et al. 1986, Seitz et al. 2005, Steck et al. 1995, Theile et al. 1995).

The concordance between the complex phenotypes of cancer cells with the over- or under-expression of thousands of genes

Examples of complex phenotypes correlating with differential expression of thousands of genes include growth autonomy, abnormal nuclear and cellular morphology, highly abnormal metabolism, invasiveness and metastasis, and acquired or inherited multi-drug resistance (Aggarwal et al. 2005, Duesberg et al. 2005, Gao et al. 2007, Nicholson and Duesberg 2009, Pollack et al. 2002, Tsafrir et al. 2006).

6.2 AUTOCATALYZED PROGRESSION OF ANEUPLOIDY *IS* CARCINOGENESIS

We do not know how progression occurs...

(Bishop 1983)

The chromosomal imbalance theory provides a more precise description of the two steps of carcinogenesis: induction and progression.

Induction of random aneuploidy

Two observations suggest that the first step in carcinogenesis is the induction of random aneuploidy: 1) Carcinogens such as the highly aneugenic SV40 activated genes, or less aneugenic chemical or physical carcinogens and defective, heritable chromosome instability syndromes, all induce aneuploidy (Section 5.3.6). 2) When tested, random aneuploidy is found to precede neoplastic transformation (Section 6.1.3).

Evolution from preneoplastic to neoplastic karyotypes

Promotion involves the evolution of quasi-stable neoplastic karyotypes from unstable initiating random aneuploid karyotypes (Section 6.1.5). Because of the inherent instability of aneuploidy (Section 5.2) and the effects of selection, the karyotypes of aneuploid cells evolve autocatalytically (Section 5.3.11) toward two stable endpoints: the quasi-stable karyotypes of immortal cancer cell populations and the lethal karyotypes of cells dying from nonviable combinations of chromosomes (Figure 6.1). Because the odds of generating the new complex functions that define cancer (Duesberg et al. 2005, Nicholson and Duesberg 2009) by random alterations of a karyotype are very low, the evolution of new neoplastic karyotypes from randomly aneuploid cells will be rare—comparable to the evolution of new species (Section 6.6). This explains the typically long neoplastic latent periods of years to decades between the induction of aneuploidy by carcinogens and human cancer (Section 6.1.3).

Recent experiments by Heng et al. lend further support to the two steps of carcinogenesis (Section 6.1.5). The authors found that randomly aneuploid fibroblasts from patients with the heritable Lie-Fraumeni aneuploidy syndrome generated clones of transformed cells *in vitro* with clonal marker chromosomes that were stable over 180 cell generations, despite the simultaneous "emergence and disappearance" of highly unstable marker chromosomes. Furthermore, independent clones from the same parental cells had distinct karyotypes, despite the same heritable mutation or mutations.

6.2.1 The Hayflick limit is due to the autocatalyzed growth of aneuploidy

Due to the work of Leonard Hayflick, the finite lifetime of diploid cells in culture has become commonly known as the Hayflick limit (Hayflick and Moorhead 1961, Hayflick 1965). After a period of active multiplication, generally less than one year (around 50 cell divisions), primary human fetal cells in culture demonstrate an increased generation time, gradual cessation of mitotic activity, accumulation of cellular debris and, ultimately, total degeneration (Hayflick 1965). Only during the degenerative phase (phase III) in cell culture do primary cells lose contact inhibition and become obviously aneuploid (Hayflick and Moorhead 1961, Hayflick 1965, Saksela and Moorhead 1963). In contrast to primary diploid cells derived from animals or humans, cell lines (immortal cells) are a heterogeneous mix of heteroploid cells.

According to Levan and Biesele, the very first mitoses of mouse cells *in vitro* show chromosomal irregularities. A zero level of numerical and structural chromosomal abnormalities "is possible only with cells in situ, and that as soon as they are explanted they start mutating.... . The genetic diversity thus induced in the tissue culture will increase steadily as new aberrations are continuously released by the mitotic mutation process. This situation prevails either until the tissue culture fades out or until some truly superior cells happen to appear..." (Levan and Biesele 1958).

As explained in Section 5.3.11, the survival advantage of hyperploid over hypoploid cells, coupled with the inherent genetic instability of aneuploid cells (Section 5.2), leads to the autocatalyzed progression of aneuploidy with each cell division. The autocatalyzed progression of aneuploidy explains the observed time-course of the Hayflick limit, including differences in spontaneous rates of transformation (Rasnick 2000). The simplest model of an autocatalyzed process is shown in Scheme 6.1.

$$D + \phi \xrightarrow{\;k\;} \phi$$

Scheme 6.1 Rate model of an autocatalyzed process

It is important to point out that Scheme 6.1 represents an autocatalyzed process, not an autocatalyzed chemical reaction, which would be written as A+B→2B (Steinfeld et al. 1989). D and ϕ do not represent chemical species undergoing a chemical reaction and, therefore, do not imply stoichiometry. D is the diploid fraction and ϕ is the aneuploid fraction of a cancer cell (Section 5.4). D and ϕ can also represent the average diploid and aneuploid fractions, respectively, of a population of cells.

Following an event that produces the initial aneuploidy (i.e., produces $\phi > 0$), the values of D and ϕ tend to change with each mitotic division because of the instability caused by chromosomal imbalance (Section 5.2). It has been shown that the aneuploid fraction, ϕ, is equivalent to the flux control coefficient or control strength of metabolic control analysis and is a measure of the extent to which a given aneuploid segment of the genome controls phenotypic transformation (see Appendix A of ref. (Rasnick and Duesberg 1999)).

The rate equation for the increase in the aneuploid fraction, ϕ, in Scheme 6.1 is given by Equation 6.1. Since ϕ appears on both sides of Scheme 6.1, the growth of the aneuploid fraction, ϕ, is autocatalyzed. In other words, the greater the level of aneuploidy, the faster the growth of ϕ. The constant k in Scheme 6.1 is a measure of the growth-rate of the aneuploid fraction ϕ.

$$\frac{d\phi}{dt} = k\phi D \tag{6.1}$$

Since the sum of the diploid fraction, D, and the aneuploid fraction, ϕ, always equals 1, the diploid fraction can be expressed in terms of ϕ, i.e., D=1-ϕ. Making this substitution for D in Equation 6.1 gives Equation 6.2.

$$\frac{d\phi}{dt} = k\phi (1-\phi) \tag{6.2}$$

Integrating Equation 6.2 yields Equation 6.3, which gives the aneuploid fraction, ϕ, as a function of time (Rasnick 2000).

$$\phi_t = \frac{1}{e^{-kt}\left(\dfrac{1}{\phi_0} - 1\right) + 1} \tag{6.3}$$

The constant ϕ_0 is the initial aneuploid fraction of an individual cell, or the average for a population of cells, at time zero, the time when aneuploidy is initiated (e.g., by a carcinogen or cell culturing). The rate constant k has units cell-cycle^{-1} when t is in cell-cycles. Since ϕ_t ranges from 0–1 (i.e., 0–100%), Equation 6.3 gives the time-course for the progression of aneuploidy, and thus the time course of any phenotypic change that depends on it.

With the growth of aneuploidy there is a corresponding reduction in the diploid fraction, and hence a reduction in the number of dividing cells since aneuploid cells are less viable than diploid cells (Section 5.3.11). If we assume that cell proliferation is due primarily to diploid cells, then the number of non-transformed dividing cells in culture at time t is proportional to the diploid fraction D_t. Using the relationship $D_t=1-\phi_t$ and the value of ϕ_t in Equation 6.3, we can derive Equation 6.4, which shows that the number of dividing cells (N_t) remaining during serial passaging is equal to the number of cells at time zero (N_0) times the diploid fraction at time t (i.e., $N_t=N_0(1-\phi_t)$).

$$N_t = N_0 \left(1 - \frac{1}{e^{-kt}\left(\dfrac{1}{\phi_0} - 1\right) + 1} \right) \qquad (6.4)$$

Figure 6.3 (solid blue line) shows that Equation 6.4 fits the data from Hayflick's figure 3 (Hayflick 1965) for the serial passaging of the human cell strain WI-44 (Rasnick 2000). The sharp decline in the cell count at around 43 passages (beginning of the degenerative phase III) is mirrored by the steep growth in the aneuploid fraction ϕ (Figure 6.3, broken red line).

The calculated value ϕ_0=0.0004 (Figure 6.3) indicates that the average aneuploid fraction at the beginning of the log-phase (phase II) of cell culture was 0.04%. However, this initial aneuploid fraction is only an estimate since it does not take into account that some aneuploid cells will be viable and divide, albeit at a reduced level. The initial aneuploidy in the WI-44 cells was almost certainly caused by the culturing process itself, especially the mechanical and enzymatic treatments used to promote proliferation *in vitro* (Hayflick and Moorhead 1961, Hayflick 1965, Huna et al. 2011, Levan and Biesele 1958, Serrano et al. 1997).

In contrast to cultured cells, the extra copy of chromosome 21 in Down syndrome individuals is present at fertilization. Trisomy of chromosome 21 represents an aneuploid fraction ϕ=0.018 for each cell (Rasnick and Duesberg 1999, Sandler and Hecht 1973, Shapiro 1983), which is substantially larger than the initial average value in Figure 6.3 for normal human explants. As a consequence, the Hayflick limit of Down syndrome cells in culture should be significantly shorter than normal cells. A perusal of the literature shows this prediction is correct. Schneider and Epstein reported in 1972 that, "Skin fibroblasts derived from patients with Down syndrome have a significantly decreased number of cumulative cell population doublings…measured from the initial passage to senescence when compared with cultures from karyotypically normal age-matched controls…" (Schneider and Epstein 1972).

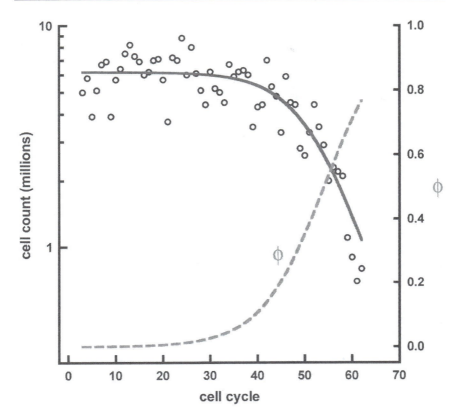

Fig. 6.3   Hayflick limit for the human cell.
The blue solid line is the best-fit curve of Equation 6.4 to the serial passaging data from Figure 3 of Hayflick for an embryonic human cell strain (Hayflick 1965). The broken red line represents the autocatalyzed progression of the aneuploid fraction ϕ for the same data using Equation 6.3. The parameters from the best fit were k=0.15 cell-cycle^{-1}, ϕ_0=3.6×10^{-4}.
Color image of this figure appears in the color plate section at the end of the book.

The Hayflick limit they measured for the Down syndrome cells was 20% shorter than for normal cells, which represents an approximate 2000-fold reduction in the exponential process of cell doubling.

More recently, Mukherjee and Costello used fluorescence in situ hybridization to study the progression of aneuploidy in cultured fibroblasts from patients with three premature aging syndromes: Cockayne, Hutchinson-Gilford, and Werner (Mukherjee and Costello 1998). "[T]he interphase aneuploidy levels of all chromosomes under study were significantly higher in cells from the syndromes as compared to those of the normal controls at both earlier and later passages. In general, the interphase aneuploidy levels of each of the chromosomes in both the control and experimental cell cultures increased with *in vitro* proliferation and aging, although to a much lesser extent in the controls..." (Mukherjee and Costello 1998).

The meticulous studies of Hayflick indicated that, "the finite lifetime of...diploid cell strains is an innate characteristic of the cells..." (Hayflick and Moorhead 1961, Hayflick 1965). Furthermore, he argued: "Cells which can be cultivated indefinitely *in vitro* (heteroploid cell lines) can only be compared with continuously cultivable cells *in vivo*, i.e., transplantable tumors. Likewise, diploid cells having a finite lifetime *in vitro* can only be compared with normal cells *in vivo*, i.e., normal somatic cells..." (Hayflick 1965). Hence, the transformation of mortal (diploid) cells in culture into immortal (aneuploid) cell lines "can be regarded as oncogenesis *in vitro*..." (Hayflick 1965).

Transformation of cells in culture can also be viewed as evolution *in vitro* (Section 6.6). Explanted cells are forced to evolve into viable single cell organisms in the laboratory or perish. Most cultured human cells stop dividing after entering phase III (the degenerative phase) and only rarely (if ever) undergo spontaneous transformation into immortal (aneuploid) cell lines (Hayflick 1965, Huna et al. 2011, Serrano et al. 1997). In contrast, cultured primary rodent cells frequently undergo spontaneous transformation to become immortal (aneuploid) cell lines (Levan and Biesele 1958, Todaro and Green 1963). The 70% shorter Hayflick limit (14 cell divisions) may be a clue as to why primary mouse fetal cells spontaneously transform into immortal (aneuploid) cell lines much more readily than human cells (Todaro and Green 1963).

While Equation 6.2 gives the continuous rate of change of the aneuploid fraction, ϕ, as a function of the instantaneous value of ϕ, the actual change in DNA content of an aneuploid cell is a discrete process that takes place at each cell division. Alternatively, then, the growth of the aneuploid fraction, ϕ, of Scheme 6.1 can be modeled by Equation 6.5, a form of the logistic rate equation that is known to reveal certain chaotic processes (Alligood et al. 1996, Posadas et al. 1996, Williams 1997).

$$\phi_{n+1} = r\phi_n \left(1 - \phi_n\right) \tag{6.5}$$

The right side of Equation 6.5 is formally identical with Equation 6.2. However, the left side of Equation 6.5 replaces the instantaneous change in ϕ with a discrete value of the aneuploid fraction for each cell division. Equation 6.5 shows that the average aneuploid fraction of a population of cells at the n+1 cell division is determined by the average level of aneuploidy at the nth cell division. The control parameter, r, in Equation 6.5 is unitless and is different from the rate constant, k, in Equation 6.2. Since ϕ_i ranges from 0–1, Equation 6.5 models the discrete growth, including chaotic, of any phenotypic change that depends on the progression of aneuploidy.

Since the non-disjunction frequency of mammalian cells in situ and short-term culture has been measured to be 10^{-4} to 10^{-5} per chromosome per mitosis (Gisselsson et al. 2010, Holliday 1989), and since there are 23 and 20 chromosome pairs in normal human and mouse cells, respectively, then the initial aneuploid fraction, ϕ_0, for the explanted cells is no larger than approximately 10^{-6} (i.e., $10^{-4}/(20 \text{ or } 23) \approx 10^{-6}$). Therefore, $\phi_0 = 10^{-6}$ was used in Equation 6.5 to model the autocatalyzed progression of aneuploidy for human and mouse primary fetal cells in culture (Figure 6.4) (Rasnick 2000).

Several values of the control parameter, r, were tried before finding the value of 1.35 (Figure 6.4) that reproduces the 35 cell-divisions of phase II and the sigmoidal growth of the aneuploid fraction, ϕ, of Figure 6.4 for human cells (Rasnick 2000). Values of the control parameter, r, greater than 1.5 were completely unrealistic since they resulted in phase III starting at about 10 cell divisions for the human cells.

In contrast to the human cells, values of r≥2.5 were necessary to reproduce the Hayflick limit of about 14 cell-cycles for primary mouse fetal cells. But what is more interesting, a value of the control parameter greater than 3.57 worked just as well. A value of r>3.57 for the logistic equation (Equation 6.5) results in chaotic growth patterns (Williams 1997). Thus, the value r=3.7 in Figure 6.4B produces a chaotic progression of the aneuploid fraction, φ, for all cell divisions beyond about 12 cycles (Figure 6.4B).

Figure 6.4A,B fails to take into account the fact that almost all of the aneuploid cells would lose the chromosome-shuffling lottery and not become transformed into an immortalized cell line. However, if a transforming karyotype did happen to appear, for example the hypothetical one circled in Figure 6.4B, and was cloned,

Fig. 6.4   Chaotic growth of aneuploidy drives transformation.
Equation 6.5 was used to model the autocatalyzed progression of aneuploidy for primary human (A) and mouse (B) fetal cells in culture. The control parameter r=1.35 and the initial average aneuploid fraction $\phi_0 = 10^{-6}$ modeled the 50 cell-cycle Hayflick limit for primary human fetal cells (A). The data points represent the average aneuploid fraction φ for the population of cells at each cell cycle. In panel B the same ϕ_0 was used for the mouse fetal cells. Although values of r>1.5 were completely unrealistic for modeling the Hayflick limit of human cells, values of the control parameter greater than 3.57 could be used to model mouse cells. A control parameter greater than 3.57 for the logistic equation (Equation 6.5) produces chaotic growth patterns (Williams 1997). Therefore, the value r=3.7 produced a chaotic progression of the aneuploid fraction φ for all cell divisions beyond around 12 cycles (B). While the aneuploid human cells would probably die out before being transformed into an immortal cell line (because so little genome space is being explored) the chaotic redistribution of the mouse genome provides a greater opportunity for the cells to hit upon a genetic combination that leads to transformation and immortalization. Panel C shows that when the transforming genome in panel B (red arrow) is cloned, its intrinsic karyotypic instability immediately leads to a heterogeneous population of heteroploid offspring.
Color image of this figure appears in the color plate section at the end of the book.

Fig. 6.4 Contd. ...

Fig. 6.4 Contd. ...

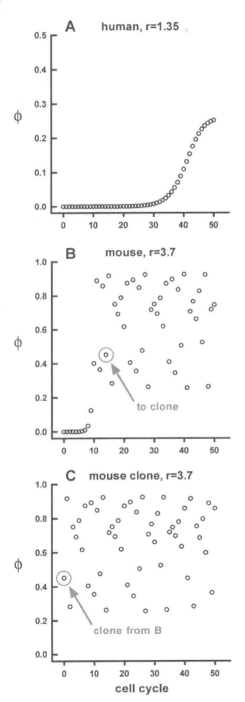

it would generate from the very first cell division a heterogeneous population of heteroploid offspring as shown in Figure 6.4C. Figure 6.4C models exactly the well-known karyotypic instability of cloned transformed cells (Duesberg et al. 1998, Hayflick 1965, Lengauer et al. 1997). As Hayflick and Moorhead have said, "The use of cloning as a means of reducing…variability in heteroploid cell lines is unfortunately limited by the rapid re-emergence of a range of chromosomal types among the progeny of a clone…" (Hayflick and Moorhead 1961). In fact, "each subpopulation can regenerate the entire range of subpopulations…" of a heteroploid population of cells (Kimmel and Axelrod 1990).

The results of Figure 6.4 show that it is possible to model the hypothesis that the transformation-prone mouse cells can exhibit a substantially more chaotic pattern of aneuploidy than the transformation-resistant primary human cells in culture. If the results of Figure 6.4 reflect reality, an interesting question presents itself. What are the physical, biochemical, genetic, or other factors responsible for the dramatically different values of the control parameter, r, that lead to the non-chaotic growth of aneuploidy in primary human cells in culture on the one hand, and a chaotic progression of aneuploidy in primary mouse cells on the other? A possible answer is presented in the next section and Section 6.4.

6.2.2   The sigmoidal age distribution of human cancer

… the time course of carcinogenesis is deeply mysterious

(Brash and Cairns 2009).

In the 1950s, log-log plots of cancer death-rates versus age were roughly linear with slopes of about 6 (Armitage and Doll 1954). That meant that cancer death-rates increased proportionally with the sixth power of age. It wasn't long before the sixth-power dependence was interpreted in light of the gene mutation hypothesis of cancer.

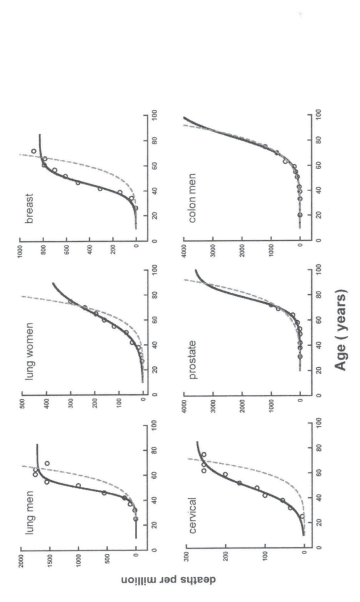

Fig. 6.5   Autocatalyzed growth of aneuploidy explains age distribution of human cancers.

The superiority of the sigmoidal curve of Equation 6.7 for the autocatalyzed progression of aneuploidy was best demonstrated by comparing it with Equation 6.6 for the multi-hit version of the gene mutation theory of human carcinogenesis. Equation 6.7 gives a good fit (solid black lines) to the number of deaths per million people for six typical cancers as a function of age (Armitage and Doll 1954). The broken red lines show the best-fit curves to the same data for the 7-successive mutation model (Equation 6.6). The only good fit for Equation 6.6 was with colon cancer deaths in men.

Color image of this figure appears in the color plate section at the end of the book.

It was hypothesized that cancer is the end-result of seven successive mutations (Kinzler and Vogelstein 1996, Nordling 1953). However, this hypothesis did not lead to the observed result in all circumstances (see Figure 6.5, broken red lines). Aware of this shortcoming, Armitage and Doll warned that the successive cellular changes leading to the development of cancer are not necessarily gene mutations (Armitage and Doll 1954). This is an important consideration since carcinogenic and mutagenic activities do not always go hand-in-hand (Section 6.3). This insight was short-lived, however. In deriving Equation 6.6 to model the incidence-rate of cancer with age, Armitage and Doll assumed that seven mutations lead to cancer, and that the mutations should be specific, discrete, stable, and proceed in a unique order (Armitage and Doll 1954).

$$\text{cancer rate}_t = k p_1 p_2 p_3 p_4 p_5 p_6 p_7 t^6 \qquad (6.6)$$

Equation 6.6 shows that the incidence-rate of cancer at age t (assumed to be proportional to death-rate (Armitage and Doll 1954)) will be proportional to the product of the probabilities (p_i) of the occurrences of each of the seven mutations and to the sixth power of age, where k is the rate constant. With the exception of colon cancer in men, Equation 6.6 is a poor model of the incidence-rate for a number of human cancers (Figure 6.5, broken red lines). In an effort to salvage their model, Armitage and Doll argued that, due to ignorance of the individual mutation probabilities (p_i) they had to combine all the probabilities into one constant. According to the authors, this combined probability was the source of the poor fit between Equation 6.6 and the data. They speculated, that if only one knew the individual mutation probabilities or could fashion a suitable weighting scheme to derive the appropriate mean probability, then Equation 6.6 should fit the real-world data. Unfortunately, the authors were not able to come up with either. Indeed, it is difficult to see how a parabolic equation could be found that fits the sigmoidal incidence-rate data of Figure 6.5.

However, it turns out that Equation 6.7, which is based on Equation 6.3 for the autocatalyzed progression of aneuploidy (Rasnick 2000), gives a good fit to all the cancer incidence-rate data of Figures 6.2 and 6.5 (solid black lines).

$$N_t = N_\infty \left(\cfrac{1}{e^{-kt}\left(\cfrac{1}{\phi_0} - 1\right) + 1} \right) \qquad (6.7)$$

In this case, the number of cancer deaths per million persons (N_t) at age t is equal to the plateau number of cancer deaths per million persons (N_∞) times the right side of Equation 6.3, which is the average aneuploid fraction, ϕ_t, for a population of cells. The only good fit for Equation 6.6 is with the incidence of colon cancer deaths in men, which may be the source of Kinzler and Vogelstein's proposal that seven gene mutations are responsible for colon cancer (Kinzler and Vogelstein 1996).

The sigmoidal nature of the mortality-rate data is more obvious in Figure 6.5 than in the log-log plots used by Armitage and Doll. The data points for lung cancer in men and women, as well as breast and cervical cancer in women, span much more of the sigmoidal region of Equation 6.7 than do the data for prostate and colon cancer in men (Figure 6.5). This difference is due to the much later onset of prostate and colon cancer in men than with the other four examples. The inflection point of the sigmoidal curves is a measure of this difference. The inflection points for lung cancer in men and women, and for breast and cervical cancer in women are around 50, 60, 45, and 50 years of age, respectively (Figure 6.5). However, the inflection points for prostate and colon cancer in men are at much older age: around 75 and 80 years, respectively (Figure 6.5).

6.2.3 Tumor formation

I am convinced that any theory of tumor formation will have to accommodate a step that has the character of a lottery.

(Boveri 1914)

There is no faster way of finding out how a chaotic system will evolve than to watch its evolution

(Jensen 1987)

Most cancers appear to begin with undifferentiated epithelial cells at the boundary between two cell types (Sell and Pierce 1994). Differentiation is a bifurcation process where the normal diploid cell "chooses", for example, to become either a squamous cell or a glandular cell (Anderson 1991). Aneuploidy introduces an additional bifurcation of the developmental course of a cell. Once a cell becomes aneuploid, its offspring irreversibly head toward either almost certain oblivion or very rarely cancer. It is the progression to cancer that is of interest here. (The spontaneous disappearance of tumors is discussed in Section 7.3.1.)

Normal tissues are made up of countless diploid cells with characteristic and reproducible properties that form an organized structure that spans the tissue. Cancers, on the other hand, are made up of a mass of autonomous aneuploid cells, no two of which are genetically alike (Fabarius et al. 2003, Fox et al. 2009, Kan et al. 2010, Lengauer et al. 1998, Levan and Biesele 1958, Rasnick and Duesberg 1999, Reshmi et al. 2004, Yosida 1983). Aneuploid tumor cells behave like a coherent beehive of single-cell organisms (Duesberg et al. 2000a, Levan 1969, Rasnick 2000). The coherence makes them pathogenic. What is the source of the coherence? How do autonomous aneuploid cells organize themselves into a tumor beehive?

Complexity theory addresses, among other things, the process of self-organization. A self-organizing system spontaneously creates a globally coherent pattern out of the local interactions of initially independent components. A self-organizing system has properties that are emergent if they are not intrinsically found within any of the parts (e.g., an individual gene or an individual aneuploid cell) and exist only at a higher level of description (e.g., an aneuploid phenotype or a tumor mass).

Typical features of self-organization include: 1) absence of centralized control, 2) evolution over time, 3) fluctuations, 4) symmetry breaking or loss of freedom, 5) instability, 6) self-reinforcing choices, 7) multiple equilibria, 8) thresholds, 9) global order, 10) energy usage and export, 11) insensitive to damage, 12) self-maintenance, 13) adaptation, 14) complexity, 15) structural hierarchies (Lucas 1999). These general features of self-organization are characteristic of carcinogenesis and tumor formation, and are completely inexplicable in terms of the gene mutation theory.

Self-organizing systems by definition organize in the absence of external direction. A dynamical system of dividing aneuploid cells can spontaneously move from a disorganized state towards the more organized state called an attractor of the system, which in this case is a tumor. An attractor is a preferred position for a system, such that if the system is started from another state it will evolve until it arrives at the attractor and will stay there in the absence of other influences. A DNA index around 1.7 is an attractor for many cancers (Kato et al. 1998, Rasnick and Duesberg 1999, Rasnick 2000, Rasnick 2002, Shackney et al. 1995b).

Equation 6.5 models the auto-catalyzed progression of aneuploidy (Section 6.2.1). The variable ϕ ranges from 0–1 and is the fraction of the genome, as described above, that is out of balance relative to the euploid state (Section 5.4). The term $1-\phi$ represents the fraction of the genome that is diploid. The control parameter, r, is a unitless measure of the strength of the non-linear growth of ϕ. Equation 6.9 shows that the average aneuploid fraction, ϕ, of a population of cells at the n+1 cell division is determined by the average level of aneuploidy at the *n*th cell division (Rasnick 2000). The growth control parameter, r, plays a key role in carcinogenesis and the self-organization of aneuploid cells into a tumor.

$$\phi_{n+1} = r\phi_n - r\phi_n^2 \qquad (6.9)$$

Equation 6.9 (expanded version of Equation 6.5) is the well-studied logistic equation (Jensen 1987, Rasnick 2000) and is used here to describe the progression of aneuploid states as the growth control parameter, r, increases (Figure 6.6). The dynamics of a self-organizing system are typically non-linear because of positive and negative feedback. Since an analytical description of the non-linear dynamics of Equation 6.9 is impossible, the best we can hope for is a statistical theory that predicts the likelihood of the variable ϕ taking on any particular value (Jensen 1987).

Following an induction period of slow growth, the positive feedback term $r\phi_n$ of Equation 6.9 eventually leads to an explosive growth in aneuploidy (Rasnick 2000), which ends when all the aneuploid cells have been absorbed into the attractor states of DNA indices between 1.5–2 that are characteristic of cancer (Kato et al. 1998, Rasnick and Duesberg 1999, Rasnick 2000, Rasnick 2002, Shackney et al. 1995b). The same process governs the progression of aneuploidy and determines the DNA index attractor values of cells at the higher multiples of genome doubling. Once in the attractor, the aneuploid cells are controlled by the negative feedback term $-r\phi_n^2$, which allows for the relatively smooth evolution towards the equilibrium state at $\phi = 0.7$ (Figure 6.6). It was shown previously that after a large number of cell divisions DNA index $= 1 + \phi$ (Rasnick and Duesberg 1999, Rasnick 2000). Therefore, the attractor converging at $\phi = 0.7$ of Figure 6.6 corresponds to DNA index $= 1.7$ (Rasnick 2002).

Figure 6.6 is a map of the probability that after numerous divisions cells will have particular values of the aneuploid fraction, ϕ, for various values of r. At relatively low values of r, the aneuploid states of Figure 6.6 bifurcate until r reaches the critical value of 3.57, beyond which the progression of aneuploidy becomes chaotic (Jensen 1987, Rasnick 2000). Nevertheless, even in the midst of chaos there are regions of order represented by the dark streaks of Figure 6.6 that mark the upper and lower boundaries and crisscross the chaotic domains. The intersections of the dark streaks correspond to crises in the chaotic dynamics, where disjoint intervals of chaotic regions collide to form larger regions (Jensen 1987). The most spectacular crisis is readily visible at r $= 3.68$ and $\phi = 0.72$ (DNA index $= 1.7$) (Rasnick 2002).

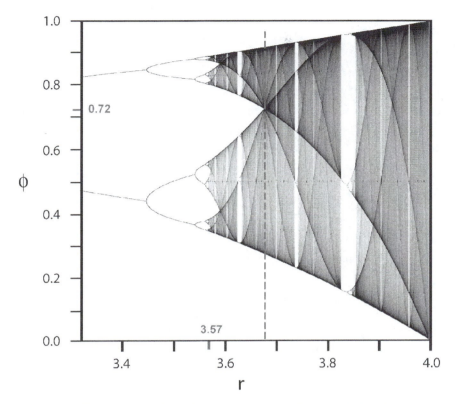

Fig. 6.6   The autocatalyzed progression of aneuploidy leads to cancer DNA indices near 1.7.

After a large number of cell divisions, DNA index = 1+φ (Rasnick and Duesberg 1999, Rasnick 2000). Equation 6.9 was iterated to generate a map of the probability that after numerous divisions cells will have particular values of the aneuploid fraction φ for various values of the growth control parameter r (Rasnick 2002). At relatively low values of r, the aneuploid states bifurcate until r reaches the critical value of 3.57 (red tick mark on abscissa), beyond which the progression of aneuploidy becomes chaotic. The denser regions of the probability map represent the more likely values of φ. Aneuploid cells evolve towards the attractor converging at r = 3.68 (broken red line) and φ = 0.72 (DNA index = 1.7) (red tick mark on the ordinate). At values of r greater than 3.68, the aneuploid cells become less coherent as their genomes become too disorganized and chaotic to sustain viability. That is why mature cancers tend to have DNA indices near 1.7 and its overtone multiples—the point of maximum disorder of the genome that still sustains life.

Color image of this figure appears in the color plate section at the end of the book.

The "order in chaos" that is apparent in Figure 6.6 plays an important role in delineating the range of the long-term viable values of the aneuploid fraction, ϕ, and the structure of the statistical descriptions (Jensen 1987). The denser regions of the probability map represent the more likely values of ϕ. The probability density is greater for values of ϕ above 0.7 than below. Therefore, one would expect near-tetraploid aneuploid cells to be the more common precursors of invasive cancer. The dark streaks represent values of ϕ that are most probable and visited more often as the cells evolve towards the DNA index = 1.7. The attractor of the next higher overtone converges at DNA index = 3.4 (i.e., 2 × 1.7 = 3.4) (Auer et al. 1980, Rasnick and Duesberg 1999, Rasnick 2002). At values of r greater than 3.68 the dark streaks diverge and the probability density thins out. Therefore, beyond r = 3.68, the aneuploid cells become less coherent as their genomes become too disorganized and chaotic to sustain viability. That is why mature cancers tend to have DNA indices near 1.7 and its overtone multiples, where they have evolved to the point of maximum disorder of the genome that still sustains life (Rasnick 2002).

It may seem paradoxical, but the basic mechanism underlying the self-organization of a population of aneuploid cells into a coherent tumor mass is the random, or entropy-driven variation inherent in each cell division. Every time an aneuploid cell divides, the genome is scrambled and becomes more disorganized than before. In other words, the entropy of the genome increases with each cell division, causing r to increase in an iterative process converging at $\phi = 0.7$. Therefore, as aneuploid cells divide, r is not really a parameter but actually a time-dependent variable that is driven by the increase in the entropy of the aneuploid cells. In the chaotic domain, the variable r governs not only the growth in the aneuploid fraction, ϕ, but also increases as the entropy of the aneuploid cells increases. As the value of r increases towards greater entropy, the dark streaks converge to the attractor at $\phi = 0.7$ (DNA index = 1.7), where the values are most dense, i.e., the probability greatest (Rasnick 2000, Rasnick 2002). One of the important consequences of the continued presence of a carcinogen is the acceleration in the growth of r (Duesberg et al. 2000b, Rasnick 2000, Rasnick 2002).

Variation on its own without further constraints produces entropy and disorder; however, disorder is generally held in check by selection, resulting in the elimination or reduction of part of the variety of configurations produced by variation (Heylighen 1999a). Albert Levan has pointed out, "The fact that the chromosome variation in tumors is never haphazard but gathers around stemlines and sidelines is compatible with the idea that the development in each tumor takes place according to an evolutionary pattern: the most viable karyotype prevails at all times" (Levan 1969). In keeping with Levan's idea, there are values of the aneuploid fraction, ϕ, that are compatible with the long-term viability of aneuploid cells and values that are not. Specifically, for values of r up to the viable limit of disorder at r = 3.68, no significant long-term populations of cells having an aneuploid fraction below $\phi \approx 0.3$ (DNA index = 1.3) or above $\phi \approx 0.9$ (DNA index = 1.9 and corresponding overtones) are expected (Figure 6.6). These limiting values of ϕ represent the window of chromosomal imbalance leading to cancer. This result is consistent with the range of 60 to 90 chromosomes ($\phi = 0.30$ to $\phi = 0.96$) observed in mature human cancer (Shackney et al. 1995a). While the limiting values of ϕ remain the same for all of the genome doubling overtones, the limiting DNA indices increase and the distributions of cells on both sides of the peaks broaden (Auer et al. 1980, Rasnick 2002).

Once a population of aneuploid cells has converged at DNA index = 1.7, the freedom of the individual aneuploid cells to act independently and evolve to a different attractor is restricted to the overtone multiples. The restriction to limited values of DNA content is equivalent to an increase of coherence, which defines self-organization (tumor beehive) and causal closure (Heylighen 1999b). Closure sets the tumor apart from the host, defining it as an autonomous new species of obligate parasite (Section 6.6).

6.2.4 Drug resistance is an inevitable consequence of aneuploidy

A second enduring principle has been that human tumours, when they do respond, contain subclones that become resistant.

(Chabner and Roberts 2005)

The simple strategy of chemotherapy is to kill as many cancer cells as possible without killing the patient. But a one gram tumor is composed of 10^8–10^9 cells. If a drug kills 99.9% of the cancer cells, that still leaves 10^5–10^6 cancer cells, which "will escape clinical and radiological detection but will be a few hundred micrometers in diameter" (Tannock 1998). Since the fractional cell survival after an entire course of adjuvant chemotherapy is about 0.01 (Tannock 1998), chemotherapy doesn't even come close to eliminating all the cancer cells.

A fundamental misconception is that there are anti-cancer drugs. The drugs used to treat cancer are actually antiproliferative drugs that target the same DNA and RNA synthesis, microtubule assembly and function, and topoisomerases required by normal cells, especially rapidly proliferating normal cells (Tannock 1998). It is not surprising that these drugs are quite toxic. To get around the high toxicity of the current crop of chemotherapeutic agents, efforts are now being directed towards developing drugs that possess greater specificity for cancer cells. Nevertheless, the problem remains of identifying cancer-specific targets. But as Hansemann said, "[cancer] displays no characters absolutely and completely lacking in the mother cell" (Whitman 1919). The only hope, then, of finding a cancer specific target for drug development would be to determine if there are essential genes expressed in cancers that are not as critical in normal tissues. None has been identified to date. Even if a new drug target is discovered, it will likely be rendered ineffective by the rapid appearance of drug resistant cancer cells (Section 5.3.5).

The gene mutation hypothesis is hard pressed to explain drug resistance, especially the appearance of multidrug resistance both before and after exposure to a chemotherapeutic drug targeting a specific gene or gene product. The chromosomal imbalance theory, on the other hand, readily explains the rapid appearance of drug resistance (Section 5.3.5). Indeed, chromosomal reassortment due to aneuploidy has been demonstrated experimentally to produce rapid drug- and multidrug-resistance in Chinese hamster cells (Duesberg et al. 2000a), mouse mammary carcinoma cell lines (Klein et al. 2010), human cancer cell lines (colon and breast) (Li et al. 2005), pathogenic yeast (Polakova et al. 2009, Selmecki et al. 2009), and the parasite Leishmania (Leprohon et al. 2009, Ubeda et al. 2008).

The collective order of a tumor's aneuploid cells (Section 6.2.3) protects it from perturbations. This robustness is achieved by the distributed or redundant control provided by the myriad of unique metabolic solutions produced by individual aneuploid cells. Thus, drug- or radiation-induced death to the susceptible part of a tumor can be replaced by the remaining, undamaged aneuploid cells.

The chromosomal imbalance theory shows that it is very unlikely that essential cancer-specific genes exist (Sections 4.4.2 & 5.4). As with normal cells, there are essential gene products for each individual cancer cell. However, there is a profound difference between normal and cancer cells. Normal cells of a particular type and from a particular tissue express a consistent ensemble of essential genes. Cancer cells, on the other hand, comprise a heterogeneous mix of heteroploid cells expressing perhaps an uncountable assortment of essential genes (Lengauer et al. 1998, Levan and Biesele 1958, Rasnick and Duesberg 1999). Many cancer cells express a common subset of essential genes, but due to the scrambling of the genome as a result of aneuploidy, other cancer cells from the same tumor will either not express or not rely on one or more of those genes. Therefore, these privileged cancer cells will not be sensitive to drugs targeted at gene products they do not express or no longer rely upon. This, at least in part, explains the phenomenon of intrinsic resistance to chemotherapeutic drugs in the absence of prior exposure. Thus, the appearance of estrogen receptor-negative breast tumors in women treated with tamoxifen (Li et al. 2001) and relapse in 72–80% of blast crisis leukemia patients on Gleevec (Druker et al. 2001, Gorre et al. 2001) is not surprising.

6.3   ANEUPLOIDY CAUSES THE WARBURG EFFECT BY INCREASING ATP DEMAND

The Warburg effect is a metabolic phenotype displayed by cancer cells and normal cells under certain conditions. It is characterized by glycolysis with lactic acid production in cancer cells even under normal oxygen saturation (aerobic glycolysis) concomitant with mitochondrial oxidative phosphorylation (Warburg 1956). The Warburg effect is also seen in rapidly proliferating normal mammalian cells, e.g., lymphocyte, endothelial, hair follicle and fibroblast (Vazquez and Oltvai 2011). Yet, the emergence of this mixed metabolic phenotype is seemingly counterintuitive, given that glycolysis produces only 2 moles of ATP per mole of glucose, 18 times less than the 36 generated by mitochondrial respiration. It turns out the Warburg effect of cancer cells is caused by increased demand for ATP production caused by aneuploidy.

Aneuploidy, in contrast to the mutation hypothesis, readily explains the tremendous alterations in the metabolic activity of cancer cells compared to their normal counterparts (Section 5.4), as well as the other unique phenotypes of cancer cells (Section 6.1.8). For example, the high DNA indices (1.5–2) (Rasnick 2000, Rasnick 2002), that are found in most malignant cancers (Wolman 1983), are directly compatible with the 10–100% increased levels of cytoplasmic RNA and protein in solid cancers (Caspersson et al. 1963, Caspersson 1950, Foley et al. 1965, Sennerstam et al. 1989). The increase in protein is limited by the space and solvating capacities of the cell. A typical normal cell contains 25% (17–40%) protein by weight (Fulton 1982). To compensate for the osmotic stress caused by the extra protein, cancer cells increase their volume (Caspersson 1950). In view of this, Caspersson et al. concluded that a typical, active cancer cell "is characterized by extreme stimulation and extreme activity of the system for protein formation... . [The] different organelles of the cytoplasmic protein-forming system are very well developed, and the cell shows all indications of excessive protein formation" (Caspersson 1950).

Elevations in cellular protein profoundly alter the physiology and biochemistry of cells. A 10% increase in cellular protein produces a 2-fold increase in membrane-bound proteins, and a 40% increase causes a 32-fold elevation in membrane-bound proteins (Minton 1994). In addition, a 10% increase in cellular protein causes a 5-fold boost in flux across membranes, and a 19% increase results in a 30-fold elevation in transmembrane fluxes (Minton 1994). Therefore, the membrane-bound tumor-associated antigens (Frati et al. 1984) and the high levels of secreted proteins (Alderman et al. 1985) that are responsible for the invasiveness and loss of contact inhibition of cancer cells are the consequence of the over-expression of protein due to aneuploidy.

Protein synthesis is a highly energy demanding process. Ten ATP/GTP are required for: 1) the synthesis of a codon, which is the nucleic acid information unit for an amino acid, 2) the charging reaction of a tRNA by its synthetase with the correct (cognate) amino acid, and 3) the subsequent incorporation of this amino acid into the nascent peptide chain (Szaflarski and Nierhaus 2007). Therefore, the energy requirements needed to produce the significantly increased protein levels of cancer cells necessarily places extra demands on ATP production (Caspersson et al. 1963, Caspersson 1950, Foley et al. 1965, Sennerstam et al. 1989).

Figure 6.7 shows that ATP production (f_{ATP}) is distributed among three fundamental pathways:

1) glycolysis via pyruvate through mitochondria (f_M, red arrow),

2) aerobic cytoplasmic glycolysis with production of lactic acid (f_L, blue arrow),

3) fatty acid oxidation through the mitochondria (f_{FA}, gold arrow).

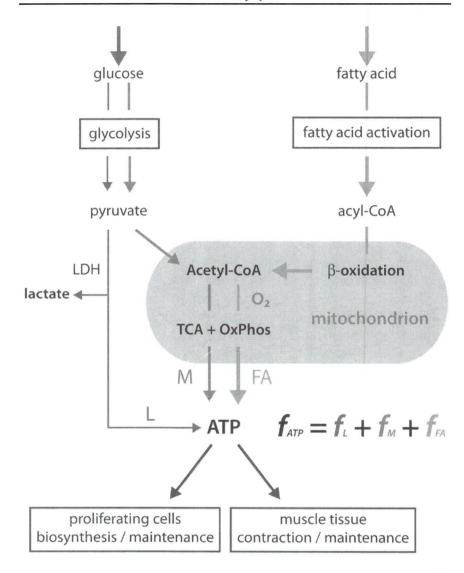

Fig. 6.7   Aneuploidy causes the Warburg effect by increasing ATP demand.

The energy requirements needed to produce the significantly increased protein levels of cancer cells necessarily places extra demands on ATP production. The production of ATP is distributed among three fundamental pathways: 1) glycolysis via pyruvate through mitochondria (f_M, red arrow), 2) aerobic cytoplasmic glycolysis with production of lactic acid (f_L, blue arrow), and 3) fatty acid oxidation through the

Fig. 6.7 Contd. ...

The normal distribution of metabolic flux along these three pathways changes dramatically as a consequence of the unavoidable physical constraints imposed by the chromosomal imbalance of cancer cells. The solvent capacity of cells for the allocation of cytosolic enzymes and mitochondria is limited. The mitochondria contribute 5 to 50 times more to molecular crowding than glycolytic enzymes and lactate dehydrogenase (LDH), and at least 1,000 times more than fatty-acyl-CoA synthase (acyl-CoA) (Figure 6.7) (Vazquez and Oltvai 2011). Aneuploidy massively disrupts the distribution of the enzymatic components of these systems.

Recently, Vazquez and Oltvai used flux balance analysis, coupled with the limited solvent capacity constraint (Vazquez et al. 2008), to explain the Warburg effect (Vazquez and Oltvai 2011). Because enzymes occupy a finite volume there is always an upper limit for the maximum rate of any metabolic pathway, which is roughly determined by the achievable maximum concentration of catalytic units, or more precisely by the solvent capacity constraint (molecular crowding) mentioned above. The authors' results for the low glucose uptake rates (or low ATP demand) of the normal cell, as expected, showed ATP was entirely produced by the mitochondrial aerobic pathway (f_M, red arrow, Figure 6.7) and there was no lactate production. This trend continued up to a threshold of ATP demand, when mitochondria occupied the entire cell volume fraction available for ATP production pathways.

Fig. 6.7 Contd. ...

mitochondria (f_{FA}, gold arrow). The normal distribution of metabolic flux along these three pathways changes dramatically as a consequence of the unavoidable physical constraints imposed by the chromosomal imbalance of cancer cells (see Section 6.3). ATP is produced entirely by the mitochondrial aerobic pathway during the low ATP demand of the normal cell (f_M, red arrow and f_{FA}, gold arrow). The production of ATP increases with demand until the mitochondrial capacity is saturated. Beyond this limiting value, the excessive ATP demand of aneuploid cancer cells leads to the abrupt turning on of the cytosolic glycolytic pathway for additional ATP production with increasing secretion of lactate (f_L, blue arrow).

Color image of this figure appears in the color plate section at the end of the book.

Beyond this threshold value the concentration of mitochondria, and therefore the rate of ATP production through oxidative phosphorylation, could not be increased further. The excessive ATP demand of aneuploid cancer cells is larger than mitochondria can satisfy. This leads to the abrupt turning on of the cytosolic glycolytic pathway—the Warburg effect—for additional ATP production with the increasing secretion of lactate (f_L, blue arrow, Figure 6.7).

For proliferating cells using both fatty acids and glucose as energy sources (e.g., prostate cells (Liu 2006)) and muscle metabolism, at low ATP demand it was more efficient to produce ATP entirely from fatty acid because this results in a higher yield of ATP than glucose catabolism (f_{FA} gold arrow, Figure 6.7). Fatty acid uptake increases until the mitochondrial capacity was saturated for ATP production from fatty acids. As demand for ATP continues to increase beyond the fatty acid limit, catabolism of glucose increases, gradually replacing the utilization of fatty acids (f_M red arrow, Figure 6.7).

When the ATP production from pyruvate in the mitochondria was saturated, cytoplasmic glycolysis and lactate production takes over as described above. From these results, Vazquez and Oltvai concluded that "ATP demand is the major driver of the Warburg effect. This observation contradicts the generally accepted notion that the need for precursor metabolites is the main driving force behind the Warburg effect in proliferating cells (Vander Heiden et al. 2009)."

6.4 BALANCED MITOTIC FORCES AND SPECIES-SPECIFIC SEQUENTIAL CHROMATID SEPARATION MAY GOVERN THE RATE OF TRANSFORMATION

The appearance of spontaneously transformed variants is very common when mouse cells are being cultured, rather rare for hamster and rat cells, and virtually unknown for human and chicken cells.

(Cairns 1978)

Understanding the role of intracellular geometry, as well as the effect of cell size and shape, on physiological processes should become an important future direction of cell biology.

(Storchova and Kuffer 2008)

It is well known that rodent cells in culture can spontaneously transform into immortal cell lines with either benign or malignant phenotypes, while human cells do not (Cairns 1978, Hayflick 1965, Levan and Biesele 1958, Todaro and Green 1963). What is not known is the reason for this qualitative difference.

Mice and humans both have about 25,000 genes (Collins et al. 2004). "Approximately 99% of mouse genes have a homologue in the human genome. For 96%, the homologue lies within a similar conserved syntenic interval [appear in the same order] in the human genome" (Waterston et al. 2002). Lisa Stubbs of the Lawrence Livermore National Laboratory said, "I know of only a few cases in which no mouse counterpart can be found for a particular human gene, and for the most part we see essentially a one-to-one correspondence between genes in the two species" (Stubbs 2010). These results suggest the profoundly different susceptibilities of mouse and human cells to transform spontaneously is almost certainly not due to differences in their genes. Here is proposed a physical and temporal explanation for why mouse cells spontaneously transform with relative ease and human cells do not.

Cells containing misaligned chromosomes are delayed an unusually long time before entering anaphase (Zirkle 1970). At metaphase, a single misaligned chromosome can inhibit further progression into anaphase (Gorbsky and Ricketts 1993, Nicklas et al. 1995). So-called checkpoint genes have been proposed to monitor various aspects of the mitotic process (Cahill et al. 1998). However, others have argued that "intelligent", supervisory genes are not required to monitor and "police" mitosis. Alternatively, it has been proposed that geometry, symmetry (Hodge et al. 1995, Nicklas et al. 1995, Rodman et al. 1978, Welter and Hodge 1985) and consequent balance of mitotic forces (Li and Nicklas 1995, McIntosh 1991) govern the metaphase to anaphase transition.

In mitotic normal diploid cells, chromosomes are always incorporated into a single rosette, and there is a remarkable tendency for chromosome homologs to be positioned in proximity (Hodge et al. 1995, Kosak et al. 2007). In plants: all centromeres are non-randomly positioned on the metaphase plate; complete haploid genomes tend to be separated in diploid and hybrid cells; and heterologous chromosomes are in a predictable mean fixed order within each of the haploid genomes (Heslop-Harrison and Bennett 1984).

The territorial organization of chromosomes in interphase (chromosome territories) are organized in a species-specific (Neusser et al. 2007) and cell type-specific (Parada et al. 2004) manner which constitutes a basic feature of nuclear architecture (Cremer et al. 2001, Cremer and Cremer 2010). The global chromosome positions of diploid mammalian cells are transmitted through mitosis preserving the chromosome arrangements throughout the cell cycle (Gerlich et al. 2003). The congression of chromosomes results in a simple linear projection of prophase chromosome positions onto the metaphase rosette, which no longer contains spatial information about their original positions along the spindle axis. However, the chromosome positions along the spindle axis are re-established by timing differences of sister chromatid separation at anaphase onset and maintained during poleward movements of chromosomes and expansion of daughter nuclei.

Synchronous synthesis of homologous α-satellite pairs is essential to maintaining diploid chromosome territories. Litmanovitch et al. "showed an association between replication timing of α-satellite sequences and centromeric function. Chromosome pairs whose homologous α-satellite loci replicated highly synchronously revealed low rates of aneuploidy, whereas chromosome pairs with a slightly asynchronous replication pattern (i.e., short intervals between early- and late-replicating loci) revealed intermediate rates of aneuploidy, and chromosome pairs exhibiting asynchrony with long-time intervals between early- and late-replicating loci showed the highest rate of aneuploidy" (Litmanovitch et al. 1998).

The timing of centromere separation is important because premature separation leads to nondisjunction and, hence, genesis of aneuploidy (Garcia-Orad et al. 2000) in human (Angell 1997) and mouse (Mailhes et al. 1998) oocytes. More recently, Gerlich et al. demonstrated that Bisbenzimide (a producer of mitotic errors and multinucleated cells) efficiently randomizes the positions of rat chromosomes by disrupting the normal sequence of chromatid separation (Gerlich et al. 2003). A relationship between premature separation and aneuploidy has been reported for a G-group chromosome in a case of chronic myelogenous leukaemia (Vig 1984) and chromosomes 1 and 3 in multiple myeloma (Vig 1987). Errors of centromere separation have also been reported for certain other types of leukemia (Gallo et al. 1984, Knuutila et al. 1981, Littlefield et al. 1985, Shiraishi et al. 1982).

The centromeres of a given genome at the metaphase-anaphase junction resolve into two units in a non-random sequence. In humans, the centromeres of chromosomes 2, 8, 17, and 18, for instance, are the ones to separate the earliest, whereas those of chromosomes 13, 14, 15, 21, and 22 are among the last to split into two sub-units. A similar situation holds for frog, Indian muntjac, wood lemming, kangaroo rat, Chinese hamster and mouse (Garcia-Orad et al. 2000).

The centromeres without any detectable heterochromatin separate first at the metaphase-anaphase junction. Generally, the centromeres flanked by increasing quantities of pericentric heterochromatin are delayed in separation in direct proportion to the quantity of heterochromatin. However, human chromosomes carrying the largest quantities of pericentric heterochromatin (numbers 1, 9, and 16) are neither the first nor the last ones to separate at the metaphase-anaphase junction. The heterochromatin in man is fairly complex and is made up of several different classes of satellites. In mice, the primary constituent of heterochromatin is the major satellite DNA (minor satellite does not appear to constitute the heterochromatin). Apparently, it is not only the quantity of pericentric heterochromatin which controls the timing of separation but also the composition of DNA and associated proteins (Garcia-Orad et al. 2000).

Not only do the centromeres in a genome separate in a nonrandom manner, but the inactive centromeres found along the dicentric and multicentric chromosomes in human aneuploid cell lines always separate ahead of active centromeres—the inactive centromeres split at pro-metaphase (Garcia-Orad et al. 2000). Like other parts of the chromosome, the centromeres do not separate immediately upon completion of DNA replication. Among the various parts of a chromosome which divide into two visible subunits in a hierarchical sequence, the centromeres separate last. Thus, there seems to be some sort of "maturation" or lag period between completion of replication and separation of the chromatids, especially at the centromere.

In the mouse, the inactive centromeres which separate early at prometaphase also replicate earlier in the S phase. Considering this relationship as well as the fact that various parts of a chromosome replicate and separate at different periods in the cell cycle, it is logical to assume that the sequence of centromere separation reflects the sequence of replication of the centric region for all centromeres. In other words, the earlier an active centromere replicates, the earlier it separates (Garcia-Orad et al. 2000).

Proposed mechanism governing the rate of transformation

During mitosis, chromosomes arrange their centromeres non-randomly in a circular ring or rosette (Kosak et al. 2007). It appears that the order of arranging the chromosomes around the metaphase ring varies widely depending on the species (Heslop-Harrison and Bennett 1984). A symmetrical ring is necessary to balance the mitotic forces permitting the transition from metaphase to anaphase. The spindle apparatus forms a physical link that mechanically separates the sister chromatids during anaphase. If a symmetrical ring does not form, then the mechanical forces that separate the chromatids are out of balance and either prevent or delay anaphase and cytokinesis. If cell division is only delayed, the daughter cells are likely to be aneuploid.

It is proposed that a species-specific sequence of chromatid separation in conjunction with a simple balance of forces resulting from a high degree of symmetry about the axis connecting the centromeres of a normal cell controls disjunction of chromatids during anaphase. As Li and Nicklas have said, "These events are tied to the accurate distribution of chromosomes in cell division by an elegant use of mitotic forces not only to move chromosomes, but also to detect errors" (Li and Nicklas 1995). Anything that disrupts this symmetry or species-specific sequence of centromere separation disrupts the normal separation of chromatids during anaphase leading to aneuploidy.

Once aneuploidy is established, it autocatalytically perpetuates the scrambling of chromosome territories. For example, the aneuploid cervical cancer HeLa (Walter et al. 2003) and fibrosarcoma HT-1080 (Thomson et al. 2004) cell lines experience major changes in the chromosome territories from one cell cycle to the next—another manifestation of chromosome instability. Likewise, Cremer et al. observed substantial differences between the chromosomal territories of 7 of 8 aneuploid cancer cell lines when compared to 6 normal cell types, indicating a loss of radial chromatin order in tumor cell nuclei (Cremer et al. 2003).

It is further proposed that the mouse has much greater flexibility than the human in arranging aneuploid chromosomes to form the circular metaphase ring and balance mitotic forces. Consequently, mouse cells are considerably more likely than human to produce combinations that lead to a growing population of aneuploid cells.

Karyograms are arranged on the basis of chromosome size and the position of the centromere. There is a greater range in the locations of centromeres and sizes of human chromosomes compared to mouse. Most human chromosomes contain two arms. In karyograms, the p-arm is shorter and is oriented above the centromere while the q-arm is below the centromere. Mouse chromosomes are more similar in size and all are acrocentric (having the centromere close to one end and thus no p-arm). Thus, the greater range of sizes and varied location of the centromeres combine to produce many more unbalanced permutations of human chromosomes arranged in the metaphase rosettes compared to mouse. In other words, the combination to unlock anaphase is much simpler in mouse compared to human. The addition or subtraction of chromosomes would be expected to disrupt the balance of mitotic forces less for the mouse compared to human. This may account for the much shorter Hayflick limit of rodent compared human cells in culture as well as the orders of magnitude greater rate of spontaneous transformation of rodent cells in culture (Cairns 1978, Rasnick 2000).

To summarize, if combinations of aneuploid and marker chromosomes can form balanced circular, hence symmetrical metaphase rings, then the metaphase to anaphase transition will take place. The multipolar prometaphase cells common in aneuploid cells tend to bi-polarize before anaphase onset, and the residual merotelic attachments would produce chromosome missegregation due to anaphase lagging chromosomes (Silkworth et al. 2009) but only if a sufficient balance of mitotic forces can be maintained. Those cells that can form a balance of spindle forces around the metaphase ring by incorporating only the normal complement of chromosomes will be the most resistant to aneuploid cell production. This is proposed to hold for normal human cells. Chromosome breaks and rearrangements may increase the likelihood of human aneuploid cells forming a force-balanced symmetrical metaphase ring. This would greatly increase the propagation of aneuploidy by balancing the mitotic forces allowing the cells to progress from metaphase to anaphase.

6.5 CANCER VACCINE IS VERY UNLIKELY

"A surprising argument used in some of the reviews dealing with immune surveillance is based on the assumption that in any information transfer system, such as somatic cell replication, there are inevitable errors, and neoplastic transformation therefore must be frequent. The argument is made that immunological surveillance must be efficacious or overt clinical neoplasia would necessarily be more frequent than it actually is. This circular argument also includes the assumption that frequent accidents of somatic cell replication produce neoplastic variants that are invariably antigenic and thus can be rapidly eliminated by the immune system."

(Stutman 1975)

The idea that the clinical course of cancer depends on whether or not a tumor's potential for unrestricted growth wins out over inherent host defenses is 200-years-old (Edinburgh 1806). A modern formulation of this view known as the immune surveillance hypothesis of cancer was advanced by Burnet (Burnet 1957) and Thomas (Thomas 1959). The main assertions of the immune surveillance hypothesis are: most tumors are antigenic and such antigenic differences can "under appropriate conditions" provoke an immune response (Burnet 1971).

Based on this thinking, in the late 1950s Jonas Salk attempted to stimulate the immune systems of terminally ill cancer patients by injecting them with what he thought were monkey heart cells. He had hoped that the patients' activated immune systems would attack the cancer cells. However, in 1978 Salk revealed that he had not injected the cancer patients with monkey heart cells but mistakenly with HeLa cancer cells (Gold 1986). The cancer patients' immune systems did indeed become activated and functioned well enough to eliminate the small tumors formed at the sites of injection of the HeLa cells within three weeks, never to return. Yet the activated immune systems of these same cancer patients were not effective against their natural tumors.

It is not the purpose here to rehash the exhaustive analysis of, and compelling arguments against, the immune surveillance hypothesis (Dawson et al. 1978, Herberman 1984, Hewitt et al. 1976, Prehn 2005, Rygaard and Povlsen 1974, Stutman 1974, Stutman 1975) but simply to add that the chromosomal imbalance theory provides additional support for the view that there is no significant connection between cancer and textbook immunity.

While the notion of immune surveillance has run out of steam, the efforts to vaccinate against cancer continue full speed (Chitale 2009, McLemore 2006), though not without criticism (Calder 2009, Sweet 2008). Hundreds of vaccine clinical trails in patients with metastatic cancer have been published (Ribas et al. 2003, Rosenberg et al. 2004) but the "anticancer vaccines don't work" (Prehn 2005). The fundamental reason for the vaccine failures is "a basic lack of tumor-specific antigenicity rather than a blocking or suppression of an immune response" (Prehn 2005).

Cancer is us because it is derived from our very own genome. What makes cancer cells not us is that they have rearranged our genome to differ from the diploid predecessors in both the number of chromosomes and the dosage of thousands of genes. Since there are no new genes, and no cancer-specific mutant genes, and no new chromosomes (except hybrid or marker chromosomes) in cancer cells (Duesberg et al. 2000a), there is little or nothing for immune surveillance to monitor. This is especially true for the earliest stages of carcinogenesis where the immune surveillance mechanism is supposed to be most effective but the aneuploid cells are least abnormal. Even if an aberrant antigenic cell happened to result from the chaotic scrambling of the genome, the immune system could be expected to eliminate it, while the vast majority of aneuploid cells remained invisible to the immune system. Therefore, even in principle, there is no possibility of an immune surveillance system guarding against the appearance of cancer cells.

6.6   "CANCERS ARE A GENUINE TYPE OF SPECIES"

Heritable genomic variation and natural selection have long been acknowledged as striking parallels between evolution and cancer. The logical conclusion, that cancer really is a form of speciation, has seldom been expounded directly.... . The implications of the "cancer as species" idea may be as important for biology as for oncology, providing as it does an endless supply of observable if accelerated examples of a phenomenon once regarded as rare.

(Vincent 2010)

Speciation is the product of nature's most definitive and far-ranging mutation: chromosome number and structure variation. The chromosomal theory of speciation has been described in detail (King 1993) and is growing in acceptance (Faria and Navarro 2010). Because a species is defined by a specific number of chromosomes and the gene sequences within (Matthey 1951, O'Brien et al. 1999, Shapiro 1983, White 1978, Yosida 1983), and not necessarily by a species-specific gene pool (O'Brien et al. 1999), cancer falls within the definition of speciation. By contrast, the number and even the function of genes is not necessarily changed in speciation. For example, among mammalian species the specific number of chromosomes and the sequences of genes within are definitive, whereas the gene pools of mice and men (Waterston et al. 2002)—indeed, mammals in general (O'Brien et al. 1999)—are basically conserved (Section 6.4).

It follows that aneuploid cells, above all cancer cells, are by definition new species that differ from their diploid predecessors in both the number of chromosomes and the dosage of thousands of genes. However, as a species of its own, aneuploid cancer is a parasite unable to function independent of its diploid host. Moreover, because of the inherent instability of aneuploid karyotypes, cancer cells are unlikely to retain acquired properties long enough to evolve phylogenetic permanence (Duesberg et al. 2011).

The view that cancer cells are a species of their own is completely compatible with Hansemann's theory of anaplasia, which postulates that cancer results from an alteration of the cell's species, "eine Artenveraenderung der Zellen" (Hansemann 1897). According to Hansemann, this alteration is not de-differentiation or trans-differentiation of a normal cell to a cancer cell, "but the cells change their character in every regard morphologically and physiologically to a new species" (Hansemann 1897). The pathologist Hauser, a contemporary of Hansemann, described cancer cells as a "new cell-race" (Bauer 1963, Braun 1969, Hauser 1903). Hauser used this term to account for the multiplicity of characters that set apart cancer cells from normal counterparts. The new species-analogy also confirms the suspicion of the geneticist Whitman, who tried to reconcile cancer with gene mutation in 1919: "The trouble is, indeed, not that the changes observed in cancer cells prove too little, but that they seem rather to prove too much" (Whitman 1919).

Boveri thought carcinogenesis "could be achieved by the loss of single chromosomes" (Harris 2007). Accordingly, he set out to induce cancer in a rabbit cornea by inducing chromosome nondisjunction. To this end, he induced tetraploidy with inhibitors of mitosis, which would then favor losses or gains of chromosomes by non-disjunctions in subsequent mitoses. But no tumors appeared "after some time" in these animals (Harris 2007). It is clear Boveri's period of observation was too short and his cells treated insufficiently with mitotic inhibitors to evolve a new autonomous cancer karyotype. Indeed, within a year after Boveri published his classic paper, Yamagiwa and Yoshikawa demonstrated in 1915, it required applying carcinogenic tar 2–3 days per week for one year to induce cancer in rabbits (Yamagiwa and Ichikawa 1915).

After World War II, Hauschka equivocated between attributing "the pathological differentiations of oncogeny" either to "differential gene activation" or to "more drastic reorganizing of the somatic karyotype in a mutation-selection sequence analogous to phylogeny" (Hauschka 1961). For this latter possibility, Hauschka relied on Julian Huxley's definition of autonomous growths as "equivalent to new biological species" (Huxley 1956). According to Huxley, "Once the neoplastic process has crossed the threshold of autonomy, the resultant tumour can be logically regarded as a new biological species, with its own specific type of self-replication and with the capacity for further evolution by the incorporation of suitable mutations. From the angle of biological classification, all tumors, whether of plants or animals, could then be regarded as constituting a special organic phylum or major taxonomic group, with the following characteristics: 1) universal parasitism, but with the parasite always originating from its host; 2) some loss of supracellular organization; 3) lack of limit to proliferation; 4) (a) in most cases each individual tumor is the equivalent of a biological species…and each species becomes extinct on the death of its host; (b) …in tumors maintained artificially…a certain amount of evolutionary divergence may occur in substrains." (Huxley 1956).

Then, in 1959, Rous confirmed Haldane's view that the gap between cancer cells and their normal predecessors is too big to be explained by known gene mutations: "The cells of the most fatal human cancers are far removed from the normal in character, and almost no growths fill the gap between, much less a graded series of them, such as one might expect were they the outcome of random somatic mutations" (Rous 1959). In other words, Rous even pointed out missing links, an evolutionary hallmark regarding the relationships among species.

The probable answer to the question of missing links in phylogeny and oncogeny is that both are based on the common mechanism of chromosome number variation, which involves coordinate changes of thousands of genes. The concept that aneuploidy defines a species also explains why mutations that cause cancer are "somatic" rather than germinal. Aneuploidy is not heritable because the product would either be non-viable (Hassold 1986, Hook 1985, Muller 1927) or it would be a new species of its own. By contrast, gene mutations, particularly those that are postulated to cause cancer, can be inherited by transgenic animals (Donehower et al. 1992, Hariharan et al. 1989, Kim et al. 1993, Purdie et al. 1994, Sinn et al. 1987), or congenitally in humans (Haber and Fearon 1998, Knudson 1985) without causing cancer, although they may increase the cancer risk.

In 1969, Foulds included "cytogenetic and biochemical individuality" in the definition of cancers (Foulds 1969). In the same year, Levan considered cancer-specific karyotypes as an alternative to "invisible genetic changes" (i.e., mutations). He said "it would be reasonable to expect a priori that each tumor type would be characterized by one karyotype, just as...a species is characterized by its karyotype" (Levan 1969). Case after case of comparisons of close species has revealed the "extraordinary" fact that chromosomal differentiation of some sort is almost always present (White 1978). By 1991, Van Valen had proposed that HeLa cells could be considered a new species. HeLa is one of the oldest and most commonly used human cell lines derived from cervical cancer cells taken from Henrietta Lacks, a patient who eventually died of her cancer on October 4, 1951. Van Valen proposed the name Helacyton *gartleri*, after Stanley M. Gartler, who was the first author to recognize "the remarkable success of this species" (Van Valen and Maiorana 1991).

The idea of cancer as new species was recently brought up to date in Mark Vincent's 2010 article "The animal within: carcinogenesis and the clonal evolution of cancer cells are speciation events *sensu stricto*" (Vincent 2010). Vincent said it is legitimate to ask: exactly what form of life is represented by a cancer? In answering this, one must consider the development of the eukaryotic cell, and the evolution of multicellular organisms (metazoa). For cancer is a problem both of the nature of the eukaryotic cell, and the multicellular organism as a whole because it has no meaning except in a metazoan context. Since metazoa are composed only of eukaryotic cells (but eukaryotic cells are not always metazoa) it is probable that the nature of cancer is to be found in the derangement of those acquired aspects of the eukaryotic cell that enable multicellular life. Metazoan existence requires cell adhesion, differentiation, and coordination (Rokas 2008), three features whose breakdown is fundamentally characteristic of cancer (Hanahan and Weinberg 2000). The net result is a life form which is not only destructive of the metazoan host, but also one that is, at most, a colony of loosely cooperating but often competing, and independently evolving individual cells (Merlo et al. 2006); or, at worst, a collection of unicellular eukaryotic organisms that are fully capable of existence independent of each other (if not from the host, or, at least, some host).

According to Vincent, knowledge that cancer is an evolutionary process is widespread (Greaves 2000, Merlo et al. 2006). It is well appreciated that, in metazoa, random mutations, heritable variation, and selection pressures all operate to both generate and further evolve malignant tumors. A strangely neglected aspect of this is that bona fide speciation seems to be involved, in fact serial speciation. Perhaps the notion of species as applied to asexual life forms was a barrier to this realization, or that we have all been educated to believe that speciation is both very rare and imperceptibly slow. Maybe the instability of the cancer genotype has tended to obscure the fact of speciation, because we naturally expect species to be stable for thousands if not millions of years (Kutschera and Niklas 2004), and it is counterintuitive to acknowledge that something that is unstable over months can be a species, or even a series of species (Vincent 2010).

Vincent said, "As a clinician, one is in a privileged position to observe evolution in action, in a way that is rarely if ever possible for other people. Cancers develop, behave in a radically different way from normal cells, move into altogether different niches, and successively evolve into even more aggressive and treatment-resistant forms. If this is not serial speciation it is hard to know what it is; a cunning, if ultimately short-sighted survival strategy, perhaps, but certainly not intelligent design, either from the viewpoint of the human host, or with respect to the eventual fate of the cancer itself."

So why do cancer species evolve so much faster than normal phylogenetic species did according to the fossil record—namely in many months to decades compared to millions of years? This seems to be a matter of complexity. Since the genetic complexity of normal sexual species consisting of many highly differentiated cells is orders of magnitude higher than that of their corresponding asexual, single-cell cancer species, the probability of forming new karyotypes of sexual species by random karyotypic variation would be exceedingly much lower and thus slower than forming a new cancer species by the same mechanism.

Using the same criteria by which bacteria and other asexual organisms are now considered to belong to bona fide species, it should be clear that there are no fundamental reasons not to consider cancers in the same way. It is true, however, that the original event (i.e., carcinogenesis) takes place "in sympatry" with the surrounding tissue of origin, that is, in the same place. This used to be considered unusual in speciation, but apparently is no longer the dominant view (de Aguiar et al. 2009, Via 2009). Subsequent episodes of speciation, resulting in clonal evolution to more aggressive or drug resistant forms of cancer, may take place away from the original parent cell, that is, "in allopatry." However, given the asexual nature of cancer cell reproduction, these terms (sympatry and allopatry), crucial in the speciation of sexual organisms, have diminished relevance. Indeed, gene flow (i.e., sex) is the biggest obstacle to sympatric speciation in sexual organisms, but this obstacle disappears in the context of carcinogenesis, which is aneuploidy-driven and, being asexual, does not require reproductive isolation.

It is likely that a certain amount of serial change takes place within a cancer species lineage (anagenesis) but eventually exceeding some significant threshold of change, so as to justify the belief that further rounds of speciation have occurred. For a while, the new species may coexist (be paraphyletic) with the older species of cancer cell, until the older ones are out-competed and go extinct. In this way, the tumor genome may be "swept clean," and homogeneity reestablished (for a while) by a form of genetic cohesion, as with bacteria.

Vincent asks whether the speciation concept goes far enough. If this was merely speciation, one would expect the generation of another, roughly similar metazoan. But what seems to be entailed in carcinogenesis is the appearance not only of a new and separate individual (and then rapidly a whole set of new individuals), but that these new individuals all seem to belong to a wholly different branch in the eukaryotic Tree of Life: a different species from the host, for sure, but also a different genus, family, order and even phylum. Huxley also made this point: "all tumors…could then be regarded as constituting a special organic phylum or major taxonomic group…" (Huxley 1956). The closest existing taxonomic group in which a cancer cell might very tentatively be placed is as a *holozoan opisthokont* (animal-like) protist (Gray et al. 2004, Rivera and Lake 2004) with facultative colonial attributes. In this view, cancer patients are literally being eaten inside out by a very primitive type of animal; primitive in the sense of being "minimized due to secondary losses of characteristics," rather than "truly primitive" in the sense of being simple and ancestral (Dacks et al. 2008, Krebs 1981). One might tentatively propose the subgroup title *dyskaryota* (abnormal karyotype) to embrace the entire malignant assemblage.

7

New Perspectives for Cancer Prevention, Diagnosis and Therapy

Industry, like academia, has placed too much emphasis on employing technology and too little on understanding the pathology of human cancer.
(Dermer 1994)

Chromosomal imbalance theory puts everything about cancer in a different light. To begin with, the "war on cancer" metaphor should be replaced. Cancer is better viewed as an ecological problem: the patient and the cancer are an environmental whole. Taking this small step would truly be one giant leap for humankind.

Six recommended changes in cancer research

1. Cancer research should dramatically shift interest away from the submicroscopic realm of individual genes and toward the domain of the pathologist and health practitioner.
2. The dynamics of the progression of aneuploidy in benign, invasive, metastasized, and recurrent human tumors should be studied.
3. Physical and chemical agents as well as drugs should be tested and cataloged for their aneuploidogenic (aneugenic) potencies and correlated with their carcinogenic potentials.
4. The detection and quantification of aneuploidy should be incorporated into the diagnosis, staging and grading of cancer.

5. Rosner warned as early as 1978 that, "in the future, cytotoxic agents would be used reluctantly to treat cancer because of the potential risk to the patient of developing acute leukemia or a second neoplasm" (Rosner 1978). Rosner's prophecy has partially come true with a vengeance: chemotherapy-induced second cancers now account for one in six of all new cancers diagnosed in the United states (Allan and Travis 2005, National Cancer Institute 2005a), but there is little reluctance to treat cancer patients with aneugenic drugs.

6. Investigating the nature of the spontaneous remission of tumors should be given high priority. The results could guide the development of non-toxic therapies.

Cancer research would benefit from de-emphasizing cell culture

Bruce Alberts said to an audience of cancer researchers at the quarterly meeting of the President's Cancer Panel in December 1990, "we have rarely seen a real human tumor" (Alberts 1990). Thirteen years later and less subtly, Horrobin asked: "Does the functioning of cells in culture bear a sufficiently strong relationship to the functioning of cells in an organ *in vivo* such that conclusions drawn from the former are useful in predicting behaviour of the latter?" He raised the question because, "what the *in vitro* system cannot do is construct a functional and valid *in vivo* biochemistry. And that is potentially a fatal flaw. For in most human diseases it is the functional biochemistry and not the anatomical biochemistry which goes wrong. When we ask cell culture to inform us about *in vivo* cell function, in most cases we ask too much" (Horrobin 2003). The situation has only gotten worse and needs to be reversed because the bulk of modern research papers are not about human cancer but reports of experiments with aneuploid cell lines.

The fact that cancers are clonal, coupled with the belief that mutant genes cause cancer, has contributed to the decades of countless *in vitro* studies using immortal cell lines. Cell culture was thought to reduce cancer to its essentials, away from the complications of tumors in patients. All one needed to do was identify and then manipulate the cancer-causing genes. Almost all of the subdivisions of cancer research owe their existence to the facility of using cell lines as research tools. Molecular biology, immunology, tumor cell biology, biochemistry, carcinogenesis, endocrinology, drug development, and virology of cancer all rely on information derived from the cell line model (Dermer 1994). Failure to recognize the fact that primary cancer cells almost always die soon after being put into culture (Atkin and Baker 1990, Dermer 1994)—signifying a fundamental difference between immortal cell lines and human cancer—has tainted virtually all cancer research.

Attempts to establish permanent cell lines from primary tumors by prolonged culture *in vitro* are successful in less than 10% of cases (Mamaeva 1998). The few cell lines obtained from primary tumors were most often highly malignant and poorly differentiated. According to Mamaeva: "It is impossible to obtain cell lines from highly differentiated cells of benign solid tumors. For instance, [a] permanent cell line has never been observed to appear spontaneously from human meningioma, a usually benign tumor characterized by a specific monosomy on chromosome 22.... . There is no permanent cell line reaching the stabilization stage in its evolution, which would have been proven to have cell clones with the near-haploid chromosome number or with nullisomies on some chromosome." The easiest ways of obtaining cell lines are: 1) spontaneous transformation of rodent cells, 2) treatment with carcinogen, and 3) most efficient of all, transfection with one or several oncogenic viruses.

The massive karyotypic heterogeneity of tumors *in vivo*, signifying substantial genomic imbalance and complexity (Hoglund et al. 2001a), contrasts sharply with karyotypes of permanent cell lines that have reached the stage of stabilization in their evolution (Mamaeva 1998). For example, the karyotypic peculiarities of leukemia and tumor cells *in vivo*, apart from specific chromosome translocations, are partial or complete monosomies in certain autosomes (Mitelman 1994). The situation is very much different in permanent cell lines.

Mamaeva et al. developed a method called generalized reconstructed karyotype (GRK). By identifying fragments of all marker chromosomes of a cell line, it was possible to determine the total amount of chromosomal material from each normal chromosome of a cell. Ideograms of entire chromosomes and fragments were reconstructed from markers and placed in a karyogram resembling the normal chromosomes. By summing up reconstruction results of chromosome sets of individual cells or cell clones, the GRK of a cell line is obtained (Mamaeva 1998).

When GRK was applied to 500 human and animal cell lines, the results demonstrated that at least 85% retained disomy of all autosomes (Mamaeva 1998). "This means," the authors said, "that to exist *in vitro*, permanent line cells must have at least [the equivalent of the] two homologs of each autosome. Establishment of such lines is achieved either by selection of a preexisting original stem cell with an increased proliferative potential and with the diploid autosome set or by a gradual evolution of the karyotype of the initial hypodiploid cells *in vitro*, with chromosomes undergoing structural and numerical rearrangements in the course of evolution." Furthermore, the karyotype evolution as a result of polyploidization of initial cells with monosomes is frequently accompanied by changes in their phenotypic and growth characteristics as well as in the degree of malignancy. For instance, a study of glioma cell lines has shown that polyploid cells that appear during prolonged culture differ from the initial near-diploid tumor cells by the absence of a number of specific proteins, by a three- to five-fold increase in the doubling time of the cell population and by the loss of malignancy (Mamaeva 1998).

Aside from aneuploidy, cell lines and human cancer have almost nothing else in common. Cell culture can never duplicate the interaction between tumor and patient that has an inevitable and fundamental effect on malignancy. Transfer from an *in vivo* to an *in vitro* environment precipitates a crisis for the cells. A cell that survives the crisis period and learns how to live forever in culture undergoes fundamental changes that render it and its descendants profoundly different from cells that live in patients.

The nature of life itself changes (Dermer 1994). Cell lines have evolved to single-cell organisms adapted to life in the laboratory and must be appreciated as autonomous living systems rather than a faithful recapitulation of cancer *in vitro* (Mamaeva 1998).

By contrast, advanced, malignant cancers are aneuploid organisms that have evolved to survive in the environment of the person (Section 6.6). This explains why the thousands of cell line studies have contributed so little to a better understanding of human cancer or more accurate diagnoses and effective treatments. "Given the potential lack of relevance of *in vitro* research, one might have expected rigorous theoretical discussion and experimental exploration of the problems of the issue of congruence of *in vitro* and *in vivo* studies. But there is almost nothing: certainly there is no general sense among medical researchers using *in vitro* systems that their work involves so many untested and unjustified assumptions that its congruence with any useful *in vivo* world must be in serious doubt. And if that congruence is unproven, there must exist the risk that not only might the *in vitro* work be useless, it may be actively misleading" (Horrobin 2003).

Acknowledging, that within clearly defined limits, cell culture could be useful, Horrobin said, "It is reasonably safe to say that if a particular biochemical step is present *in vitro*, then that particular biochemical step is also likely to be present in at least some form *in vivo*. We can therefore construct a network of all possible biochemical events *in vivo* by examining all possible biochemical events *in vitro*." Thus, if one sticks to specific scientific questions while mindful of the limitations of *in vitro* experiments, cell culture can be a useful tool to investigate the production and consequences of chromosomal imbalance.

7.1 INTERNATIONAL REGULATION OF ANEUPLOIDY-INDUCING AGENTS

The wider and closer one looks the more clearly does one see that chemical and physical agents start off nearly all human tumors, and that these latter are occupational diseases resulting from the exceedingly hazardous occupation of living out a life in this world.

(Rous 1967)

Sources of environmental exposures to potentially aneugenic agents are many and include occupational and therapeutic exposures, and exposures associated with lifestyle habits.

(Pacchierotti and Eichenlaub-Ritter 2011)

The initial focus of genetic toxicity testing in the early and mid 1970s concerned the detection of point mutations and structural chromosome aberrations. This was largely due to the use of easy and inexpensive assays for induced mutations (e.g., Ames Salmonella Assay) because of the expressed belief that "Carcinogens are Mutagens" (Ames et al. 1973). Concern for the potential of chemicals to induce aneuploidy grew in the late 1970s and has continued to the present. Aneuploidy-inducing agents are called aneugens (aneuploidogens). Over the years, interest in aneuploidy led regulatory, academic, and industrial scientists around the world (Aardema et al. 1998) to develop a wide range of testing methods capable of detecting and assessing the aneugenic potential of chemicals (Phillips and Venitt 1995). A summary of the subsequent regulatory guidelines and testing schemes that address the assessment of numerical chromosome changes is given in Table 7.1.

For almost 30 years, a major effort of a number of research programs supporting regulatory requirements of the European Union has been the determination of the DNA reactivity of newly developed chemicals and the development of appropriate methods for the detection and assessment of aneugenic chemicals (Parry and Sors 1993). Recent reassessments of the strategies for the testing of chemicals within the European Union have resulted in the increasing recognition of the importance of evaluating the induction of aneuploidy in both somatic and germ cells. The UK Department of Health's Advisory Committee on the Mutagenicity of Chemicals (COM) recommended in 2000 a requirement for the measurement of aneugenic potential in its revised guidelines for the testing of chemicals (Parry 2000) and it is expected that similar requirements will be introduced throughout Europe.

Table 7.1 Overview of regulatory guidelines and testing schemes as of 1998 for assessing numerical chromosome changes.*

Governing Body	*Current Guidelines*	*Future*
OECD	Report polyploidy and endoreduplication when seen. *In vivo* micronucleus test can detect whole chromosome loss.	None
ICH	*In vivo* micronucleus tests have the potential to detect some aneuploidy inducers; there may be aneuploidy inducers that act preferentially during meiosis, but there is no conclusive evidence for these chemicals.	Same
CEC	For chemicals consider aneuploidy-induction or spindle inhibition.	Harmonize with OECD and ICH
Japan	Routine assessment of polyploidy in *in vitro* cytogenetic assay.	Harmonize with OECD and ICH
Canada	*In vitro* chromosome assay should include some chromosome counts to gain information on potential aneuploidy induction.	Harmonize with OECD and ICH
USA-EPA	None	Harmonize with OECD
USA-FDA	None	Harmonize with OECD and ICH

*Adapted from Table 9 on page 23 of Aneuploidy: a report of an ECETOC task force (Aardema et al. 1998). OECD—Organization for Economic Co-operation and Economic Development; ICH—International Conference on Harmonization of Technical Requirements for Registration of Pharmaceuticals for human Use; CEC—Commission of the European Communities; US-FDA—US Food and Drug Administration; US-EPA—US Environmental Protection Agency.

The World Health Organization sponsored an International Drafting Group Meeting of experts in 2007 to update the 1996 IPCS Harmonized Scheme for Mutagenicity Testing. The results of that meeting were published in 2009 (Eastmond et al. 2009). Safety assessments of substances with regard to genotoxicity were based on a combination of tests to assess effects on three major end points of genetic damage associated with human disease: 1) gene mutation (i.e., point mutations or deletions/insertions that affect single or blocks of genes), 2) clastogenicity (i.e., structural chromosome changes), 3) aneuploidy (i.e., numerical chromosome aberrations).

Not surprisingly, the results of international collaborative studies and the large databases demonstrated no single assay can detect all genotoxic substances. For detection of the aneuploidy, only clastogenic and aneugenic assays are meaningful. With regards to cancer, the gene mutation assays will generally do more harm than good because of misleading results (Pradhan et al. 2010, Rennert et al. 2007, Secretary's Advirory Committee on Genetics 2008).

7.2 CANCER DETECTION

Personalized medicine is the application of genomic and molecular data to better target the delivery of health care, facilitate the discovery and clinical testing of new products, and help determine a person's predisposition to a particular disease or condition.

(Obama et al. 2007)

Genes Show Limited Value in Predicting Disease. The era of personal genomic medicine may have to wait.

(Wade 2009)

Legally, the histological diagnosis of cancer is the responsbility of pathologists. Unfortunately, the diagnosis of cancer is based on interpretations that are unavoidably subjective. As Crum et al. have stated, much of practice in cytology and histology involves evaluating abnormal smears and biopsies under suboptimal circumstances or rendering diagnoses that are frequently based more on instinct than objective criteria (Crum et al. 1997). Consequently, false positive and false negative diagnoses are common (Anthony 1998).

More and more, however, molecular biology is claiming to do a better job under the rubric of personalized medicine. The

US Department of Health and Human Services, which funded the human genome project, is a major promoter of personalized healthcare, saying it consists of "medical practices that are targeted to individuals on the basis of their specific genetic code in order to provide a tailored approach" (Obama et al. 2007). Thus, personalized medicine is an extension of the idea genes are the cause of disease, in particular cancer, and can be used for diagnosis and as targets of therapy.

Currently, the big push is to look for gene mutations that "predispose" a person to cancer based on the belief these mutations are causative in some way. Screening for "predisposing" gene mutations has, unfortunately, led to the growth industry of prophylactic surgery. "The primary interventions for mutations carriers for highly penetrant syndromes such as multiple endocrine neoplasias, familial adenomatous polyposis, hereditary nonpolyposis colon cancer, and hereditary breast and ovarian cancer syndromes are primarily surgical" (Guillem et al. 2006).

However, diagnoses based on gene mutation have so far not demonstrated any real benefits to patients (Maher 2008). The much ballyhooed *BRCA1* and *BRCA2* mutations have not been proved to cause breast cancer (Secretary's Advisory Committee on Genetics 2008) and are at best uninformative (Pradhan et al. 2010, Rennert et al. 2007). A 2007 report in the *New England Journal of Medicine* described the results of a national population-based study of Israeli women to determine the influence, if any, of a *BRCA1* or a *BRCA2* mutation on the prognosis in breast cancer. The authors concluded, "Breast cancer-specific rates of death among Israeli women are similar for carriers of a *BRCA* founder mutation and noncarriersî (Rennert et al. 2007). In short: *BRCA* mutations are irrelevant to breast cancer in Israeli women.

Likewise, the reported association between a particular gene mutation of *NuMA* and familial breast cancer turned out not to hold up. Women with this mutation were reported to have more than a four-fold increased risk of breast cancer compared to women with "benign disease" (Kammerer et al. 2005). Kilpivaara et al. attempted to confirm these results but concluded instead:

"Our results do not support the role of *NuMA* variants as breast cancer susceptibility alleles" (Kilpivaara et al. 2008).

Faced with the poor results of using a handful of gene mutations to predict susceptibility to cancer, high throughput searches for single nucleotide polymorphisms (SNPs) have grown in popularity. SNPs represent a colossal number of point mutations throughout the whole genome. Microarray techniques make it possible to detect thousands, even millions of point mutations present in a person's DNA. According to the National Cancer Institute, cancer researchers across the country are looking for correlations between SNPs and precancerous conditions, SNPs and drug resistance in chemotherapy, SNPs and cancer susceptibility, and SNPs and drug response (National Cancer Institute 2005b).

The theory of chromosomal imbalance supports the predictions of others that disappointment awaits (Terwilliger and Weiss 2003, Terwilliger and Hiekkalinna 2006, Weiss and Terwilliger 2000). Indeed, a very recent critique of SNPs said, "the biggest effects that exist for this class of genetic variant…is packing much less of a phenotypic punch than expected" (Goldstein 2009). The ever-growing emphasis on detecting gene mutations as a means of identifying future risk of cancer is misguided and will ultimately do more harm than good if widely implemented. A much more accurate means of detection and prognosis would be to incorporate quantification of aneuploidy into the diagnosis, staging and grading of tumors.

7.2.1 Quantification of aneuploidy for diagnosis and prognosis

The earliest systematic study of cell division in malignant tumors was made in 1890 by the German pathologist David Hansemann. He drew attention to the frequent occurrence of aberrant mitoses in carcinoma biopsies and suggested that this phenomenon could be used as a criterion for diagnosing the malignant state (Heim and Mitelman 2009). Since 1952, preneoplastic aneuploidy of hyperplastic, dysplastic and carcinogen-exposed cells has been used to forecast and prevent human cancer with remarkable clinical success, first by Papanicolau et al. (Mellors et al. 1952) and then by several others (Auersperg et al. 1967, Böcking and Chatelain 1989, Hammarberg et al. 1984, Menke-Pluymers et al. 1994, Osterheld et al. 2004, Reid et al. 2000, Rennstam et al. 2001, Spriggs 1974, Steinbeck 1997). Although there was no consistent underlying theory guiding these efforts, the success of the tests lends unbiased clinical support for the chromosomal imbalance theory of carcinogenesis (Duesberg et al. 2004a). The early presence of random aneuploidy in hyperplastic, dysplastic or asymptomatic cells exposed to carcinogens (Section 5.3.6), and the late appearance of cancers with individual clonal karyotypes long after aneuploidization (Sections 5.3.8 & 6.1.5), also support this theory.

In the mid 1980s, Aldaz et al. demonstrated that progressive aneuploidy and dysplasia constitute general phenomena in the life of every papilloma. According to those data, aneuploidy could be considered as an early pre-malignant marker in mouse skin papillomas (Aldaz et al. 1987, Conti et al. 1986). "These findings seem to correlate with those of human pre malignant lesions of different epithelia, e.g., cervical dysplasias, colonic adenomas, and laryngeal dysplasias. In those lesions, studies using different DNA quantitative techniques such as flow cytometry, cytophotometry, and cytogenetics usually agree that the higher the degree of aneuploidy, the worse the prognosis of the patient. [I]t appears that aneuploidy is an early premalignant marker, and…is probably the earliest in a sequence that includes keratin modifications and GGT expression. It is…possible that the degree of aneuploidy in a premalignant epithelial lesion could become a valuable prognostic factor in clinical practice" (Aldaz et al. 1988a).

Clinical data have established the diagnostic and prognostic power of measuring the degree of aneuploidy. Comparisons between different cancers have shown "Numerous associations between genomic abnormalities and clinical behavior" (Albertson et al. 2003)—the more aneuploid the karyotype the more malignant is the cancer (Duesberg et al. 2005, Foulds 1969, Tsafrir et al. 2006, Wolman 1986). Indeed, Castro et al. analyzed 79 solid tumor types with at least 30 karyotypes from the Mitelman Database of Chromosome Aberrations (Mitelman 2011) and found this to be true (Castro et al. 2006). Using the survival data from population-based registries in the Surveillance, Epidemiology and End Results (SEER) Program data base they obtained the Kaplan-Meier distributions of the cumulative survival estimates of 60-month cohorts for each of the 79 solid tumor types (Surveillance Epidemiology and End Results (SEER) 2009). Their results showed "that [chromosomal] aberration spread is specific for each tumor type, with high degree of diversity for those tumor types with worst survival indices. Those tumor types with preferential variants (e.g., high proportion of a given karyotype) have shown better survival statistics, indicating that aberration recurrence is a good prognosis".

For half-a-century pathologists have used the Papanicolaou stain to detect large, irregular nuclei indicative of all levels of neoplasia and invasive cancer. Because near-diploid cells are indistinguishable to the eye from normal cells, only near triploid and higher DNA containing cells are noticeable to pathologists. This has led to the general (though not absolute) rule that in order to diagnose a neoplastic cell its nucleus must be at least 3 times larger than obviously normal nuclei (Crum et al. 1997). Since all levels of neoplasia must satisfy this rule, pathologist use numerous visual clues to discriminate among the various stages of cancer progression (Table 7.2).

Looking to improve upon the highly subjective cytomorphological criteria of Table 7.2, Bulten led a group of Dutch scientists using in situ hybridization to measure the changes in 8 chromosomes of cervical samples (Bulten et al. 1998). Since abnormal numbers of chromosomes appear at the very earliest stages of cancer, the authors had hoped to find the different progressive stages of cervical intraepithelial neoplasia were characterized by different aneusomies. What they observed, however, was a quantum jump in all chromosomes in all of the stages of cervical cancer (Figure 7.1A). Similar results have been reported by Gebhart and Liehr in

Table 7.2 Cytologic criteria of low (LSIL) and high (HSIL) grade squamous intraepithelial lesions.*

Parameter	LSIL	HSIL
Architecture	Isolated cells or sheets	Isolated cells or pseudosyncytia
Cell size	Mature & rarely immature	Mature & immature
Nuclear size	Greater than 3 times area of normal intermediate nucleus (with some exceptions)	Greater than 3 times area of normal intermediate nucleus (with some exceptions)
N/C ratio	Increased	Increased + +
Hyperchromasia	+	+ +
Chromatin	Smooth to finely granular	Coarsely granular
Nuclear outlines	Round, regular to slightly irregular	Irregular

*Adapted from Table 4–2 on page 57 of *Pathology of Early Cervical Neoplasia* (Crum et al. 1997)

Germany for a large number of cancers including cancer of the cervix (Gebhart and Liehr 2000) and by Mertens et al. for over 3,000 cancers in Sweden (Mertens et al. 1997). By the time nuclei have evolved to the point of satisfying the morphological criteria of Table 7.2, even low grade cells are highly aneuploid.

Even though Bulten et al. failed to identify diagnostic aneusomies, their results can still be used to advantage. Consistent with a growing number of observations that chromosome instability is predictive of clinical outcome (Carter et al. 2006, Doak 2008, Lee et al. 2011, Walther et al. 2008), a plot of data from Table 2 of Bulten et al. (Figure 7.1B) shows a direct correlation between the standard deviation of the chromosome indices of all chromosomes (a measure of chromosome instability) and progression towards invasive cancer.

The fact that aneuploid chromosomes are present at all stages of carcinogenesis can be used as the basis of a universal and definitive approach to detecting cancer and pre-cancerous lesions. Measuring the level of aneusomy of any single chromosome (or

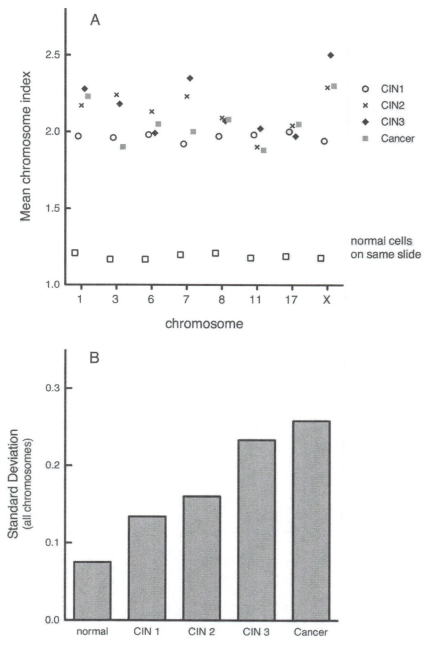

Fig. 7.1   Aneusomy of any chromosome or combination of chromosomes can detect the presence of neoplasia at all grade levels.

Fig. 7.1 Contd. ...

combination of chromosomes for better statistics and reliability) could provide an objective means of determining the presence or absence of neoplasia at all the grade levels pathologists use. This has been tested experimentally for similar levels of aneusomy of chromosomes 7 and 17 present in the same cervical samples (Figure 7.2A, author's unpublished results) and for equivalent fractions of cells with aneusomy for chromosomes 1 and 17 in the same grades of cervical neoplasia (Figure 7.2B) (Hariu and Matsuta 1996). An immensely important benefit of using chromosomal analysis to detect neoplasia would be the virtual elimination of indeterminate diagnoses (Walther et al. 2008).

There are commercial tests approved for the detection of aneuploidy in certain cancers. UroVysion, for example, uses fluorescence in situ hybridization of three chromosomes and one chromosomal segment for the specific detection of bladder cancer recurrence (Food and Drug Administration 2005). The test was developed empirically on the assumption that copy number changes in chromosomes 3, 7, 17 and chromosomal segment p16 were "specific" to bladder cancer. However, chromosomal imbalance theory and results such as those of Bulten et al., show

Fig. 7.1 Contd. ...

Graph of data from Table 2 on page 501 of reference (Bulten et al. 1998). Bulten et al. prepared serial sections of cervical samples to stain for pathologist grading followed by in situ hybridization to detect numerical aberrations of eight chromosomes in the progressive stages of cervical intraepithelial neoplasia (CIN). Artifacts due to serial sectioning led to slightly elevated chromosome index (CI) values for the normal diploid cells above the expected value of 1 (average CI=1.19). Aneusomy was present only in regions of morphologically dysplastic epithelium. (A) Quantum jump in all chromosomes in all of the stages: CIN1, CIN2, CIN3, invasive cervical cancer. Dividing the average CI value of the aneusomic chromosomes (CI=1.97, 2.14, 2.17, 2.06, for CIN1, CIN2, CIN3, invasive cancer respectively) by the average CI value of normal cells (CI=1.19) yielded DNA indices=1.7 for CIN1, 1.8 for CIN2, 1.8 for CIN3, and 1.7 for invasive cancer. (B) Graph of the standard deviation data (a measure of chromosome instability) for the eight chromosomes from Table 2 of (Bulten et al. 1998) shows a direct correlation with progression towards invasive cancer.

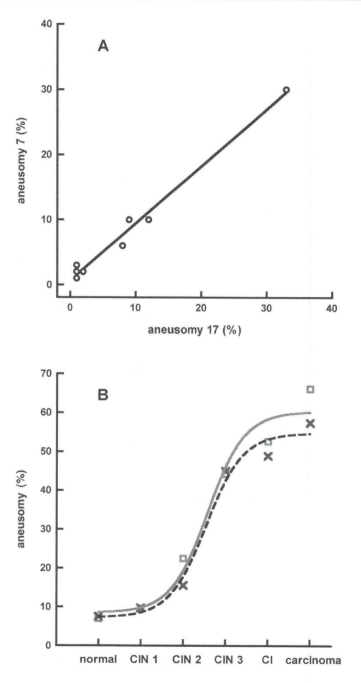

Fig. 7.2   Chromosomes are equally efficient at detecting neoplasia.

Fig. 7.2 Contd. ...

that any combination of chromosomes could be used to diagnose all cancers. An appreciation of chromosomal imbalance theory could unleash the power of UroVysion and similar tests for the detection and prognosis of any cancer. For example, a Swedish group recently adapted UroVysion to measure aneuploidy in patients with mesothelioma to distinguish malignant from reactive (non-malignant) cells, particularly when cytology is inconclusive (Flores-Staino et al. 2010).

Singh et al. demonstrated a significant correlation between degree of aneuploidy and progression of cervical cancer (Singh et al. 2008), concluding "that aberrant DNA content reliably predicts the occurrence of squamous cell carcinoma in cervical smear. Walther et al. recently published an analysis of 63 colorectal cancer (CRC) studies that correlated ploidy status with outcome in 10,126 patients (Walther et al. 2008). The authors concluded that the published data support the view that chromosomal instability (determined as aneuploidy/polyploidy) is associated with a worse prognosis in CRC and can stratify CRC patients further after standard pathological staging. But Professor Alfred Böcking at the Institute for Cytopathology in Germany leads the field in using aneuploidy in the diagnosis and prognosis of all types of cancer. His group has developed and commercialized in Europe an automated system (Böcking 2008) to diagnose cancers in 19 different tissues by detecting and quantifying aneuploidy (Böcking 2009).

Fig. 7.2 Contd. ...

Measuring the level of aneusomy of any single chromosome (or combination of chromosomes for better statistics and reliability) could provide an objective means of determining the presence or absence of neoplasia at all the grade levels pathologists use. (A) This has been tested experimentally for similar levels of aneusomy of chromosomes 7 and 17 present in the same cervical samples (author's unpublished results). (B) Equivalent fractions of cells with aneusomy for chromosomes 1 (red) and 17 (black) in the same grades of cervical neoplasia (Hariu and Matsuta 1996).

Color image of this figure appears in the color plate section at the end of the book.

7.2.2   Chromosomal imbalance theory applied to transcript microarray data

Microarray-based prognostic tests are irremediably moving to the clinics, but their clinical utility might never be formally established. The conclusions of most microarray-based prognostic studies have so far been overoptimistic. The performances of published signatures are much poorer than stated, the predictions are quite discordant, and the list of genes is very unstable. The concept of a unique prognostic signature is open to question.

(Koscielny 2008)

Microarray technology is used around the world to quantify the expression of thousands of genes in tumor samples. There is literally a mountain of expression data available for analysis. Analyzing large gene expression datasets is a relatively new area of data analysis with its own unique challenges (Macgregor and Squire 2002, Simon et al. 2003, Terwilliger and Hiekkalinna 2006, Weiss and Terwilliger 2000).

Various statistical methods are used to sift through tens of thousands of data points searching for stable subsets of genes that are correlated with specific normal and abnormal phenotypes. The supervised and unsupervised statistical algorithms produce annotated lists of genes according to differences in expression (Ochs and Godwin 2003). The lists of genes are then assembled into genetic roadmaps that are thought to govern the specific phenotypes being investigated. This strategy comes from the general belief that a relatively small number of specific genes control certain normal and disease phenotypes. In spite of the promising initial results, this approach has not lived up to expectations, particularly with respect to cancer (Dunkler et al. 2007, Dupuy and Simon 2007, Eden et al. 2004, Koscielny 2008, Massague 2007, Michiels et al. 2005, Michiels et al. 2007, Ntzani and Ioannidis 2003, Pan et al. 2005, Reid et al. 2005, Shi et al. 2008, Simon et al. 2003).

The results to date indicate that the genetic roadmaps are not providing the rules governing the dynamic interplay between genotype and phenotype (Bains 2001). Gene knockout experiments, for example, have repeatedly shown that the whole animal, down to the cellular phenotype, is usually unaffected by the loss of one or a few genes (Newman 2002, Shastry 1995) and when there are phenotypic consequences, consistent phenotypes are rarely obtained by modification of the same gene even in mice (Pearson 2002, Sigmund 2000). "The disruption of a gene in one strain of mice may be lethal, whereas disruption of exactly the same gene in another strain of mice may have no detectable phenotypic effect. If this is true of the impact on one gene of the rest of the mouse genome, how much more is it likely to be true of the impact of the rest of the genes in the human genome?" (Horrobin 2003).

Adding to the problem of associating subsets of genotype with phenotype, the "genetic signatures" generated from microarray experiments are highly unstable (Massague 2007, Miklos and Maleszka 2004b, Shi et al. 2008), particularly for genomically unstable cancers (Dupuy and Simon 2007, Ein-Dor et al. 2005, Koscielny 2008, Massague 2007, Michiels et al. 2007, Pan et al. 2005). A partial explanation for the instability was recently offered by Shi et al: "reproducibility has seldom been, but in the future should be, used as a crucial criterion to judge the validity of data analysis procedures" (Shi et al. 2008). However, aneuploidy and normal genetic variance (Lucito et al. 2003) are the major sources of the instability plaguing the genetic signatures derived from conventional data mining. According to Elser and Hamilton, "It seems that the only hope for creatively interrogating new data is to develop new, integrated theoretical frameworks to inform strategies for that interrogation" (Elser and Hamilton 2007).

Transcript microarrays measure values that are approximately proportional to the numbers of copies of different mRNA molecules in samples. It has long been recognized that variability can exist between arrays, some of biological interest and other of non-biological interest. These two types of variation are classified as either interesting or obscuring. It is this obscuring variation researchers seek to remove when normalizing array data. Therefore, normalization of the raw data is one of the most important (and often contentious) steps in analyzing microarray data (Bolstad et al. 2003, Huettel et al. 2008).

An important consideration when applying a normalization method to data from a typical comparative microarray experiment, is how many genes are expected to change between conditions and what those changes will be. Two important assumptions are generally made with regards to microarrays. Specifically, most normalization methods require that either the number of genes changing in expression between conditions be small or that an equivalent number of genes increase and decrease in expression. For this reason, all of the arrays in a particular experiment are typically normalized together as a single group (Bolstad et al. 2003). This approach is clearly not suited to analyzing aneuploid cancer cells.

Transcriptional changes resulting from aneuploidy must be described in terms of chromosomes and chromosome regions that are numerically altered and whether changes in expression are in cis or trans regions (on the same or other chromosomes, respectively). Clearly, the choice of microarray data analysis methods has a substantial impact on results and, in particular, normalization methods that are robust to large-scale shifts in gene expression need to be applied in studies of aneuploidy. Unfortunately, many of the popular normalization transforms are not appropriate for microarray data sets with large-scale expression level shifts as seen with aneuploid cancer because they violate the underlying assumptions of the methods (Huettel et al. 2008).

For example, normalization *via* division by the median signal per sample assumes that the median expression level of different samples should be the same. However, large fractions of the genome of aneuploid cells are expressed at higher levels compared to diploid precursors violating the assumption of a constant median expression level of all samples.

Another common practice adversely affecting microarray results from aneuploid cancer samples is aggressive normalization, which is prone to over-fitting. Over-fitting in this context means that the transform may consider true data features as technical bias. Removing the wrongly estimated technical trend then not only subtracts biological signal but may also introduce signal artifacts. In general, one should aim for a transform as conservative as possible.

Huettel et al. have described in detail how popular normalization methods (e.g., quantile and loess) violate their stated assumptions when applied to aneuploid cells. The underlying assumption of quantile is that the signal distribution for all samples is the same. This assumption clearly doesn't hold when aneuploid cells are compared to normal cells. Thus, when quantile is applied to aneuploid cancer samples, there is substantial loss of the differential expression, even to the point of removing it entirely.

Loess assumes that the differential expression signal should be independent of the average gene expression levels and average to zero. Again, aneuploid samples violate this assumption. A large fraction of genes are more highly expressed in aneuploid samples whereas no corresponding strong down-regulation effect can be observed. As a result, the conditions for the loess normalization are not satisfied. Because of this, Huettel et al. concluded that inappropriate normalization can drastically affect the power of an analysis to detect even the large-scale expression differences caused by aneuploidy (Huettel et al. 2008).

A recent example using microarray experiments designed specifically to analyze the expression of aneuploid cells illustrates the difficulty in attempting to reverse the problems introduced by a standard normalization procedure. Torres et al. applied the Agilent normalization method to the microarray data from aneuploid yeast even though it assumes a ratio of 1 between experimental strains and wild type reference (Torres et al. 2007). The authors realized this assumption does not hold for strains carrying extra chromosomes. Therefore, their expression data, that had already been normalized by the loess method, were renormalized to account for the extra chromosomes. With this correction, genes contained on all aneusomic chromosomes over several replicate experiments increased in expression by an average of 1.8-fold. Torres et al. assumed this result was reliable.

However, when Huettel et al. performed the average bias subtraction fix of Torres et al. to their previously loess normalized data from trisomic 5 *Arabidopsis thaliana*, the results were worse than before normalization (Huettel et al. 2008). "The trend of genes on other chromosomes clearly deviated from the zero axis, showing bias for up-regulation for lowly and highly expressed genes and also a bias for down-regulation for genes with moderate expression levels."

As an aid to those considering doing microarray experiments on cancer samples, here is a technical summary of the approach used by Huettel et al. to normalize data from chromosomally unbalanced cells (Huettel et al. 2008). (See their paper for the complete details of the procedure.) To get around the problems inherent in standard normalization methods used in the microarray analysis of aneuploid cells, the probe level signals were conservatively normalized for different backgrounds and overall hybridization intensities of individual chips using an iterative 20%-trimmed least squares fit of a generative model with additive-multiplicative noise. This approach is robust both to outliers and to systemic large-scale shifts, as demonstrated when estimating transform parameters from all data or only from genes not on aneusomic chromosomes. The generalized log transform for the model used to stabilize variation was calibrated for asymptotic equivalence to a standard $\log_2$ transformation. Refrain from further transforms in a first examination of data characteristics.

DATE analysis applied to breast cancer microarray data

Metabolic control analysis (MCA) is a well-established integrated theoretical foundation upon which to construct a general theory of biological change (Fell 1992). A modification of MCA called DATE (Differentiation, Adaptation, Transformation, Evolution) analysis was recently applied to microarray expression data to investigate aneuploidy in primary cancer cells (Rasnick 2009).

DATE analysis differs from MCA in that its essential task lies in the *comparison* of related phenotypes rather than in the precise definition or description of each. In place of tracking the kinetic particulars of thousands of individual cellular components, DATE analysis uses, instead, two biological equations of state to calculate their aggregate effects (Cornish-Bowden 2004, Kacser and Burns 1981, Rasnick and Duesberg 1999). This approach makes it possible to analyze the phenotypic changes of whole cells, organs and organisms (Rasnick 2009).

For large datasets, the state variables F, RNA index, and ϕ (Section 5.4) are intensive variables, thus not dependent on a defined subset of specific genes (Rasnick 2009). This result is of fundamental importance and distinguishes DATE analysis from conventional data mining, which seeks a stable and unique set of genes as either diagnostic tools or targets of drug therapy. In contrast to data mining for genetic signatures, DATE analysis provides a robust strategy of correlating specific phenotypes with the state variables F, RNA or DNA index, and ϕ along with the measures of data dispersion D and γ (Rasnick 2009).

Occasionally one can find useful microarray data for human cancer. The best data contain replicate runs (Lee et al. 2000, Novak et al. 2002, Pavlidis et al. 2003) and primary diploid reference samples from the same tissue type as the cancer (Lucito et al. 2003). Analogous to flow cytometry, histograms of cancer transcript microarray data visualize aneuploidy. The six cancers shown in Figure 7.3—pancreas (Iacobuzio-Donahue et al. 2003), colon (Notterman et al. 2001), lymphoma (Bohen et al. 2003), breast (Perou et al. 2000), stomach (Chen et al. 2003), kidney (Barrett et al. 2005)—were compared to normal tissue of the same type from which the cancer originated. By way of comparison, it was possible to compare the background spread in transcripts from normal tonsil and skin (graphs at top of Figure 7.3) because data from more than one sample were available (Bohen et al. 2003, Storz et al. 2003). However, these kinds of data are rarely part of micorarray results (Lucito et al. 2003, Rasnick 2009).

Figure 7.3 shows that the normal tissues were characterized by a tight distribution of transcripts centered at DNA index = 1. In contrast to normal tonsil and normal skin, the distributions of cancer transcripts were all decidedly different and irregular compared to the respective normal tissues. The aneuploid fractions ϕ and DNA indices (Section 5.4) were characteristically large for all the cancers, indicating advanced malignancies (Rasnick and Duesberg 1999, Rasnick 2000, Rasnick 2002). For determination of ϕ, the normal range of π was set to $0.5 \leq \pi \leq 1.5$ (Rasnick 2009).

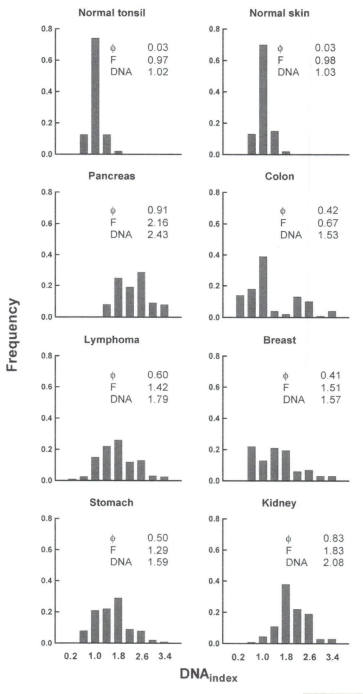

Fig. 7.3 Contd. ...

DATE analysis was performed on the microarray data from 36 invasive ductal carcinomas of the breast for which there were clinical data (Zhao et al. 2004). The stated purpose of the Zhao et al. study was to determine if there were distinct genetic signatures distinguishing invasive ductal carcinoma from invasive lobular carcinoma. The authors did not correlate their results with clinical grade of the tumors. Because there was only one Grade 1 and no Grade 3 invasive lobular carcinoma patients, only the ductal carcinoma data were analyzed (Rasnick 2009).

Since chromosomal imbalance causes the genetic instability characteristic of invasive cancer (Section 5.2), the ductal carcinoma patients (represented by the black squares in Figure 7.4) were sorted along the horizontal axis by increasing values of D and γ, objective measures of genomic imbalance (hence chromosomal instability) of cancer cells (Rasnick 2009). Grade 3 tumors were concentrated at high values of D and γ. With a notable exception, the few examples of Grade 1 favored low values of D and γ. The Grade 1 tumor circled at the lower right of Figure 7.4A,B had the highest distribution entropy D of all the cancers. This patient's tumor was likely misclassified and probably highly malignant.

The Grade 2 tumors were disperse but tended to the left side of the graphs, with low and intermediate values of D and γ. It is likely intermediate Grade 2 is so uninformative as to be of little value

---

Fig. 7.3 Contd. ...

Fig. 7.3   Six advanced human cancers.
Cancer cells are aneuploid and this genomic imbalance determines their properties. Normal tonsil (Bohen et al. 2003) and skin (Storz et al. 2003) were characterized by a tight distribution of transcripts centered at DNA index=1 (top graphs). The six cancers— pancreas (Iacobuzio-Donahue et al. 2003), colon (Notterman et al. 2001), lymphoma (Bohen et al. 2003), breast (Perou et al. 2000), stomach (Chen et al. 2003), kidney (Barrett et al. 2005)—were compared to normal tissue of the same type from which each originated. In contrast to normal tonsil and skin, the distribution of cancer transcripts were all decidedly different and irregular compared to normal tissues (Rasnick 2009). The aneuploid fractions φ and DNA indices were characteristically large for all the cancers, indicating advanced malignancies.

(Ellsworth et al. 2007, Ivshina et al. 2006). This was recognized some years ago for cervical cancer when the intermediate category CIN-2 was eliminated (Bollmann et al. 2001, National Cancer Institute 1989). Now there are only low and high grade cervical lesions. This simplified classification scheme has also been recommended for neoplastic lesions of esophagus, stomach, colon and rectum (Schlemper et al. 2000).

7.3 CANCER THERAPY

It is noteworthy that chemotherapy often provides a long period of remission of symptoms and control of growth without substantially lengthening life, the rapidity of the terminal phase counter balancing the remission due to treatment.

(Foulds 1954)

After decades of intensive clinical research and development of cytotoxic drugs, there is no evidence for the vast majority of cancers that chemotherapy

---

Fig. 7.4   Thirty-six invasive ductal carcinomas of the breast correlated with two measures of chromosomal instability.
The solid squares represent a patient's tumor that had been graded 1, 2, or 3, for increasing severity. It is generally accepted that the most malignant cancers are the most genetically unstable. Both D and γ are measures of the genomic instability of cancer cells (See reference (Rasnick 2009) for detailed descriptions of D (the distribution entropy of histogram data) and γ). A) Patients along the horizontal axis were sorted by increasing D (broken blue line). B) Patients were sorted by increasing γ (broken red line). Grade 3 tumors were concentrated at high values of D and γ (indicating high levels of chromosome instability). The few examples of Grade 1 favored low values of D and γ. The Grade 1 tumor circled at the lower right of both graphs was likely misclassified and probably highly malignant. Grade 2 tumors were disperse but tended to the left side of the graphs, with low and intermediate values of D and γ. It is likely intermediate Grade 2 is so uninformative as to be of little value. This was recognized some years ago for cervical cancer when the intermediate category CIN-2 was eliminated (Bollmann et al. 2001, National Cancer Institute 1989). Now there

Fig. 7.4 Contd. ...

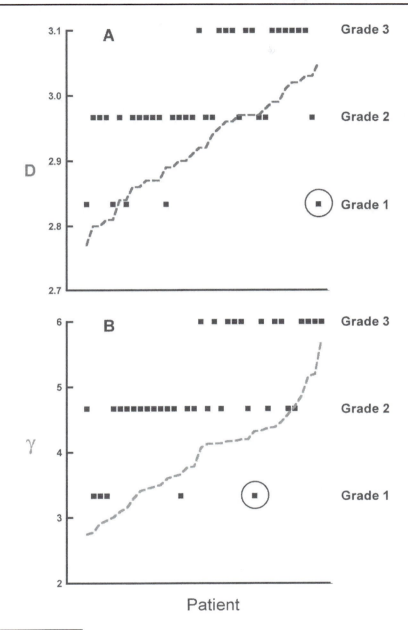

Fig. 7.4 Contd. ...

are only low and high grade cervical lesions. This simplified classification scheme has also been recommended for neoplastic lesions of esophagus, stomach, colon and rectum (Schlemperet al. 2000).

Color image of this figure appears in the color plate section at the end of the book.

exerts any positive influence on survival or quality of life in patients with advanced disease.

(Epstein 1998)

Throughout the clinical development of anticancer drugs, researchers repeatedly encountered significant problems because of the acute and long-term toxicities of chemotherapies, which affected virtually every organ of the body.

(Chabner and Roberts 2005)

Meredith Wadman covered the 2011 Annual Meeting of the American Association for Cancer Research for *Naturenews*. She published an article on the results of Mathew Ellis et al. of the Siteman Cancer Center (Wadman 2011). The authors sequenced 50 breast cancers alongside matching DNA from the same patients' healthy cells in order to identify genomic alterations present only in the cancerous cells. Their findings revealed that the "cancers' genetic fingerprints" were highly diverse. Of the 1,700 gene mutations they found in total, most were unique to individual patients' tumors, and only three occurred in 10% or more. The genomic changes were also of all kinds, from single nucleotide variations and frame shifts to translocations and deletions.

Ellis said, "The results are complex and somewhat alarming, because the problem does make you sit down and rethink what breast cancer is." Further, the complexity of the results indicates that when it comes to developing therapeutics "very clearly the only way forward is the genome-first approach. No single blockbuster drug will answer the problem of drug resistance." In spite of the disquieting result, he said careful analysis of the data, combined with what is already known about the functions of the affected genes, yields a wealth of new therapeutic possibilities. Undaunted, Ellis "has already begun his next step: to repeat the experiment on at least 1,000 more tumors."

The disappointing picture is not new. Herceptin is a case in point. In March, 2002, as part of the fast-track approval process, the National Institute for Health and Clinical Excellence in the UK recommended Herceptin for use in women with *HER2*-positive advanced breast cancer, either alone or in combination with paclitaxel. An editorial in the *New England Journal of Medicine* said

Herceptin was "maybe even a cure" for breast cancer (Hortobagyi 2005). However, an editorial in the *Lancet* countered with: "the best that can be said about Herceptin's efficacy and safety for the treatment of early breast cancer is that the available evidence is insufficient to make reliable judgments. It is profoundly misleading to suggest, even rhetorically, that the published data may be indicative of a cure for breast cancer" (2005).

A 2002 study assessed the value of a number of the new treatments. It reported that drugs approved in the first six years of the new European Medicines Evaluation Agency did not meet the expectations generated by the gains in basic knowledge on cancer (Garattini and Bertele 2002). The authors stressed, in order to reach the market swiftly new drugs are often candidates for second or third line treatment of rare cancers, and they are evaluated in small phase II studies which assess their equivalence or non-inferiority (rather than superiority) to standard treatments. In spite of not improving survival, quality of life, or safety, these new drugs cost much more than the standard treatments.

As for safety, most drugs caused the usual signs of cytotoxicity, including neutropenia, thrombocytopenia, fever, infections, and gastrointestinal toxicity. In no instance did comparisons show a clear cut advantage, in terms of adverse reactions, over the reference drugs or analogous agents. The study also evaluated monoclonal antibodies, a completely new class of anticancer agents, and said the efficacy of this class of drugs has not yet been confirmed by appropriate studies (Garattini and Bertele 2002), and their safety seems unfavorable, contrary to all expectations (White et al. 2001). The authors concluded, "From these results over the past six years there is little to justify some of the promises made to the public."

It is clear that the long-standing strategy of trying to kill cancer cells before killing the patient with radiation and cytotoxic drugs has just not worked for the vast majority of lethal cancers (Tannock 1998). So what are the prospects for prevention and therapeutic intervention in cancer? Does the theory of chromosomal imbalance suggest new strategies? The answer is yes.

7.3.1 Spontaneous tumor disappearance

That cancer cells are often sick cells and die young is known to every pathologist.

(Rous and Kidd 1941)

One of the most stubborn misconceptions is that cancer cells are rapidly dividing super cells, "the enemy within" that is bent on our destruction (Nowell et al. 1998). Hence the military metaphors and the "war on cancer". However, observationally, experimentally, and theoretically, cancer cells are aneuploid cells (Chapters 5 & 6). Aneuploidy damages cells—the more severe the chromosomal imbalance, the greater the damage (Lindsley et al. 1972, Liu et al. 1998). Being damaged, aneuploid cells typically divide at slower rates than normal diploid cells (Hayflick 1965) and "progression does not necessarily lead to dominance of the tumor over its host" (Foulds 1954). Being damaged, aneuploid cells tend to die at high rates (Steel and Lamerton 1969), one of the "liabilities of the neoplastic state" (Rous and Kidd 1941). It is only the "successful" tumors that attract attention; the "unsuccessful" ones escape notice (Foulds 1954). Herein lies the key to prevention and much more effective and less toxic therapeutic approaches to cancer.

Cancer cells are not super cells but damaged aneuploid cells, which for the most part spontaneously die. Since aneuploid cells typically lose in competition with normal diploid cells (Atkin and Baker 1990, Rous and Kidd 1941), the new strategy is to stop devising poisons to kill cancer cells and to focus more on the interactions between tumor and host. The fact that propagation of primary human cancer cells *in vitro* requires finally-tuned, stable environments (see *Cancer research would benefit from de-emphasizing cell culture*, p. 196) implies that non-toxic perturbations of the host may be sufficient to nudge the tumor out of its stable, comfortable environment into a different attractor that leads to the death of the cancer cells.

As Upender et al. said, "[T]he normalization of the complex dysregulation of transcriptional activity in carcinomas requires a more general, less specific, and hence more complex interference" (Upender et al. 2004). Chromosomal imbalance theory demonstrates that in order to alter a complex phenotype non-toxically requires only moderate changes in the activities of hundreds or even thousands of genes and their products (Section 5.4). While changes in the activities of large numbers of genes merely exercise normal cells, the resulting physical and metabolic perturbations stress and destabilize cancer cells, reducing or terminating their viability within the host (Niakan 1998, Pettigrew et al. 1974, Rous and Kidd 1941). Such perturbations may be responsible for the 741 documented examples of spontaneous remission from more than 45 different types of cancer (Challis and Stam 1990, Kleef et al. 2001). (Some advocate using the term "remission" for permanent disappearance of cancer and "regression" for temporary or partial improvement. In practice, however, spontaneous regression and remission are commonly used as synonyms for unexpected transient or final improvements in cancer.)

The most striking feature of virtually all cases of spontaneous remission was a prior fever-inducing infection: diphtheria, gonorrhea, hepatitis, influenza, malaria, measles, smallpox, syphilis and tuberculosis, as well as various other pyogenic and non-pyogenic infections (Nauts 1980). The Remission Project of the Institute of Noetic Sciences surveyed the literature and found the common factor in all the infections associated with spontaneous regression was high fever, usually 40°C for 3–5 days (O'Regan and Hirschberg 1993). It should be noted that these infections do not always produce fevers that high or for that long. Less common global perturbations associated with spontaneous remission are pregnancy, severe dietary changes, and operative trauma with subsequent infection (Challis and Stam 1990, Hoption Cann et al. 2003). There are also the cases of spontaneous remission that seem to happen for no apparent reason (Kappauf et al. 1997), perhaps because as Rous and Kidd have observed, many neoplasms "require continual aid for their survival" (Rous and Kidd 1941) and without it they perish.

It is generally accepted that spontaneous remission is a natural phenomenon whose causes remain unknown at the present. Only when we begin to address the most basic questions can we start to determine the epidemiology of remission. The Medline database shows between 1966–1992, the terms "spontaneous remission" or "spontaneous regression" appeared 10,603 times as a descriptor and 718 times in titles (O'Regan and Hirschberg 1993). Of these 718 papers, more than 80% of them have appeared in the period 1975–1992 and over 40% appeared during 1985–1992. There have been literally no comprehensive reviews of spontaneous remission of diseases other than cancer (O'Regan and Hirschberg 1993).

One way to determine an overall epidemiology of remission would be to establish a National Remission Registry modeled after the National Tumor Registry. In that way, information on spontaneous remission could be collected and cases of remission tracked in a systematic manner. The building of such an epidemiology could lead to increased understanding of cancer and treatment and the ability to advise patients more precisely regarding prevention and outcome.

Spontaneous remission probably has more to do with changes in the person than changes in the tumor. The former emphasis on controlling and preventing cancer through diet, exercise, avoidance of carcinogens and similar nontoxic strategies needs to be revived and vigorously investigated. The more remission is recognized as legitimate and the more it is understood, the more likely it is we can understand how to stimulate the natural self-repair capacities that exist in everyone to some degree.

7.3.2   Induction of fever as cancer treatment

Spontaneous tumour regression has followed bacterial, fungal, viral, and protozoal infections. This phenomenon inspired the development of numerous rudimentary cancer immunotherapies, with a history spanning thousands of years. Coley took advantage of this natural phenomenon, developing a killed bacterial vaccine for cancer in the late 1800s. He observed that inducing a fever was crucial for tumour regression. Unfortunately, at the present time little credence is given to the febrile response in fighting infections—no less cancer.

(Hoption Cann et al. 2003)

The treatment of cancer by injection of bacterial products is based on the fact that for over two hundred years neoplasms have been observed to regress following acute infections, principally streptococcal (Nauts et al. 1953) (Section 7.3.1). If these cases were not too far-advanced and the infections were of sufficient severity or duration, the tumors completely disappeared and the patients remained free from recurrence. If the infections were mild, or of brief duration, and the neoplasms were extensive or of histological types which were less sensitive to infections or their toxins, only partial or temporary regressions occurred. The first known observation of this phenomenon was published in 1866 by Wilhelm Busch (Busch 1866). But it was William Coley, Chief of the Bone Service at Memorial Hospital in New York, who is best remembered for devoting a lifetime to the subject. He had an international reputation and was an honorary fellow of the Royal College of Surgeons in London.

Prompted in 1891 by the loss of one of his first patients to a sarcoma on her arm, Coley searched for information that might help him treat other patients with similar conditions. This led him to study all cases of sarcoma treated in a New York Hospital during the preceding 15 years (Hoption Cann et al. 2002). His interest in the possible therapeutic value of infections or their toxins was aroused by the strange case of Fred K. Stein. Five operations on Stein's sarcoma of the cheek had failed to control the disease. However, according to hospital records, he recovered completely after two attacks of erysipelas (an acute, sometimes recurrent disease caused by a bacterial infection). Stein was released from the hospital in 1885. In 1892, after a long search, Coley finally located Stein who agreed to be examined and was found to be still free of cancer 7 years after leaving the hospital (Hall 1997).

Coley immediately attempted to produce erysipelas in a patient with twice-recurrent inoperable myxosarcoma of the tonsil and neck. After repeated trials, using four different bacterial cultures, he succeeded. The resulting severe erysipelas caused a spectacular regression of the tumors, leaving only the scar tissue from the former operations (Coley 1906). This patient lived well for 8 more years, but eventually died from a local recurrence.

After attempting to induce erysipelas in 12 other patients, Coley recognized the difficulties—eight were successfully infected through live bacteria and developed a tumor response (two complete remissions), but two died of erysipelas—leading Coley to abandon the use of living cultures. Starting 1893, he settled on a mixture of two heat-killed bacteria, *Streptococcus pyogenes* and *Serratia marcescens*, that successfully induced remission in a number of cancer patients (Hobohm 2009). This mixture became known as Coley's toxins, or Mixed Bacterial Toxins as it is now called.

There were many strong supporters of Coley's toxins, even among the highest levels of the medical establishment, including the Mayo brothers, Joseph Lister and Henri Matagne (Coley 1936, Matagne 1953). But he also had notable opponents. One of the best known was James Ewing, chief pathologist at the New York Hospital where Coley worked. Ewing became famous for describing the sarcoma that would soon to be named after him (Ewing's Sarcoma). The fact that Coley claimed some of his best results in Ewing's sarcoma may have softened Ewing's opposition (Coley 1910), because he acknowledged in his internationally famous textbook *Neoplastic Diseases* that, "In some recoveries from endothelioma of the bone, there is substantial evidence that [Coley] toxins played an essential part" (Ewing 1941).

William Coley's daughter, Helen Coley Nauts, organized and published all of her father's collected papers. In 1997, she won the National Institute of Social Sciences' Gold Medal for Distinguished Service to Humanity. In collaboration with George Fowler and Louis Pelner, she wrote 18 monographs on different cancers treated with Coley Toxins by her father and his contemporaries, including Sarcoma (Nauts 1969), Colorectal Cancer (Fowler 1969b), Melanoma (Fowler 1969a), and Neuroblastoma (Fowler and Nauts 1969). In all, they reported around 2000 cases, of these, 896 were microscopically confirmed. The overall 5-year survival was 51% in operable cases and 46% in inoperable ones (Nauts 1982). For example, in 104 cases of soft tissue sarcomas, 50% of those injected with the toxins lived 5–20 years (O'Regan and Hirschberg 1993).

While tumor regression was often noted within hours of injection with Coley's toxins (Nauts et al. 1953), primary adaptive immune responses were often delayed by several days to a week (Medzhitov 2001). In fact, Coley's experience (Nauts et al. 1953) and an exploratory evaluation of case reports of spontaneous regression (Hoption Cann et al. 2002, Nauts 1980), support the concept that infection-stimulated tumor regression generally results from a "non-specific" innate immune response (Hoption Cann et al. 2003).

In cases where the regression was partial and the acute or febrile phase of the infection subsided, residual tumor generally re-grew (Nauts 1980). Similarly, if the infection recurred or was reintroduced, tumor regression proceeded as before (Nauts 1980). Coley stated that daily injections should be given, if the patient could bear it, as discontinuing the vaccine even for a few days would often lead to re-growth of residual tumor (Coley 1906)—again suggesting that specific anti-tumor immunity was not a primary mechanism of this vaccine. The broad diversity of organisms capable of eliciting spontaneous regression coupled with its speed are consistent with the observed general, non-specific disruption of the chromosomally unbalanced and metabolically damaged cancer cells.

One unexpected observation by Coley of no small importance was the salutary effect of fever on cancer pain (Nauts et al. 1953). This beneficial property had been observed by others in association with infection-induced tumor regression (Hoption Cann et al. 2003). In fact, patients would often reduce or discontinue their use of narcotic pain medications while receiving treatment. This phenomenon appears to be independent of tumor regression, as it often occurred immediately after toxin injection, preceding such regressions. Lagueux, after many years of experience using Coley's toxins, commented that, "pain always disappeared after the first injections" (Nauts et al. 1953). Actually, this remarkable analgesic effect had long been noted. The well known description of inflammation by Celsus is followed by a largely unappreciated observation on the benefits of fever: "Now the signs of an inflammation are four: redness and swelling with heat and pain

... if there is pain without inflammation, nothing is to be put on: for the actual fever at once will dissolve the pain" (Hoption Cann et al. 2003).

Researchers in the 1960s and 1970s saw things differently. They believed Coley's toxin worked by stimulating the patient's immune system to defeat cancer. It was proposed that specific cell mediated (type 1) or humoral (type 2) immune responses were the key mediators of cancer regression. This led eventually to the discovery of small cell-signaling proteins called cytokines as the cancer-fighting agents of the immune system. However, the cytokines did not turn out to be very effective against cancer in spite of considerable efforts to commercialize them for that purpose (Hoption Cann et al. 2003).

Apparently, the simple fact of elevating temperature was not seriously considered for its anti-cancer effects. This is unfortunate since it is known that slightly elevated temperature causes general non-specific disruption that is damaging to aneuploid cells (Torres et al. 2007). It is also known that high temperature (>39°C) kills cancer cells both *in vivo* and *in vitro*, in a temperature and time dependent manner (the higher the temperature and the longer the exposure, the higher the number of killed cells) (Keech and Wills 1979, Mackey et al. 1992, Pettigrew et al. 1974). Indeed, Coley's toxins produced violent febrile reactions, which he duly noted was the symptom most associated with tumor regression. The fevers he induced usually did not last more than 24 to 48 hours. However, it would seem from subsequent data that a temperature of at least 40°C maintained for 48 to 96 hours is more likely to produce remission than lower temperatures for shorter duration. A retrospective study of patients with inoperable soft tissue sarcomas treated with Coley's toxins found a superior 5-year survival in patients whose fevers averaged 38–40°C, compared with those having little or no fever (<38°C) during treatment (60% v 20%) (Hoption Cann et al. 2003).

Unfortunately, with the widespread use of antibiotics to treat infections and antipyretics to "manage" symptoms of an infection, the critical part played by fever is often overlooked. In hospital settings, fever is frequently suppressed as a matter of routine (Edwards et al. 2001, Isaacs et al. 1990, Thomas et al. 1994). Many modern immunology texts make little mention of fever, e.g., (Delves et al. 2006), and may disregard it as being "insignificant" (Parslow et al. 2001) or refer to it as a "mystery" (Rosenberg and Gallin 1999).

As Hoption Cann said, "Nature exists in a delicate balance, the immune system being no exception. Attempts to create an increasingly sterile environment may further reduce our innate cancer curing ability, until we may finally convince ourselves that it never existed at all" (Hoption Cann et al. 2003). Four medical advances have systemically eroded Coley's fever-inducing therapy for cancer:

1. Cancer surgery, like any other operation, became a sterile procedure after acceptance of Lister's aseptic techniques in the late 1800s (Lister 1906). In fact, in a 1909 discussion paper on cancer treatment, one surgeon suggested that the postoperative infections that were common in the past improved survival and should be encouraged (Thiery 1909). Yet in this new era, his suggestion was harshly criticized as "a doctrine that would make surgery go backwards".

2. By the time of Coley's death in 1936, radiotherapy had become an established cancer treatment, and chemotherapy was rapidly gaining acceptance. These treatments could be more easily standardized than Coley's approach and the hope that these therapies would eventually lead to a cure for cancer was high. Such therapies, however, ran counter to Coley's "immunotherapy", as they are highly immunosuppressive.

3. Following World War II, antibiotic use during and after surgery became commonplace. Thus, post-surgical infection rates were reduced even further, in addition to diminishing the severity and duration of those infections that did occur.

4. Once the immune system became "redundant" in fighting infections, antipyretics came into routine use to eliminate

the discomforting symptoms of an immune response. Hence, reports of spontaneous regression have become less commonplace, although an association with acute infections is often noted when it occurs (Bowles and Perkins 1999, Delmer et al. 1994, Fassas et al. 1991, Frick and Frick 1993, Garcia-Rayo et al. 1996, Ifrah et al. 1985, Maekawa et al. 1989, Marcos Sanchez et al. 1991, Mitterbauer et al. 1996, Rebollo et al. 1990, Ruckdeschel et al. 1972, Sureda et al. 1990, Tzankov et al. 2001) (Section 7.3.1). Table 7.3 summarizes the literature on spontaneous remission of neoplastic diseases associated with infection and/or fever.

Table 7.3   Selected references supporting infection and/or fever in association with spontaneous remission of neoplastic diseases.*

Cancer	*Source*
Bone	(Callan et al. 1975, Cole and Ferguson 1959, Copeland et al. 1985, Eisenbud et al. 1987, Levin 1957)
Brain	(Kapp 1983, Margolis and West 1967)
Breast	(Larsen and Rose 1999)
Burkitt's lymphoma	(Bluming and Ziegler 1971, Ziegler 1976)
Colorectal	(Fucini et al. 1985, Nowacki and Szymendera 1983)
Gastric	(Rebollo et al. 1990, Zambrana Garcia et al. 1996)
Gynecological	(Friedrich 1972)
Head and neck	(Temesrekasi 1969, Woods 1975)
Leukemia (AML, ALL, CML, CLL)	(Barton and Conrad 1979, Bassen and Kohn 1952, Burgess and de Gruchy 1969, Jono et al. 1994, Kizaki et al. 1988, Lefrere et al. 1994, Maekawa et al. 1989, Matzker and Steinberg 1976, Treon and Broitman 1992, Vladimirskaia 1962, Wiernik 1976, Wyszkowski et al. 1969, Zhu and Qian 1986)
Liver	(Chien et al. 1992, Grossmann et al. 1995, Markovic et al. 1996, Tarazov 1996)
Lung	(Greentree 1973, Marcos Sanchez et al. 1991, Mentzer 1995, Ruckdeschel et al. 1972, Takita 1970)

Table 7.3 Contd. ...

Table 7.3 Contd. ...

Lymphoma & Non-Hodgkin	(De Berker et al. 1996, Drobyski and Qazi 1989, Gattiker et al. 1980, Grem et al. 1986, Rao et al. 1995, Sawada et al. 1994, Sureda et al. 1990, Wolf 1989, Zygiert 1971)
Melanoma	(Cook 1992, Grafton 1994, Gunale and Tucker 1975, Motofei 1996, Wagner and Nathanson 1986, Wormald and Harper 1983)
Multiple myeloma	(London 1955)
Prostate	(Katz and Schapira 1982, Schurmans et al. 1996)
Kidney	(Edwards et al. 1996, Mangiapan et al. 1994)
Retinoblastoma	(Hunter 1968, Jain and Singh 1968, Verhoeff 1966)
Sarcoma	(Berner and Laub 1965, Lei et al. 1997, Penner 1953, Weintraub 1969)

*Adapted from Table 1 on page 58 of (Kleef et al. 2001). AML, Acute Myeloid Leukemia; ALL, Acute Lymphoblastic Leukemia; CML, Chronic Myeloid Leukemia; CLL, Chronic Lymphoblastic Leukemia.

A small sign things may be changing is a 2011 report funded by the National Cancer Institute. Huang et al. published the first study to examine the relationship between severity of menopausal symptoms and breast cancer risk among postmenopausal women (Huang et al. 2011). The authors observed that, "increasing intensity of hot flushes was associated with progressively lower risks of all 3 histologic subtypes of breast cancer studied. In particular, women who experienced severe hot flushes with awakening had lower risks of breast cancer compared with women who experienced menopausal symptoms other than hot flushes with awakening and also compared with women who had hot flushes without perspiration." Unfortunately, the authors considered hot flashes as merely "a surrogate marker for hormonal changes that are relevant to the etiology of breast cancer." Hopefully, follow-up studies will investigate the intrinsic importance of hot flashes in warding off cancer.

Hyperthermia

A 2001 report sponsored by the Office of Alternative Medicine at the National Institutes of Health in Bethesda concluded that, "Pyrogenic substances and the recent use of whole-body hyperthermia to mimic the physiologic response to fever have been successfully administered in palliative and curative treatment protocols for metastatic cancer. Further research in this area is warranted" (Kleef et al. 2001).

The National Cancer Institute's website says: "Hyperthermia (also called thermal therapy or thermotherapy) is a type of cancer treatment in which body tissue is exposed to high temperatures (up to 45°C). Research has shown that high temperatures can damage and kill cancer cells, usually with minimal injury to normal tissues (van der Zee 2002). By killing cancer cells and damaging proteins and structures within cells (Hildebrandt et al. 2002), hyperthermia may shrink tumors. Hyperthermia is under study in clinical trials (research studies with people) and is not widely available." (http://www.cancer.gov/cancertopics/factsheet/Therapy/hyperthermia)

There are several kinds of hyperthermia treatments: local, regional and whole body. Whole body hyperthermia is being used to treat cancer but unfortunately only in combination with radiation and chemotherapy (van der Zee 2002, Wust et al. 2002). Nevertheless, hyperthermia treatments are becoming more and more common and available in the US, while Germany is leading the use of this promising therapy.

The therapeutic benefit due to elevated temperature is understandable in light of the theory of chromosomal imbalance. As we have seen, aneuploid cells are particularly temperature sensitive to elevated temperatures that are harmless to normal human cells (Keech and Wills 1979, Mackey et al. 1992, Torres et al. 2007). This can be explained (at least in part) by the abnormal metabolism of cancer cells (e.g. the Warburg effect, Section 6.3) caused by aneuploidy. When the temperature raises, there is an increase in the entropy of the cancer cells, which were already at the maximum level of disorganization consistent with viability (Section 6.2.3), leading to rapid and substantial tumor regression.

8

Conclusion

[A] reductionistic approach to organization in general, and to life in particular, [is] as follows: throw away the organization and keep the underlying matter. The relational alternative...says the exact opposite, namely: when studying an organized material system, throw away the matter and keep the underlying organization.

(Rosen 1991)

Commenting on the clinical implications of basic research, Lupski said, "the molecular medicine model that is promulgated in every medical school is based on sickle cell disease, in which the predominant type of mutation is a base-pair change, which alters the coding sequence and results in the synthesis of a mutant protein" (Lupski 2007). Virtually all the resources continue to be committed to this model of finding the focal lesions (mutations) in DNA that cause cancer. While that colossal effort has consumed many billions of dollars it has failed to find one or even a combination of mutations that have been demonstrated to cause cancer.

Richard Strohman was an outspoken critic and one of the most insightful commentators on the significance and shortcomings of the molecular medicine model. "The revolution that came with the DNA double helix," Strohman said, "has largely been a technological one in which a huge research effort has developed to further our understanding of 'the nature of the genetic code, the mechanism of protein synthesis, and the manner of gene replication' " (Wilkins 1996). While this is a correct statement of what Watson and Crick were after, he said, "it is also an exceedingly narrow view of what was later adopted as a paradigm by those biologists who were so quickly recruited to the molecular cause. And it is certainly a narrow and distorted view of what actually happened... . It has been made to appear that a theory that works so well for an understanding of how genes code for proteins also works just as well, and as simply, in explaining how genes cause common cancer or excessive TV viewing...; we are mixing our levels in biology and it doesn't work" (Strohman 1997).

The attraction of the gene mutation theory was its promise of simplicity; that cancer resulted from a manageable number of specific mutations. A manageable number was the hoped-for key to unlocking the mysteries of cancer that should lead to the taming of the ever growing modern scourge. Instead, we find every cancer cell has a unique karyotype that changes with each division. Far from providing insights into the nature of cancer, and hence insights into prevention and more effective treatments, the gene mutation theory is now so burdened with the complexity of its details—driver mutation, passenger mutation, gain-of-function mutation, loss-of-function mutation, haploinsufficiency, caretaker-gatekeeper-landscaper genes, gradualism, chromothripsis, synthetic lethality, synthetic sickness, synthetic viability, hard and soft synthetic lethalities, non-cell-autonomous synthetic lethality, functional buffering, gene addiction, functional redundancy, induced essentiality, genetic capacitor, network compatibility (Ashworth et al. 2011)—that it has become an empirical exercise devoid of theoretical and explanatory power.

A 1997 study provides a typical example. Comparing the genomes of normal and cancer cells (colon and pancreas), Zhang et al. acknowledged that, "most of the transcripts could not have been predicated to be differentially expressed in cancers" (Zhang et al. 1997). The authors were also surprised to discover that, "two widely studied oncogenes, c-fos and c-erbb3, were expressed at much higher levels in normal colon epithelium than in [colorectal] cancers, in contrast to their up-regulation in transformed cells [in culture]." The expression levels of mutated K-*ras*, which was said to be a "dominant" oncogene of colon cancer in 1987

(Bos et al. 1987), was not even listed. According to a personal communication from the authors, K-*ras* was not included in the analysis because it contributed fewer than five copies per cell (Personal communication, Lin Zhang, 1997).

The simple fact that chromosomal imbalance theory answers fundamental questions inexplicable in terms of the gene mutation theory of cancer (Table 8.1) accounts for its growing popularity among researchers around the world: First & Second International Conferences on Aneuploidy and Cancer: Clinical & Experimental Aspects, Oakland, CA, 23-26 January 2004; 31 January–3 February 2008 (Reith A. Special Issue Editor 2004).

Table 8.1   Questions unanswered by the gene mutation theory of cancer, yet resolved by the theory of chromosomal imbalance.

* How would non-mutagenic carcinogens cause cancer?

   As Boveri said, anything that causes aneuploidy can cause cancer.
* What kind of mutation would cause cancer only after delays of several decades and many cell generations?

   Point mutations cannot. Autocatalyzed progression of aneuploidy does.
* What kind of mutation would continually alter the phenotype of mutant cells, despite the absence of further mutagens?

   Point mutations cannot. Autocatalyzed progression of aneuploidy does.
* What kind of mutation would be able to alter phenotypes at rates that exceed conventional gene mutations by 4–11 orders of magnitude?

   Point mutations cannot. Autocatalyzed progression of aneuploidy does.
* What kind of mutation would generate resistance against many more drugs than the one used to select it?

   Point mutations cannot. Autocatalyzed progression of aneuploidy does.

Table 8.1 Contd. ...

Table 8.1 Contd. ...

- What kind of mutations would continually change the cellular and nuclear morphologies within the same "clonal" cancer?

  Point mutations cannot. Autocatalyzed progression of aneuploidy does.

- What kind of mutation would alter the expressions and metabolic activities of thousands of genes, which is the hallmark of cancer cells?

  Point mutations cannot. Aneuploidy does.

- What kind of mutation would consistently coincide with aneuploidy, although conventional gene mutations do not alter the karyotype?

  Point mutations do not. Aneuploidy makes cancer a species of its own.

- Why would cancer not be heritable via conventional mutations by conventional Mendelian genetics?

  Aneuploidy is not heritable.

We have before us the facts to explain the entire multi-step sequence of carcinogenesis. This sequence begins with a random aneuploidy, which is caused either by a carcinogen or arises spontaneously. Since aneuploidy unbalances conserved teams of proteins that segregate, synthesize and repair chromosomes, the karyotypes of aneuploid cells are perpetually at risk of autocatalytic variations. The risk and rates of karyotype variation were found to be proportional to the degree of aneuploidy. The basis for the somatic evolution of cancer cells from randomly aneuploid precursor cells is selection for advantages in growth. Thus aneuploidy is necessary and—if progresses—sufficient for carcinogenesis.

Since cancer cells are generated from precursor cells by rearranging old genes into new sets of chromosomes—just like new species are generated in phylogenesis (O'Brien et al. 1999)—cancer cells are new species of their own, rather than mutants of any precursor cells. However, due to the inherent instability of aneuploidy, aneuploid cell species are unable to achieve phylogenetic autonomy because they are too unstable to maintain the many genetic investments that are necessary for autonomy. But as parasites, aneuploid cells are able to progress from bad to worse autocatalytically, from preneoplastic phenotypes to highly malignant cancer cells.

The many idiosyncratic phenotypes of cancer cells, such as immortality, invasiveness, drug-resistance, metastasis, abnormal gene expression, etc., represent distinct subspecies from within the inherent 'polyphyletic' diversity (Hauschka and Levan 1958) of individual cancers. Since aneuploidy and genomic diversity are incompatible with the identity and survival of an autonomous, diploid species (Hassold 1986, Hernandez and Fisher 1999), the idiosyncratic phenotypes of cancer cells are never observed in diploid biology. As chromosomal imbalance theory becomes widely understood and accepted, it will alter the course of cancer research, prevention, diagnosis and treatment, with profound consequences for how governments, departments of health, clinicians and patients think about and deal with cancer.

References

Aardema MJ, Albertini S, Arni P, Henderson LM, Kirsch-Volders M, Mackay JM, Sarrif AM, Stringer DA, Taalman RD (1998) Aneuploidy: a report of an ECETOC task force. Mutat Res 410: 3-79.

Aggarwal A, Leong SH, Lee C, Kon OL, Tan P (2005) Wavelet transformations of tumor expression profiles reveals a pervasive genome-wide imprinting of aneuploidy on the cancer transcriptome. Cancer Res 65: 186-194.

Ahuja D, Saenz-Robles MT, Pipas JM (2005) SV40 large T antigen targets multiple cellular pathways to elicit cellular transformation. Oncogene 24: 7729-7745.

Ai H, Barrera JE, Pan Z, Meyers AD, Varella-Garcia M (1999) Identification of individuals at high risk for head and neck carcinogenesis using chromosome aneuploidy detected by fluorescence in situ hybridization. Mutat Res 439: 223-232.

Ai H, Barrera JE, Meyers AD, Shroyer KR, Varella-Garcia M (2001) Chromosomal aneuploidy precedes morphological changes and supports multifocality in head and neck lesions. The Laryngoscope 111: 1853-1858.

Akagi T, Sasai K, Hanafusa H (2003) Refractory nature of normal human diploid fibroblasts with respect to oncogene-mediated transformation. Proc Natl Acad Sci U S A 100: 13567-13572.

Al-Hajj M, Wicha MS, Benito-Hernandez A, Morrison SJ, Clarke MF (2003) Prospective identification of tumorigenic breast cancer cells. Proc Natl Acad Sci U S A 100: 3983-3988.

Al-Mefty O, Kadri PA, Pravdenkova S, Sawyer JR, Stangeby C, Husain M (2004) Malignant progression in meningioma: documentation of a series and analysis of cytogenetic findings. J Neurosurg 101: 210-218.

Al-Mulla F, Going JJ, Sowden ET, Winter A, Pickford IR, Birnie GD (1998) Heterogeneity of mutant versus wild-type Ki-ras in primary and metastatic colorectal carcinomas, and association of codon-12 valine with early mortality. J Pathol 185: 130-138.

Al-Mulla F, Keith WN, Pickford IR, Going JJ, Birnie GD (1999) Comparative genomic hybridization analysis of primary colorectal carcinomas and their synchronous metastases. Genes Chromosomes Cancer 24: 306-314.

Alberts B. (1990) Fundamental research on cancer: issues and progress; December 7; San Francisco. Eberlin Reporting Service. 90 pp.

Alberts B, Bray D, Lewis J, Raff M, Roberts K, Watson JD (1994) Molecular Biology of the Cell. New York: Garland Publishing, Inc.

Albertson DG, Collins C, McCormick F, Gray JW (2003) Chromosome aberrations in solid tumors. Nat Genet 34: 369-376.

Albino AP, Le Strange R, Oliff AI, Furth ME, Old LJ (1984) Transformating ras genes from human melanoma: a manifestation of tumor heterogeneity? Nature 308: 69-72.

Aldaz CM, Conti CJ, Klein-Szanto AJ, Slaga TJ (1987) Progressive dysplasia and aneuploidy are hallmarks of mouse skin papillomas: relevance to malignancy. Proc Natl Acad Sci U S A 84: 2029-2032.

Aldaz CM, Conti CJ, Larcher F, Trono D, Roop DR, Chesner J, Whitehead T, Slaga TJ (1988a) Sequential development of aneuploidy, keratin modifications, and gamma-glutamyltransferase expression in mouse skin papillomas. Cancer Res 48: 3253-3257.

Aldaz CM, Conti CJ, Yuspa SH, Slaga TJ (1988b) Cytogenetic profile of mouse skin tumors induced by the viral Harvey-ras gene. Carcinogenesis 9: 1503-1505.

Alderman EM, Lobb RR, Fett JW (1985) Isolation of tumor-secreted products from human carcinoma cells maintained in a defined protein-free medium. Proc Natl Acad Sci U S A 82: 5771-5775.

Alexander P (1983) Dormant metastases—studies in experimental animals. J Pathol 141: 379-383.

Alexander P, Senior PV, Murphy P, Clarke R (1985) Role of growth stimulatory factors in determing the sites of metastasis. In: Honn K, Powers WE, Sloane BF, editors. Mechanisms of Cancer Metastasis: Potential Therapeutic Implications. Hingham: Martinus Nijhoff Publishing.

Alexander P (1985) Do cancers arise from a single transformed cell or is monoclonality of tumours a late event in carcinogenesis? Br J Cancer 51: 453-457.

Allan JM, Travis LB (2005) Mechanisms of therapy-related carcinogenesis. Nat Rev Cancer 5: 943-955.

Allen JD, Brinkhuis RF, van Deemter L, Wijnholds J, Schinkel AH (2000) Extensive contribution of the multidrug transporters P-glycoprotein and Mrp1 to basal drug resistance. Cancer Res 60: 5761-5766.

Alligood KT, Sauer TD, Yorke JA (1996) Chaos: an introduction to dynamical systems; Banchoff TF, Devlin K, Gonnet G, Marsden J, Wagon S, editors. Textbooks in Mathematical Sciences. New York: Springer. pp. 1-42.

American Type Culture Collection (1992) Catalog of cell lines and hybridomas. Rockville, MD: The collection, 1985-1992.

Ames B, Durston WE, Yamaski E, Lee FD (1973) Carcinogens are mutagens: a simple test system combining liver homogenates for activation and bacteria for detection. Proc Natl Acad Sci U S A 70: 2281-2285.

Andalis AA, Storchova Z, Styles C, Galitski T, Pellman D, Fink GR (2004) Defects arising from whole-genome duplications in *Saccharomyces cerevisiae*. Genetics 167: 1109-1121.

Anderson GH (1991) Chapter 3: Cytology screening programs; Bibbo M, editor. Comprehensive Cytopathology. Philadelphia: W. B. Saunders Company. pp. 48-58.

Anderson GR, Stoler DL, Brenner BM (2001) Cancer: the evolved consequence of a destabilized genome. Bioessays 23: 1037-1046.

Anderson K, Lutz C, van Delft FW, Bateman CM, Guo Y, Colman SM, Kempski H, Moorman AV, Titley I, Swansbury J, Kearney L, Enver T, Greaves M (2011) Genetic variegation of clonal architecture and propagating cells in leukaemia. Nature 469: 356-361.

Andrechek ER, Hardy WR, Laing MA, Muller WJ (2004) Germ-line expression of an oncogenic erbB2 allele confers resistance to erbB2-induced mammary tumorigenesis. Proc Natl Acad Sci U S A 101: 4984-4989.

Angell R (1997) First-meiotic-division nondisjunction in human oocytes. Am J Genet 61: 23-32.

Anthony PP, editor (1998) Diagnostic pitfalls in histology and cytopathology practice. London: Greenwich Medical Media. 125 p.

Antonarakis SE, Lyle R, Dermitzakis ET, Reymond A, Deutsch S (2004) Chromosome 21 and Down syndrome: from genomics to pathophysiology. Nat Rev Genet 5: 725-738.

Aragane H, Sakakura C, Nakanishi M, Yasuoka R, Fujita Y, Taniguchi H, Hagiwara A, Yamaguchi T, Abe T, Inazawa J, Yamagishi H (2001) Chromosomal aberrations in colorectal cancers and liver metastases analyzed by comparative genomic hybridization. Int J Cancer 94: 623-629.

Armitage P, Doll R (1954) The age distribution of cancer and the multi-stage theory of carcinogenesis. British Journal of Cancer 8: 1-12.

Arnold K (2001) Study Results Help Define HPV's Role as Diagnostic Tool. J Natl Cancer Inst 93: 259-260.

Artandi SE, Chang S, Lee SL, Alson S, Gottlieb GJ, Chin L, DePinho RA (2000) Telomere dysfunction promotes non-reciprocal translocations and epithelial cancers in mice. Nature 406: 641-645.

Ashby J, Purchase IF (1988) Reflections on the declining ability of the Salmonella assay to detect rodent carcinogens as positive. Mutat Res 205: 51-58.

Ashworth A, Lord CJ, Reis-Filho JS (2011) Genetic interactions in cancer progression and treatment. Cell 145: 30-38.

Atkin NB (1964) Nuclear Size in Premalignant Conditions of the Cervix Uteri. Nature: 201.

Atkin NB, Baker MC (1966) Chromosome abnormalities as primary events in human malignant disease: evidence from marker chromosomes. J Natl Cancer Inst 36: 539-557.

Atkin NB (1974) Chromosomes in human malignant tumors: a review and assessment. In: German J, editor. Chromosomes and Cancer. New York: John Wiley and Sons. pp. 375-422.

Atkin NB (1986) Chromosome 1 aberrations in cancer. Cancer Genet Cytogenet 21: 279-285.

Atkin NB, Baker MC (1990) Are human cancers ever diploid–or often trisomic? Conflicting evidence from direct preparations and culture. Cytogenet Cell Genet 53: 58-60.

Atkin NB (1991) Non-random chromosomal changes in human neoplasia. In: Sobti R, Obe G, editors. Eukaryotic Chromosomes: Structural and Functional Aspects. New Dehli: Narosa Publishing House. pp. 153-164.

Atkin NB (1997) Cytogenetics of carcinoma of the cervix uteri: a review. Cancer Genet Cytogenet 95: 33-39.

Auer G, Kronenwett U, Roblick UJ, Franzen JK, Habermann JK, Sennerstam R, Ried T (2004) Human breast adenocarcinoma: DNA content, chromosomes, gene expression and prognosis. Cell Oncol 26: 171-172.

Auer GU, Caspersson TO, Wallgren AS (1980) DNA content and survival in mammary carcinoma. Anal Quant Cytol Histol 2: 161-165.

Auersperg N, Corey MJ, Worth A (1967) Chromosomes in preinvasive lesions of the human uterine cervix. Cancer Res 27: 1394-1401.

Augenlicht LH, Wahrman MZ, Halsey H, Anderson L, Taylor J, Lipkin M (1987) Expression of cloned sequences in biopsies of human colonic tissue and in colonic carcinoma cells induced to differentiate in vitro. Cancer Res 47: 6017-6021.

Ault JG, Cole RW, Jensen CG, Jensen LCW, Bachert LA, Rieder CL (1995) Behavior of Crocidolite Asbestos during Mitosis in Living Vertebrate Lung Epithelial Cells. Cancer Res 55: 792-798.

Awa AA (1974) Cytogenetic and oncogenic effects of the ionizing radiations of the atomic bombs. In: German J, editor. Chromosomes and Cancer. New York: John Wiley & Sons. pp. 637-674.

Bailar JC, 3rd, Smith EM (1986) Progress against cancer? N Engl J Med 314: 1226-1232.

Bailar JC, 3rd, Gornik HL (1997) Cancer undefeated. N Engl J Med 336: 1569-1574.

Bailey JE (2000) Life is complicated. In: Cornish-Bowden A, Cardenas ML, editors. Technological and Medical Implications of Metabolic Control Analysis. Dordrecht: Kluwer. pp. 41-47.

Bains W (2001) The parts list of life. Nat Biotechnol 19: 401-402.

Baker DJ, van Deursen JM (2010) Chromosome missegregation causes colon cancer by APC loss of heterozygosity. Cell Cycle 9.

Baker SJ, Markowitz S, Fearon ER, Willson JK, Vogelstein B (1990) Suppression of human colorectal carcinoma cell growth by wild-type p53. Science 249: 912-915.

Balaban GB, Herlyn M, Clark WH, Jr., Nowell PC (1986) Karyotypic evolution in human malignant melanoma. Cancer Genet Cytogenet 19: 113-122.

Balmain A (2001) Cancer genetics: from Boveri and Mendel to microarrays. Nat Reviews 1: 77-82.

Bamford S, Dawson E, Forbes S, Clements J, Pettett R, Dogan A, Flanagan A, Teague J, Futreal PA, Stratton MR, Wooster R (2004) The COSMIC (Catalogue of Somatic Mutations in Cancer) database and website. Br J Cancer 91: 355-358.

Bardi G, Johansson B, Pandis N, Bak-Jensen E, Orndal C, Heim S, Mandahl N, Andren-Sandberg A, Mitelman F (1993) Cytogenetic aberrations in colorectal adenocarcinomas and their correlation with clinicopathologic features. Cancer 71: 306-314.

Bardi G, Parada LA, Bomme L, Pandis N, Willen R, Johansson B, Jeppsson B, Beroukas K, Heim S, Mitelman F (1997) Cytogenetic comparisons of synchronous carcinomas and polyps in patients with colorectal cancer. Br J Cancer 76: 765-769.

Bardi G, Fenger C, Johansson B, Mitelman F, Heim S (2004) Tumor karyotype predicts clinical outcome in colorectal cancer patients. J Clin Oncol 22: 2623-2634.

Barrett JC (1980) A preneoplastic stage in the spontaneous neoplastic transformation of Syrian hamster embryo cells in culture. Cancer Res 40: 91-94.

Barrett T, Suzek TO, Troup DB, Wilhite SE, Ngau W-C, Ledoux P, Rudnev D, Lash AE, Fujibuchi W, Edgar R (2005) NCBI GEO: mining millions of expression profiles—database and tools. Nucl Acids Res 33: d562-566.

Barton JC, Conrad ME (1979) Beneficial effects of hepatitis in patients with acute myelogenous leukemia. Ann Intern Med 90: 188-190.

Bassen FA, Kohn JL (1952) Multiple spontaneous remissions in a child with acute leukemia. The occurrence of agranulocytosis and aplastic anemia in acute leukemia and their relationship to remissions. Blood 7: 37-46.

Baudis M (2007) Genomic imbalances in 5918 malignant epithelial tumors: an explorative meta-analysis of chromosomal CGH data. BMC Cancer 7: 226.

Bauer K-H (1963) Das Krebsproblem; Verlag S, editor. Berlin, Goettingen, Heidelberg: Springer Verlag. 1099 pp.

Bauer KH (1928) Mutationstheorie der Geschwulstentstehung. Berlin: Springer

Bauer KH (1949) Das Krebsproblem. Berlin: Springer-Verlag

Becker FF, Fox RA, Klein KM, Wolman SR (1971) Chromosome patterns in rat hepatocytes during N-2-fluorenylacetamide carcinogenesis. J Natl Cancer Inst 46: 1261-1269.

Bell RH, Jr., Memoli VA, Longnecker DS (1990) Hyperplasia and tumors of the islets of Langerhans in mice bearing an elastase I-SV40 T-antigen fusion gene. Carcinogenesis 11: 1393-1398.

Benedict WF (1972) Early changes in chromosomal number and structure after treatment of fetal hamster cultures with transforming doses of polycyclic hydrocarbons. J Natl Cancer Inst 49: 585-590.

Benedict WF, Banerjee A, Mark C, Murphree AL (1983) Non-random retinoblastoma gene is a recessive cancer gene. Cancer Genet Cytogenet 10: 311-333.

Berenblum I, Shubik P (1947) The role of croton oil applications, associated with a single painting of a carcinogen, in tumour induction of the mouse's skin. Br J Cancer 1: 379-382.

Berenblum I, Shubik P (1949) An experimental study of the initiating state of carcinogenesis, and a re-examination of the somatic cell mutation theory of cancer. Br J Cancer 3: 109-118.

Berman J (2010) Evolutionary genomics: When abnormality is beneficial. Nature 468: 183-184.

Bernards R, Weinberg RA (2002) A progression puzzle. Nature 418: 823.

Berner RE, Laub DL (1965) The Spontaneous Cure of Massive Fibrosarcoma. Plast Reconstr Surg 36: 257-262.

Beroukhim R, Getz G, Nghiemphu L, Barretina J, Hsueh T, Linhart D, Vivanco I, Lee JC, Huang JH, Alexander S, Du J, Kau T, Thomas RK, Shah K, Soto H, Perner S, Prensner J, Debiasi RM, Demichelis F, Hatton C, et al. (2007) Assessing the significance of chromosomal aberrations in cancer: methodology and application to glioma. Proc Natl Acad Sci U S A 104: 20007-20012.

Bialy H (2004) Oncogenes, Aneuploidy and AIDS: A scientific life and times of Peter H. Duesberg. Cueranavaca, Mexico: Institute of Biotechnology, Autonomous National University of Mexico. 318 pp.

Biernaux C, Loos M, Sels A, Huez G, Stryckmans P (1995) Detection of major bcr-abl gene expression at a very low level in blood cells of some healthy individuals. Blood 86: 3118-3122.

Binder RL, Johnson GR, Gallagher PM, Stockman SL, Sundberg JP, Conti CJ (1998) Squamous cell hyperplastic foci: precursors of cutaneous papillomas induced in SENCAR mice by a two-stage carcinogenesis regimen. Cancer Res 58: 4314-4323.

Birkenkamp-Demtroder K, Christensen LL, Olesen SH, Frederiksen CM, Laiho P, Aaltonen LA, Laurberg S, Sorensen FB, Hagemann R, Orntoft, TF (2002) Gene expression in colorectal cancer. Cancer Res 62: 4352-4363.

Bishop JM (1981) Enemies within: genesis of retrovirus oncogenes. Cell 23: 5-6.

Bishop JM (1983) Cellular oncogenes and retroviruses. Annu Rev Biochem 52: 301-354.

Bishop JM (1987) The molecular genetics of cancer. Science 235: 305-311.

Bishop JM (1991) Molecular themes in oncogenesis. Cell 64: 235-248.

Bishop JM (1995) Cancer: the rise of the genetic paradigm. Genes Dev 9: 1309-1315.

Bloch-Shtacher N, Sachs L (1977) Identification of a chromosome that controls malignancy in Chinese hamster cells. J Cell Physiol 93: 205-212.

Bluming AZ, Ziegler JL (1971) Regression of Burkitt's lymphoma in association with measles infection. Lancet 2: 105-106.

Böcking A, Chatelain R (1989) Diagnostic and prognostic value of DNA cytometry in gynecologic cytology. Analytical and Quantitative Cytology and Histology 11: 177-186.

Böcking A (2008) DNA image cytometry-assisted early diagnosis and grading of malignant tumors by detection and quantification of DNA aneuploidy. Wetzlar: Motic Deutschland GmbH (GERMANY)accessed Sept 14, 2009, http://www.knittkuhl.com/pdf/Broschuere.pdf

Böcking A (2009) This is an interactive, web-based course in diagnostic DNA-Image-Cytometry (DNA-ICM). accessed Sept 14, http://www.cytopathologie-dna-icm.uni-duesseldorf.de/

Bockmuhl U, Schluns K, Schmidt S, Matthias S, Petersen I (2002) Chromosomal alterations during metastasis formation of head and neck squamous cell carcinoma. Genes Chromosomes Cancer 33: 29-35.

Bohen SP, Troyanskaya OG, Alter O, Warnke R, Botstein D, Brown PO, Levy R (2003) Variation in gene expression patterns in follicular lymphoma and the response to rituximab. Proc Natl Acad Sci U S A 100: 1926-1930.

Boice JD, Monson RR (1977) Breast cancer in woman after repeated fluoroscopic examinations of the chest. J Natl Cancer Inst 59: 823-832.

Boland CR, Ricciardiello L (1999) How many mutations does it take to make a tumor? Proc Natl Acad Sci U S A 96: 14675-14677.

Bollmann R, Bollmann M, Henson DE, Bodo M (2001) DNA cytometry confirms the utility of the Bethesda System for the classification of Papanicolaou Smears. Cancer Cytopathol 93: 222-228.

Bolstad BM, Irizarry RA, Astrand M, Speed TP (2003) A comparison of normalization methods for high density oligonucleotide array data based on variance and bias. Bioinformatics (Oxford, England) 19: 185-193.

Bomme L, Bardi G, Pandis N, Fenger C, Kronborg O, Heim S (1998) Cytogenetic analysis of colorectal adenomas: karyotypic comparisons of synchronous tumors. Cancer Genet Cytogenet 106: 66-71.

Bonassi S, Hagmar L, Stromberg U, Montagud AH, Tinnerberg H, Forni A, Heikkila P, Wanders S, Wilhardt P, Hansteen IL, Knudsen LE, Norppa H (2000) Chromosomal aberrations in lymphocytes predict human cancer independently of exposure to carcinogens. European Study Group on Cytogenetic Biomarkers and Health. Cancer Res 60: 1619-1625.

Bonetta L (2005) Going on a cancer gene hunt. Cell 123: 735-737.

Bookstein R, Shew JY, Chen PL, Scully P, Lee WH (1990) Suppression of tumorigenicity of human prostate carcinoma cells by replacing a mutated RB gene. Science 247: 712-715.

Borek C (1982) Radiation oncogenesis in cell culture. Adv Cancer Res 37: 159-232.

Bos JL, Fearon ER, Hamilton SR, Verlaan-de Vries M, van Boom JH, van der Eb AJ, Vogelstein B (1987) Prevalence of *ras* gene mutations in human colorectal cancers. Nature 327: 293-297.

Bose S, Deininger M, Gora-Tybor J, Goldman JM, Melo JV (1998) The presence of typical and atypical BCR-ABL fusion genes in leukocytes of normal individuals: biologic significance and implications for the assessment of minimal residual disease. Blood 92: 3362-3367.

Boveri T (1902/1964) On multipolar mitoses as a means of analysis of the cell nucleus. In: Willer BH, Oppenheimer JM, editors. Foundations of Experimental Embryology. Englewood Cliffs, NJ: Prentice Hall. pp. 74-97.

Boveri T (1914) Zur Frage der Entstehung maligner Tumoren. Jena, Germany: Fischer.

Bowles AP, Jr., Perkins E (1999) Long-term remission of malignant brain tumors after intracranial infection: a report of four cases. Neurosurgery 44: 636-642; discussion 642-633.

Bozic I, Antal T, Ohtsuki H, Carter H, Kim D, Chen S, Karchin R, Kinzler KW, Vogelstein B, Nowak MA (2010) Accumulation of driver and passenger mutations during tumor progression. Proc Natl Acad Sci U S A 107: 18545-18550.

Brash D, Cairns J (2009) The mysterious steps in carcinogenesis. Br J Cancer 101: 379-380.

Braun AC (1969) The cancer problem. A critical analysis and modern synthesis. New York: Columbia University Press.

Breivik J, Gaudernack G (1999) Genomic instability, DNA methylation, and natural selection in colorectal carcinogenesis. Seminars in Cancer Biology 9: 245-254.

Bremner R, Balmain A (1990) Genetic changes in skin tumor progression: correlation between presence of a mutant ras gene and loss of heterozygosity on mouse chromosome 7. Cell 61: 407-417.

Breslow RE, Goldsby RA (1969) Isolation and Characterization of Thymidine Transport Mutants of Chinese Hamster Cells. Exptl Cell Res 55: 339-346.

Brewer C, Holloway S, Zawalnyski P, Schinzel A, FitzPatrick D (1999) A chromosomal duplication map of malformations: regions of suspected haplo- and triplolethality—and tolerance of segmental aneuploidy—in humans. Am J Hum Genet 64: 1702-1708.

Brinkley BR, Goepfert TM (1998) Supernumerary centrosomes and cancer: Boveri's hypothesis resurrected. Cell Motility and the Cytoskeleton 41: 281-288.

Brookes P, Lawley PD (1964) Evidence for the binding of polynuclear aromatic hydrocarbons to the nucleic acids of mouse skin: relation between carcinogenic potential of hydrocarbons and their binding to deoxyribonucleic acid. Nature 202: 781-784.

Brú A (1998) Super-rough dynamics on tumour growth. Phys Rev Lett 81.

Brú A, Albertos S, Luis Subiza J, Garcia-Asenjo JL, Bru I (2003) The universal dynamics of tumor growth. Biophys J 85: 2948-2961.

Bryan WR (1960) A reconsideration of the nature of the neoplastic reaction in the light of recent advances in cancer research. J Natl Cancer Inst 24: 221-251.

Bulten J, Poddighe PJ, Robben JCM, Gemmink JH, de Wilde PCM, Hanselaar AGJM (1998) Interphase cytogenetic analysis of cervical intraepithelial neoplasia. Am J Pathol 152: 495-503.

Burdette WJ (1955) The significance of mutation in relation to the origin of tumors: a review. Cancer Res 15: 201-226.

Burgess MA, de Gruchy GC (1969) Septicaemia in acute leukaemia. Med J Aust 1: 1113-1117.

Burnet FM (1957) Cancer—a biological approach. Brit Med J 1: 779-786.

Burnet FM (1971) Immunological surveillance in neoplasia. Transplant Rev 7: 3-25.

Busch W (1866) Ueber den Einfluss, welchen heftige Erysipele zuweilen auf organisirte Neubildungen ausiiben. Berliner klin Wochenschr 13: 245-246.

Cahill DP, Lengauer C, Yu J, Riggins GJ, Willson JK, Markowitz SD, Kinzler KW, Vogelstein B (1998) Mutations of mitotic checkpoint genes in human cancers. Nature 392: 300-303.

Cahill DP, Kinzler KW, Vogelstein B, Lengauer C (1999) Genetic instability and Darwinian selection in tumours. Trends in Biological Sciences (TIBS) 24: M57-M60.

Cairns J (1978) Cancer: Science and Society. A Series of Books in Biology. San Francisco: W. H. Freeman and Company. pp. 35-61.

Cairns J (1981) The origin of human cancers. Nature 289: 353-357.

Cairns J (2002) Somatic stem cells and the kinetics of mutagenesis and carcinogenesis. Proc Natl Acad Sci U S A 99: 10567-10570.

Calder L (2009) Gardasil: holy grail or false idol? Nurs N Z 15: 20-21.

Callan JE, Wood VE, Linda L (1975) Spontaneous resolution of an osteochondroma. J Bone Joint Surg Am 57: 723.

Camps J, Ponsa I, Ribas M, Prat E, Egozcue J, Peinado MA, Miro R (2005) Comprehensive measurement of chromosomal instability in cancer cells: combination of fluorescence in situ hybridization and cytokinesis-block micronucleus assay. Faseb J 19: 828-830.

Camps J, Armengol G, del Rey J, Lozano JJ, Vauhkonen H, Prat E, Egozcue J, Sumoy L, Knuutila S, Miro R (2006) Genome-wide differences between microsatellite stable and unstable colorectal tumors. Carcinogenesis 27: 419-428.

Capasso LL (2005) Antiquity of cancer. Int J Cancer 113: 2-13.

Caratero C, Hijazi A, Caratero A, Mazerolles C, Rischmann P, Sarramon JP (1990) Flow cytometry analysis of urothelial cell DNA content according to pathological and clinical data on 100 bladder tumors. Eur Urol 18: 145-149.

Carter SL, Eklund AC, Kohane IS, Harris LN, Szallasi Z (2006) A signature of chromosomal instability inferred from gene expression profiles predicts clinical outcome in multiple human cancers. Nat Genet 38: 1043-1048.

Caspersson O (1964a) Quantitative cytochemical studies on normal, malignant, premalignant and atypical cell populations from the human uterine cervix. Acta Cytol 8: 45-60.

Caspersson T, Foley GE, Killander D, Lomakka G (1963) Cytochemical differences between mammalian cell lines of normal and neoplastic origins: correlation with heterotransplantability in Syrian hamsters. Exp Cell Res 32: 553-565.

Caspersson T (1964b) Chemical variability in tumor cell populations. Acta Unio Int Contra Cancrum 20: 1275-1279.

Caspersson TO (1950) Disturbed systems for protein formation in the metazoan cell. Cell growth and cell function: A cytochemical study. New York: W. W. Norton & Company. pp. 141-151.

Castro MA, Onsten TT, de Almeida RM, Moreira JC (2005) Profiling cytogenetic diversity with entropy-based karyotypic analysis. J Theor Biol 234: 487-495.

Castro MA, Onsten TG, Moreira JC, de Almeida RM (2006) Chromosome aberrations in solid tumors have a stochastic nature. Mutat Res 600: 150-164.

Cavenee WK, Scrable HJ, James CD (1989) Molecular genetics of human cancer predisposition and progression. In: Cavenee WK, Hastie N, Stanbridge EJ, editors. Recessive Oncogenes and Tumor Suppression. Cold Springs Harbor: Cold Springs Harbor Press. pp. 67-72.

Cetin B, Cleveland DW (2010) How to survive aneuploidy. Cell 143: 27-29.

Cha RS, Thilly WG, Zarbl H (1994) N-nitroso-N-methylurea-induced rat mammary tumors arise from cells with preexisting oncogenic H-ras1 gene mutations. Proc Natl Acad Sci U S A 91: 3749-3753.

Chabner BA, Roberts TG, Jr. (2005) Timeline: Chemotherapy and the war on cancer. Nat Rev Cancer 5: 65-72.

Chakraborty AK, Cichutek K, Duesberg PH (1991) Transforming function of proto-ras genes depends on heterologous promoters and is enhanced by specific point mutations. Proc Natl Acad Sci U S A 88: 2217-2221.

Challis GB, Stam HJ (1990) The spontaneous regression of cancer: a review of cases from 1900 to 1987. Acta Oncologica 29: 545-550.

Chandhok NS, Pellman D (2009) A little CIN may cost a lot: revisiting aneuploidy and cancer. Curr Opin Genet Dev. 19: 74-81.

Chang EH, Furth ME, Scolnick EM, Lowy DR (1982) Tumorigenic transformation of mammalian cells induced by a normal human gene homologous to the oncogene of Harvey murine sarcoma virus. Nature 297: 479-483.

Chang WP, Little JB (1992) Delayed reproductive death as a dominant phenotype in cell clones surviving X-irradiation. Carcinogenesis 13: 923-928.

Chaum E, Ellsworth RM, Abramson DH, Haik BG, Kitchin FD, Chaganti RS (1984) Cytogenetic analysis of retinoblastoma: evidence for multifocal origin and in vivo gene amplification. Cytogenet Cell Genet 38: 82-91.

Chen X, Leung SY, Yuen ST, Chu KM, Ji J, Li R, Chan AS, Law S, Troyanskaya OG, Wong J, So S, Botstein D, Brown PO (2003) Variation in gene expression patterns in human gastric cancers. Mol Biol Cell 14: 3208-3215.

Chi H, Kawachi Y, Otsuka F (1994) Xeroderma pigmentosum variant: DNA ploidy analysis of various skin tumors and normal-appearing skin in a patient. Int J Dermatol 33: 775-778.

Chiba S, Okuda M, Mussman JG, Fukasawa K (2000) Genomic convergence and suppression of centrosome hyperamplification in primary p53-/- cells in prolonged culture. Exp Cell Res 258: 310-321.

Chien RN, Chen TJ, Liaw YF (1992) Spontaneous regression of hepatocellular carcinoma. Am J Gastroenterol 87: 903-905.

Chitale R (2009) Merck hopes to extend gardasil vaccine to men. J Natl Cancer Inst 101: 222-223.

Choma D, Daures JP, Quantin X, Pujol JL (2001) Aneuploidy and prognosis of non-small-cell lung cancer: a meta-analysis of published data. Br J Cancer 85: 14-22.

Cichutek K, Duesberg PH (1986) Harvey ras genes transform without mutant codons, apparently activated by truncation of a 5' exon (exon -1). Proc Natl Acad Sci U S A 83: 2340-2344.

Cichutek K, Duesberg PH (1989) Recombinant BALB and Harvey sarcoma viruses with normal proto-ras-coding regions transform embryo cells in culture and cause tumors in mice. J Virol 63: 1377-1383.

Clawson GA, Blankenship LJ, Rhame JG, Wilkinson DS (1992) Nuclear enlargement induced by hepatocarcinogens alters ploidy. Cancer Res 52: 1304-1308.

Coffin JM, Varmus HE, Bishop JM, Essex M, Hardy WD, Jr., Martin GS, Rosenberg NE, Scolnick EM, Weinberg RA, Vogt PK (1981) Proposal for naming host cell-derived inserts in retrovirus genomes. J Virol 40: 953-957.

Cole RL, Ferguson MR (1959) Spontaneous regression of reticulum-cell sarcoma of bone; a case report. J Bone Joint Surg Am 41-A: 960-965.

Cole RW, Ault JG, Hayden JH, Rieder CL (1991) Crocidolite asbestos fibers undergo size-dependent microtubule-mediated transport after endocytosis in vertebrate lung epithelial cells. Cancer Res 51: 4942-4947.

Coley WB (1906) Late results of the treatment of inoperable sarcoma by the mixed toxins of erysipelas and *Bacillus prodigiousus*. Am J Med Sci 131: 376-430.

Coley WB (1910) The Treatment of Inoperable Sarcoma by Bacterial Toxins (the Mixed Toxins of the *Streptococcus erysipelas* and the *Bacillus prodigiosus*). Proc R Soc Med 3: 1-48.

Coley WB (1936) The diagnosis and treatment of bone sarcoma. Glasgow Med J 126: 49-86, 128-169.

Collins FS, Lander ES, Rogers J, Waterson RH (2004) International Human Genome Sequencing Consortium: Finishing the euchromatic sequence of the human genome. Nature 431: 931-945.

Connell JR, Ockey CH (1977) Analysis of karyotype variation following carcinogen treatment of Chinese hamster primary cells. Int J Cancer 20: 768-779.

Connell JR (1984) Karyotype analysis of carcinogen-treated Chinese hamster cells in vitro evolving from a normal to a malignant phenotype. Br J Cancer 50: 167-177.

Conti JC, Aldaz CM, O'Connell J, Klen-Szanto AJ-P, Slaga TJ (1986) Aneuploidy, an early event in mouse skin tumor development. Carcinogenesis 7: 1845-1848.

Cook MG (1992) The significance of inflammation and regression in melanoma. Virchows Arch A Pathol Anat Histopathol 420: 113-115.

Cooper GM (1990) Oncogenes. Boston: Jones and Bartlett Publishers

Cooper HL, Black PH (1963) Cytogenetic studies of hamster kidney cell cultures transformed by the Simian vacuolating virus (SV40). J Natl Cancer Inst 30: 1015-1025.

Cooper PD, Marshall SA, Masinello GR (1982) Properties of cell lines derived from altered cell foci in baby mouse skin cultures. J Cell Physiol 113: 344-349.

Copeland RL, Meehan PL, Morrissy RT (1985) Spontaneous regression of osteochondromas. Two case reports. J Bone Joint Surg Am 67: 971-973.

Cornish-Bowden A (1995) Kinetics of multi-enzyme systems. In: H.-J. Rehm and G. Reed, editors. Biotechnology. 2nd ed. New York: VHC. pp. 122-136.

Cornish-Bowden A (1999) Metabolic control analysis in biotechnology and medicine. Nature Biotechnology 17: 641-643.

Cornish-Bowden A (2004) Fundamentals of enzyme kinetics. London: Portland Press, Biochemical Society.

Costa D, Queralt R, Aymerich M, Carrio A, Rozman M, Vallespi T, Colomer D, Nomdedeu B, Montserrat E, Campo E (2003) High levels of chromosomal imbalances in typical and small-cell variants of T-cell prolymphocytic leukemia. Cancer Genet Cytogenet 147: 36-43.

Cowell JK (1981) Chromosome abnormalities associated with salivary gland epithelial cell lines transformed in vitro and in vivo with evidence of a role for genetic imbalance in transformation. Cancer Res 41: 1508-1517.

Cram LS, Bartholdi MF, Ray FA, Travis GL, Kraemer PM (1983) Spontaneous neoplastic evolution of Chinese hamster cells in culture: Multistep progression of karyotype. Cancer Res 43: 4828-4837.

Crawford LV (1980) Transforming genes of DNA tumor viruses. Cold Spring Harb Symp Quant Biol 44 Pt 1: 9-11.

Cremer M, Schermelleh L, Solovei I, Cremer T (2001) Cell Cycle: Chromosomal Organization: John Wiley & Sons, Ltd http://dx.doi.org/10.1038/npg.els.0005769.

Cremer M, Kupper K, Wagler B, Wizelman L, von Hase J, Weiland Y, Kreja L, Diebold J, Speicher MR, Cremer T (2003) Inheritance of gene density-related higher order chromatin arrangements in normal and tumor cell nuclei. J Cell Biol 162: 809-820.

Cremer T, Cremer M (2010) Chromosome territories. Cold Spring Harb Perspect Biol 2: a003889.

Crnalic S, Panagopoulos I, Boquist L, Mandahl N, Stenling R, Lofvenberg R (2002) Establishment and characterisation of a human clear cell sarcoma model in nude mice. Int J Cancer 101: 505-511.

Crum CP, Cibas ES, Lee KR (1997) Pathology of early cervical neoplasia; Roth LM, editor. Contemporary Issues in Surgical Pathology. New York: Churchill Livingstone. 288 pp.

Dacks JB, Walker G, Field MC (2008) Implications of the new eukaryotic systematics for parasitologists. Parasitol Int 57: 97-104.

Daley GQ, Van Etten RA, Baltimore D (1990) Induction of chronic myelogenous leukemia in mice by the P210bcr/abl gene of the Philadelphia chromosome. Science 247: 824-830.

Danielsen HE, Brogger A, Reith A (1991) Specific gain of chromosome 19 in preneoplastic mouse liver cells after diethylnitrosamine treatment. Carcinogenesis 12: 1777-1780.

Das SK, Kunkel TA, Loeb LA (1985) Effects of altered nucleotide concentrations on the fidelity of DNA replication. Basic Life Sci 31: 117-126.

Dawson PJ, Fieldsteel AH, McCusker J (1978) Incidence of spontaneous tumours in neonatally thymectomized rats. Br J Cancer 38: 476-478.

de Aguiar MA, Baranger M, Baptestini EM, Kaufman L, Bar-Yam Y (2009) Global patterns of speciation and diversity. Nature 460: 384-387.

De Berker D, Windebank K, Sviland L, Kingma DW, Carter R, Rees JL (1996) Spontaneous regression in angiocentric T-cell lymphoma. Br J Dermatol 134: 554-558.

de Lapeyriere O, Hayot B, Imbert J, Courcoul M, Arnaud D, Birg F (1984) Cell lines derived from tumors induced in syngeneic rats by FR 3T3 SV40 transformants no longer synthesize the early viral proteins. Virology 135: 74-86.

Dellarco VL, Voytek PE, Hollaender A, editors (1985) Aneuploidy—Etiology and Mechanisms. New York and London: Plenum Press.

Dellas A, Torhorst J, Jiang F, Proffitt J, Schultheiss E, Holzgreve W, Sauter G, Mihatsch MJ, Moch H (1999) Prognostic value of genomic alterations in invasive cervical squamous cell carcinoma of clinical stage IB detected by comparative genomic hybridization. Cancer Res 59: 3475-3479.

Delmer A, Heron E, Marie JP, Zittoun R (1994) Spontaneous remission in acute myeloid leukaemia. Br J Haematol 87: 880-882.

Delves PJ, Martin SJ, Burton DR, Roitt IM (2006) Roit's essential immunology. Oxford: Blackwell Publishing. 474 pp.

Dermer G (1994) The immortal cell: why cancer research fails. Garden City Park, New York: Avery Publishing Group Inc.

Dermer GB (1983) Human cancer research. Science 221: 318.

Dietrich H, Dietrich B (2001) Ludwig Rehn (1849-1930)—pioneering findings on the aetiology of bladder tumours. World J Urol 19: 151-153.

Doak SH (2008) Aneuploidy in upper gastro-intestinal tract cancers—a potential prognostic marker? Mutat Res 651: 93-104.

Doi M, Yukutake M, Tamura K, Watanabe K, Kondo K, Isobe T, Awaya T, Shigenobu T, Oda Y, Yanmakido K, Koyama K, Kohno N. (2002) A retrospective cohort study on respiratory tract cancers in the workers of the Japanese army poison-gas-factory operated from 1929 to 1945; May 18-21; Orlando, FL. pp. 439a, Abstract 1754.

Doll R (1970) Practical steps towards the prevention of bronchial carcinoma. Scott Med J 15: 433-447.

Donehower LA, Harvey M, Siagle BL, McArthur MJ, Montgomery CA, Jr., Butel JS, Bradley A (1992) Mice deficient for p53 are developmentally normal but susceptible to spontaneous tumors. Nature 356: 215-221.

Doolittle RF, Hunkapiller MW, Hood LE, Devare SG, Robbins KC, Aaronson SA, Antoniades HN (1983) Simian sarcoma virus onc gene, v-sis, is derived from the gene (or genes) encoding a platelet-derived growth factor. Science 221: 275-277.

Doubre H, Cesari D, Mairovitz A, Benac C, Chantot-Bastaraud S, Dagnon K, Antoine M, Danel C, Bernaudin JF, Fleury-Feith J (2005) Multidrug resistance-associated protein (MRP1) is overexpressed in DNA aneuploid carcinomatous cells in non-small cell lung cancer (NSCLC). Int J Cancer 113: 568-574.

Doxsey S, Zimmerman W, Mikule K (2005) Centrosome control of the cell cycle. Trends Cell Biol 15: 303-311.

Drobyski WR, Qazi R (1989) Spontaneous regression in non-Hodgkin's lymphoma: clinical and pathogenetic considerations. Am J Hematol 31: 138-141.

Druker BJ, Sawyers CL, Kantarjian H, Resta DJ, Reese SF, Ford JM, Capdeville R, Talpaz M (2001) Activity of a specific inhibitor of the BCR-ABL tyrosine kinase in the blast crisis of chronic myeloid leukemia and acute lymphoblastic leukemia with the Philadelphia chromosome. N Engl J Med 344: 1038-1042.

Duesberg P, Vogt PK (1970) Differences between the ribonucleic acids of transforming and nontransforming avian tumor viruses. Proc Natl Acad Sci U S A 67: 1673-1680.

Duesberg P, Vogt PK, Beemon K, Lai M (1974) Avian RNA Tumor Viruses: Mechanism of Recombination and Complexity of the Genome. Cold Spring Harb Symp Quant Biol 39: 847-857.

Duesberg P (1980) Transforming genes of retroviruses. Cold Spring Harb Symp Quant Biol 44 Pt 1: 13-29.

Duesberg P (1983) Retroviral transforming genes in normal cells? Nature 304: 219-226.

Duesberg P (1985) Activated proto-onc genes: sufficient or necessary for cancer? Science 228: 669-677.

Duesberg P (1987) Cancer genes: rare recombinants instead of activated oncogenes (a review). Proc Natl Acad Sci U S A 84: 2117-2124.

Duesberg P, Zhou RP, Goodrich D (1989) Cancer genes by illegitimate recombination. Ann N Y Acad Sci 567: 259-273.

Duesberg P, Schwartz JR (1992) Latent viruses and mutated oncogenes: no evidence for pathogenicity. Prog Nucleic Acid Res Mol Biol 43: 135-204.

Duesberg P (1995) Oncogenes and cancer. Science 267: 1407-1408.

Duesberg P, Rausch C, Rasnick D, Hehlmann R (1998) Genetic instability of cancer cells is proportional to their degree of aneuploidy. Proc Natl Acad Sci U S A 95: 13692-13697.

Duesberg P (1999) Are centrosomes or aneuploidy the key to cancer? Science 284: 2091-2092.

Duesberg P, Stindl R, Hehlmann R (2000a) Explaining the high mutation rates of cancer cells to drug and multidrug resistance by chromosome reassortments that are catalyzed by aneuploidy. Proc Natl Acad Sci U S A 97: 14295-14300.

Duesberg P, Li R, Rasnick D, Rausch C, Willer A, Kraemer A, Yerganian G, Hehlmann R (2000b) Aneuploidy precedes and segregates with chemical carcinogenesis. Cancer Genet Cytogenet 119: 83-93.

Duesberg P, Li R, Rausch C, Willer A, Kraemer A, Yerganian G, Hehlmann R, Rasnick D (2000c) Mechanism of carcinogenesis by polycyclic aromatic hydrocarbons: aneuploidy precedes malignant transformation and occurs in all cancers. In: Cornish-Bowden A, Cárdenas ML, editors. Technological and Medical Implications of Metabolic Control Analysis. Dordrecht: Kluwer Academic Publishers. pp. 83-98.

Duesberg P, Rasnick D (2000) Aneuploidy, the somatic mutation that makes cancer a species of its own. Cell Motil Cytoskeleton 47: 81-107.

Duesberg P, Stindl R, Li R, Hehlmann R, Rasnick D (2001a) Aneuploidy versus gene mutation as cause of cancer. Current Science (India) 81: 490-500.

Duesberg P, Stindl R, Hehlmann R (2001b) Origin of multidrug resistance in cells with and without multidrug resistance genes: Chromosome reassortments catalyzed by aneuploidy. Proc Natl Acad Sci U S A 98: 11283-11288.

Duesberg P, Li R (2003) Multistep carcinogenesis: a chain reaction of aneuploidizations. Cell Cycle 2: 202-210.

Duesberg P (2003) Are cancers dependent on oncogenes or on aneuploidy? Cancer Genet Cytogenet 143: 89-91.

Duesberg P, Li R, Rasnick D (2004a) Aneuploidy approaching a perfect score in predicting and preventing cancer: highlights from a conference held in Oakland, CA in January, 2004. Cell Cycle 3: 823-828.

Duesberg P, Fabarius A, Hehlmann R (2004b) Aneuploidy, the primary cause of the multilateral genomic instability of neoplastic and preneoplastic cells. IUBMB Life 56: 65-81.

Duesberg P (2005) Does aneuploidy or mutation start cancer? Science 307: 41.

Duesberg P, Li R, Fabarius A, Hehlmann R (2005) The chromosomal basis of cancer. Cell Oncol 27: 293-318.

Duesberg P, Li R, Fabarius A, Hehlmann R (2006) Aneuploidy and cancer: from correlation to causation. Contrib Microbiol 13: 16-44.

Duesberg P (2007) Chromosomal chaos and cancer. Sci Am 296: 52-59.

Duesberg P, Li R, Sachs R, Fabarius A, Upender MB, Hehlmann R (2007) Cancer drug resistance: The central role of the karyotype. Drug Resist Updat 10: 51-58.

Duesberg P, Mandrioli D, McCormack A, Nicholson JM (2011) Is carcinogenesis a form of speciation? Cell Cycle 10.

Dulbecco R (1976) Francis Peyton Rous. Biogr Mem Natl Acad Sci 48: 275-306.

Dunkler D, Michiels S, Schemper M (2007) Gene expression profiling: does it add predictive accuracy to clinical characteristics in cancer prognosis? Eur J Cancer 43: 745-751.

Dunn JM, Phillips RA, Zhu X, Becker A, Gallie BL (1989) Mutations in the RB1 gene and their effects on transcription. Mol Cell Biol 9: 4596-4604.

Dupuy A, Simon RM (2007) Critical review of published microarray studies for cancer outcome and guidelines on statistical analysis and reporting. J Natl Cancer Inst 99: 147-157.

Eastmond DA, Hartwig A, Anderson D, Anwar WA, Cimino MC, Dobrev I, Douglas GR, Nohmi T, Phillips DH, Vickers C (2009) Mutagenicity testing for chemical risk assessment: update of the WHO/IPCS Harmonized Scheme. Mutagenesis 24: 341-349.

Eden P, Ritz C, Rose C, Ferno M, Peterson C (2004) "Good Old" clinical markers have similar power in breast cancer prognosis as microarray gene expression profilers. Eur J Cancer 40: 1837-1841.

Edinburgh (1806) Report of the medical committee of the society for investigating the nature and cure of cancer. Edinburgh Medical Surgery Journal 2: 382.

Editorial (2005) Herceptin and early breast cancer: a moment for caution. Lancet 366: 1673.

Edwards HE, Courtney MD, Wilson JE, Monaghan SJ, Walsh AM (2001) Fever management practises: what pediatric nurses say. Nurs Health Sci 3: 119-130.

Edwards MJ, Anderson JA, Angel JR, Harty JI (1996) Spontaneous regression of primary and metastatic renal cell carcinoma. J Urol 155: 1385.

Ein-Dor L, Kela I, Getz G, Givol D, Domany E (2005) Outcome signature genes in breast cancer: is there a unique set? Bioinformatics (Oxford, England) 21: 171-178.

Eisenbud L, Kahn LB, Friedman E (1987) Benign osteoblastoma of the mandible: fifteen year follow-up showing spontaneous regression after biopsy. J Oral Maxillofac Surg 45: 53-57.

Ellsworth RE, Ellsworth DL, Love B, Patney HL, Hoffman LR, Kane J, Hooke JA, Shriver CD (2007) Correlation of Levels and Patterns of Genomic Instability With Histological Grading of DCIS. Ann Surg Oncol 14: 3070-3077.

Elser JJ, Hamilton A (2007) Stoichiometry and the new biology: the future is now. PLoS Biol 5: e181.

Epstein CJ (1986) The consequences of chromosome imbalance: principles, mechanisms, and models. In: Barlow P, Green PB, Wylie CC, editors. Developmental and Cell Biology Series. New York: Cambridge University Press. 475 pp.

Epstein SS (1998) The Politics of Cancer Revisited. New York: East Ridge Press. 475 pp.

Epstein SS (2005) Cancer-Gate: How to Win the Losing Cancer War. Amityville, N.Y.: Baywood

Era T, Witte O (2000) Regulated expression of P210 Bcr-Abl during embryonic stem cell differentiation stimulates multipotential progenitor expansion and myeloid cell fate. Proc Natl Acad Sci U S A 97: 1737-1742.

Erenpreisa J, Cragg MS (2010) MOS, aneuploidy and the ploidy cycle of cancer cells. Oncogene 29: 5447-5451.

Erenpreisa J, Salmina K, Huna A, Kosmacek EA, Cragg M, Ianzini F, Anisimov A (2011) Polyploid tumour cells elicit para-diploid progeny through de-polyploidising divisions and regulated autophagic degradation. Cell Biol Int 35: 687-695.

Eshleman JR, Casey G, Kochera ME, Sedwick WD, Swinler SE, Veigl ML, Willson JK, Schwartz S, Markowitz SD (1998) Chromosome number and structure both are markedly stable in RER colorectal cancers and are not destabilized by mutation of p53. Oncogene 17: 719-725.

Ewald D, Li M, Efrat S, Auer G, Wall RJ, Furth PA, Hennighausen L (1996) Time-sensitive reversal of hyperplasia in transgenic mice expressing SV40 T antigen. Science 273: 1384-1386.

Ewing J (1941) Neoplastic diseases: a textbook on tumors. Philadelphia: W. B. Saunders. 578 pp.

Fabarius A, Willer A, Yerganian G, Hehlmann R, Duesberg P (2002) Specific aneusomies in Chinese hamster cells at different stages of neoplastic transformation, initiated by nitrosomethylurea. Proc Natl Acad Sci U S A 99: 6778-6783.

Fabarius A, Hehlmann R, Duesberg PH (2003) Instability of chromosome structure in cancer cells increases exponentially with degrees of aneuploidy. Cancer Genet Cytogenet 143: 59-72.

Fabarius A, Li R, Yerganian G, Hehlmann R, Duesberg P (2008) Specific clones of spontaneously evolving karyotypes generate individuality of cancers. Cancer Genet Cytogenet 180: 89-99.

Faguet GB (2005) The war on cancer: an anatomy of failure, a blueprint for the future. New York: Springer. 227 pp.

Faria R, Navarro A (2010) Chromosomal speciation revisited: rearranging theory with pieces of evidence. Trends Ecol Evol 25: 660-669.

Fassas A, Sakellari I, Anagnostopoulos A, Saloum R (1991) Spontaneous remission of acute myeloid leukemia in a patient with concurrent *Pneumocystis carinii* pneumonia. Nouv Rev Fr Hematol 33: 363-364.

Fearon ER, Vogelstein B (1990) A genetic model for colorectal tumorigenesis. Cell 61: 759-767.

Fell D (1992) Metabolic control analysis: a survey of its theoretical and experimental development. Biochem J 286: 313-330.

Fell D, Thomas S (1995) Physiological control of metabolic flux: the requirement for multisite modulation. Biochem J 311: 35-39.

Fell D (1997) Understanding the control of metabolism; Snell K, editor. Frontiers in Metabolism 2. London: Portland Press. 301 pp.

Fibiger JAG (1926) *Spiroptera carcinoma*. The Nobel Prize in Physiology or Medicine 1926. Stockholm: Nobel.org http://nobelprize.org/nobel_prizes/ medicine/laureates/1926/.

Fidler IJ, Hart IR (1982) Biological diversity in metastatic neoplasms: origins and implications. Science 217: 998-1003.

Finkel T, Der CJ, Cooper GM (1984) Activation of ras genes in human tumors does not affect localization, modification, or nucleotide binding properties of p21. Cell 37: 151-158.

Finlay CA, Hinds PW, Levine AJ (1989) The p53 proto-oncogene can act as a suppressor of transformation. Cell 57: 1083-1093.

Flagiello D, Gerbault-Seureau M, Padoy E, Dutrillaux B (1998) Near haploidy in breast cancer: a particular pathway of chromosome evolution. Cancer Genet Cytogenet 102: 54-58.

Flores-Staino C, Darai-Ramqvist E, Dobra K, Hjerpe A (2010) Adaptation of a commercial fluorescent in situ hybridization test to the diagnosis of malignant cells in effusions. Lung Cancer 68: 39-43.

Fogh J (1986) Human tumor lines for cancer research. Cancer Invest 4: 157-184.

Foley GE, Handler AH, Lynch PM, Wolman SR, Stulberg CS, Eagle H (1965) Loss of neoplasatic properties *in vitro*. II. Observations on KB sublines. Cancer Research 25: 1254-1261.

Food and Drug Administration (2005) UroVysion Bladder Cancer Kit approval letter. Center for Devices and Radiological Health, Rlckville, http://www.accessdata.fda.gov/cdrh_docs/pdf3/p030052a.pdf.

Forbes SA, Bindal N, Bamford S, Cole C, Kok CY, Beare D, Jia M, Shepherd R, Leung K, Menzies A, Teague JW, Campbell PJ, Stratton MR, Futreal PA (2011) COSMIC: mining complete cancer genomes in the Catalogue of Somatic Mutations in Cancer. Nucleic Acids Research 39: D945-950.

Foulds L (1954) The experimental study of tumor progression: a review. Cancer Research 14: 327-339.

Foulds L (1965) Multiple etiologic factors in neoplastic development. Cancer Res 25: 1339-1347.

Foulds L (1969) Neoplastic Development. London, New York, San Francisco: Academic Press.

Foulds L (1975) Neoplastic Development. London, New York, San Francisco: Academic Press.

Fowler GA, Nauts HC (1969) The Apparently Beneficial Effects of Concurrent Infections, Inflammation or Fever, and of Bacterial Toxin Therapy on Neuroblastoma. New York: Cancer Research Institute.

Fowler GA (1969a) Enhancement of natural resistance to malignant melanoma with special reference to the beneficial effects of concurrent infections and bacterial toxin therapy. New York: Cancer Research Institute.

Fowler GA (1969b) Beneficial effects of acute bacterial infections or bacterial toxin therapy on cancer of the colon or rectum. New York: Cancer Research Institute.

Fox EJ, Salk JJ, Loeb LA (2009) Cancer genome sequencing—an interim analysis. Cancer Res 69: 4948-4950.

Frank SA (2004) Age-specific acceleration of cancer. Curr Biol 14: 242-246.

Frankfurt OS, Chin JL, Englander LS, Greco WR, Pontes JE, Rustum YM (1985) Relationship between DNA ploidy, glandular differentiation, and tumor spread in human prostate cancer. Cancer Res 45: 1418-1423.

Frati L, Cortesi E, Ficorella C, Manzari V, Verna R (1984) Phenotypic and genetic markers of cancer: turning point in research. In: Aaronson SA, Frati L, Verna R, editors. Genetic and Phenotypic Markers of Tumors. New York: Plenum Press. pp. 1-20.

Freeman AE, Lake RS, Igel HJ, Gernand L, Pezzutti MR, Malone JM, Mark C, Benedict WF (1977) Heteroploid conversion of human skin cells by methylcholanthrene. Proc Natl Acad Sci U S A 74: 2451-2455.

Frick S, Frick P (1993) [Spontaneous remission in chronic lymphatic leukemia]. Schweiz Med Wochenschr 123: 328-334.

Frieben A (1902) Demonstration eines Cancroid der rechten Handruckens, das sich nach langdauernder Einwirkung von Rontgenstrahlen entwickelt hat. Fortschr Roentgenstr 6: 106-111.

Fried M (1965) Cell-Transforming Ability of a Temperature-Sensitive Mutant of Polyoma Virus. Proc Natl Acad Sci U S A 53: 486-491.

Friedrich EG, Jr. (1972) Reversible vulvar atypia. A case report. Obstet Gynecol 39: 173-181.

Friend C (1957) Cell-free transmission in adult Swiss mice of a disease having the character of a leukemia. J Exp Med 105: 307-318.

Friend SH, Bernards R, Rogelj S, Weinberg RA, Rapaport JM, Albert DM, Dryja TP (1986) A human DNA segment with properties of the gene that predisposes to retinoblastoma and osteosarcoma. Nature 323: 643-646.

Fucini C, Bandettini L, D'Elia M, Filipponi F, Herd-Smith A (1985) Are postoperative fever and/or septic complications prognostic factors in colorectal cancer resected for cure? Dis Colon Rectum 28: 94-95.

Fujimaki E, Sasaki K, Nakano O, Chiba S, Tazawa H, Yamashiki H, Orii S, Sugai T (1996) DNA ploidy heterogeneity in early and advanced gastric cancers. Cytometry 26: 131-136.

Fujimura JH (1996) Crafting Science: A sociohistory of the quest for the genetics of cancer. Cambridge: Harvard University Press.

Fujiwara T, Bandi M, Nitta M, Ivanova EV, Bronson RT, Pellman D (2005) Cytokinesis failure generating tetraploids promotes tumorigenesis in p53-null cells. Nature 437: 1043-1047.

Fulton AB (1982) How crowded is the cytoplasm? Cell 30: 345-347.

Furge KA, Lucas KA, Takahashi M, Sugimura J, Kort EJ, Kanayama HO, Kagawa S, Hoekstra P, Curry J, Yang XJ, Teh BT (2004) Robust classification of renal cell carcinoma based on gene expression data and predicted cytogenetic profiles. Cancer Res 64: 4117-4121.

Furth J, Sobel H (1947) Neoplastic transformation of granulosa cells in grafts of normal ovaries into spleens of gonadectomized mice. J Natl Cancer Inst 8: 7-16.

Fusenig NE, Boukamp P (1998) Multiple stages and genetic alterations in immortalization, malignant transformation, and tumor progression of human skin keratinocytes. Mol Carcinog 23: 144-158.

Futakawa N, Kimura W, Ando H, Muto T, Esaki Y (1997) Heterogeneity of DNA Ploidy Pattern in Carcinoma of the Gallbladder: Primary and Metastatic Sites. Cancer science 88: 886-894.

Futreal PA, Coin L, Marshall M, Down T, Hubbard T, Wooster R, Rahman N, Stratton MR (2004) A census of human cancer genes. Nat Rev Cancer 4: 177-183.

Gabor Miklos GL (2005) The human cancer genome project—one more misstep in the war on cancer. Nat Biotechnol 23: 535-537.

Gale RP, Canaani E (1984) An 8-kilobase *abl* RNA transcript in chronic myelogenous leukemia. PNAS 81: 5648-5652.

Gallo JH, Misawa S, Testa JR (1984) Centromere spreading in acute nonlymphocytic leukemia. Cancer Genet Cytogenet 12: 105-109.

Ganem NJ, Godinho SA, Pellman D (2009) A mechanism linking extra centrosomes to chromosomal instability. Nature 460: 278-282.

Ganmore I, Smooha G, Izraeli S (2009) Constitutional aneuploidy and cancer predisposition. Hum Mol Genet 18: R84-93.

Gao C, Furge K, Koeman J, Dykema K, Su Y, Cutler ML, Werts A, Haak P, Vande Woude GF (2007) Chromosome instability, chromosome transcriptome, and clonal evolution of tumor cell populations. Proc Natl Acad Sci U S A 104: 8995-9000.

Garattini S, Bertele V (2002) Efficacy, safety, and cost of new anticancer drugs. Brit Med J 325: 269-271.

Garcia-Orad A, Vig BK, Aucoin D (2000) Separation vs. replication of inactive and active centromeres in neoplastic cells. Cancer Genet Cytogenet 120: 18-24.

Garcia-Rayo S, Gurpide A, Vega F, Brugarolas A (1996) Spontaneous tumor regression in a patient with multiple myeloma: Report of another case. Rev Med Univ Navarra 40: 41-42.

Gardner HA, Gallie BL, Knight LA, Phillips RA (1982) Multiple karyotypic changes in retinoblastoma tumor cells: presence of normal chromosome No. 13 in most tumors. Cancer Genet Cytogenet 6: 201-211.

Gattiker HH, Wiltshaw E, Galton DA (1980) Spontaneous regression in non-Hodgkin's lymphoma. Cancer 45: 2627-2632.

Gebhart E, Liehr T (2000) Patterns of genomic imbalances in human solid tumors (Review). International Journal of Oncology 16: 383-399.

Geiger T, Cox J, Mann M (2010) Proteomic changes resulting from gene copy number variations in cancer cells. PLoS Genet 6.

Geigl JB, Obenauf AC, Schwarzbraun T, Speicher MR (2008) Defining 'chromosomal instability'. Trends Genet 24: 64-69.

Gerlich D, Beaudouin J, Kalbfuss B, Daigle N, Eils R, Ellenberg J (2003) Global chromosome positions are transmitted through mitosis in mammalian cells. Cell 112: 751-764.

German J (1974) Bloom's syndrome. II. The prototype of genetic disorders predisposing to chromosome instability and cancer. In: German J, editor. Chromosomes and Cancer. New York: John Wiley & Sons. pp. 601-617.

Ghadimi BM, Sackett DL, Difilippantonio MJ, Schrock E, Neumann T, Jauho A, Auer G, Ried T (2000) Centrosome amplification and instability occurs exclusively in aneuploid, but not in diploid colorectal cancer cell lines, and correlates with numerical chromosomal aberrations. Gene Chromosome Canc 27: 183-190.

Giaretti W, Santi L (1990) Tumor progression by DNA flow cytometry in human colorectal cancer. International Journal of Cancer 45: 597-603.

Giaretti W (1994) A model of DNA aneuploidization and evolution in colorectal cancer. Lab Invest 71: 904-910.

Giaretti W, Monaco R, Pujic N, Rapallo A, Nigro S, Geido E (1996) Intratumor heterogeneity of K-ras2 mutations in colorectal adenocarcinomas: association with degree of DNA aneuploidy. Am J Pathol 149: 237-245.

Gibbs WW (2001) Dissident or Don Quixote? Sci Am 265: 30-32.

Gibbs WW (2003) Untangling the roots of cancer. Sci Am 289: 56-65.

Gibson DP, Aardema MJ, Kerckaert GA, Carr GJ, Brauninger RM, LeBoeuf RA (1995) Detection of aneuploidy-inducing carcinogens in the Syrian hamster embryo (SHE) cell transformation assay. Mutat Res 343: 7-24.

Gisselsson D (2005) Mitotic instability in cancer: is there method in the madness? Cell Cycle 4: 1007-1010.

Gisselsson D, Jin Y, Lindgren D, Persson J, Gisselsson L, Hanks S, Sehic D, Mengelbier LH, Ora I, Rahman N, Mertens F, Mitelman F, Mandahl N (2010) Generation of trisomies in cancer cells by multipolar mitosis and incomplete cytokinesis. Proc Natl Acad Sci U S A 107: 20489-20493.

Goddard AD, Balakier H, Canton M, Dunn J, Squire J, Reyes E, Becker A, Phillips RA, Gallie BL (1988) Infrequent genomic rearrangement and normal expression of the putative RB1 gene in retinoblastoma tumors. Mol Cell Biol 8: 2082-2088.

Goff SP, Tabin CJ, Wang JY, Weinberg R, Baltimore D (1982) Transfection of fibroblasts by cloned Abelson murine leukemia virus DNA and recovery of transmissible virus by recombination with helper virus. J Virol 41: 271-285.

Goh K, Lee H, Miller G (1978) Down's syndrome and leukemia: mechanism of additional chromosomal abnormalities. Am J Ment Defic 82: 542-548.

Gold M (1986) A conspiracy of cells. New York: State University of New York Press. pp. 125-126.

Goldfarb MP, Weinberg RA (1981) Structure of the provirus within NIH 3T3 cells transfected with Harvey sarcoma virus DNA. J Virol 38: 125-135.

Goldie JH (2001) Drug resistance in cancer: a perspective. Cancer Metastasis Rev 20: 63-68.

Goldstein DB (2009) Common genetic variation and human traits. N Engl J Med 360: 1696-1698.

Gollin SM (2005) Mechanisms leading to chromosomal instability. Seminars in Cancer Biology 15: 33-42.

Gong JK (1971) Anemic stress as a trigger of myelogenous leukemia in rats rendered leukemia-prone by x-ray. Science 174: 833-835.

Gorbsky GJ, Ricketts WA (1993) Differential expression of a phosphoepitope at the kinetochores of moving chromosomes. J Cell Biol 122: 1311-1321.

Gorre ME, Mohammed M, Ellwood K, Hsu N, Paquette R, Rao PN, Sawyers CL (2001) Clinical resistance to STI-571 cancer therapy caused by BCR-ABL gene mutation or amplification. Science 293: 876-880.

Gorringe KL, Chin SF, Pharoah P, Staines JM, Oliveira C, Edwards PA, Caldas C (2005) Evidence that both genetic instability and selection contribute to the accumulation of chromosome alterations in cancer. Carcinogenesis 26: 923-930.

Grade M, Hormann P, Becker S, Hummon AB, Wangsa D, Varma S, Simon R, Liersch T, Becker H, Difilippantonio MJ, Ghadimi BM, Ried T (2007) Gene expression profiling reveals a massive, aneuploidy-dependent transcriptional deregulation and distinct differences between lymph node-negative and lymph node-positive colon carcinomas. Cancer Res 67: 41-56.

Grafton WD (1994) Regressing malignant melanoma. J La State Med Soc 146: 535-539.

Gray MW, Lang BF, Burger G (2004) Mitochondria of protists. Annu Rev Genet 38: 477-524.

Greaves M (2000) Cancer: the evolutionary legacy. New York: Oxford University Press. 276 pp.

Greentree LB (1973) Anaplastic lung cancer with metastases. Case report of a 15-year survival. Ohio State Med J 69: 841-843.

Grem JL, Hafez GR, Brandenburg JH, Carbone PP (1986) Spontaneous remission in diffuse large cell lymphoma. Cancer 57: 2042-2044.

Griffiths AJF, Miller JH, Suzuki DT, Lewontin RC, Gelbart WM (2000) An introduction to genetic analysis. New York: W. H. Freeman.

Grigorova M, Staines JM, Ozdag H, Caldas C, Edwards PA (2004) Possible causes of chromosome instability: comparison of chromosomal abnormalities in cancer cell lines with mutations in BRCA1, BRCA2, CHK2 and BUB1. Cytogenet Genome Res 104: 333-340.

Grimwade D, Walker H, Harrison G, Oliver F, Chatters S, Harrison CJ, Wheatley K, Burnett AK, Goldstone AH (2001) The predictive value of hierarchical cytogenetic classification in older adults with acute myeloid leukemia (AML): analysis of 1065 patients entered into the United Kingdom Medical Research Council AML11 trial. Blood 98: 1312-1320.

Grosovsky AJ, Parks KK, Giver CR, Nelson SL (1996) Clonal analysis of delayed karyotypic abnormalities and gene mutations in radiation-induced genetic instability. Mol Cell Biol 16: 6252-6262.

Gross L (1951) "Spontaneous" leukemia developing in C3H mice following inoculation in infancy, with AK-leukemic extracts, or AK-embryos. Proc Soc Exp Biol Med 76: 27-32.

Gross L (1957) Development and serial cell-free passage of a highly potent strain of mouse leukemia virus. Proc Soc Exp Biol Med 94: 767-771.

Grossmann M, Hoermann R, Weiss M, Jauch KW, Oertel H, Staebler A, Mann K, Engelhardt D (1995) Spontaneous regression of hepatocellular carcinoma. Am J Gastroenterol 90: 1500-1503.

Gruszka-Westwood AM, Horsley SW, Martinez-Ramirez A, Harrison CJ, Kempski H, Moorman AV, Ross FM, Griffiths M, Greaves MF, Kearney L (2004) Comparative expressed sequence hybridization studies of high-hyperdiploid childhood acute lymphoblastic leukemia. Gene Chromosome Canc 41: 191-202.

Guillem JG, Wood WC, Moley JF, Berchuck A, Karlan BY, Mutch DG, Gagel RF, Weitzel J, Morrow M, Weber BL, Giardiello F, Rodriguez-Bigas MA, Church J, Gruber S, Offit K (2006) ASCO/SSO review of current role of risk-reducing surgery in common hereditary cancer syndromes. J Clin Oncol 24: 4642-4660.

Gunale S, Tucker WG (1975) Regression of metastatic melanoma. Mich Med 74: 697-698.

Gusev Y, Kagansky V, Dooley WC (2001) Long-term dynamics of chromosomal instability in cancer: a transition probability model. Math Comput Model 33: 1253-1273.

Haber DA, Fearon ER (1998) The promise of cancer genetics. Lancet 351: SII 1-8.

Hagmar L, Bonassi S, Stromberg U, Brogger A, Knudsen L, Norppa H, Reuterwall C (1998) Chromosomal aberrations in lymphocytes predict human cancer: a report from the European Study Group on Cytogenetic Biomarkers and Health. Cancer Res 58: 4117-4121.

Hahn WC, Counter CM, Lundberg AS, Beijersbergen RL, Brooks MW, Weinberg RA (1999) Creation of human tumour cells with defined genetic elements. Nature 400: 464-468.

Hahn WC, Weinberg RA (2002a) Rules for making human tumor cells. N Engl J Med 347: 1593-1603.

Hahn WC, Weinberg RA (2002b) Modelling the molecular circuitry of cancer. Nat Rev Cancer 2: 331-341.

Haigis KM, Caya JG, Reichelderfer M, Dove WF (2002) Intestinal adenomas can develop with a stable karyotype and stable microsatellites. Proc Natl Acad Sci U S A 99: 8927-8931.

Hajdu SI (2004) Greco-Roman thought about cancer. Cancer 100: 2048-2051.

Hall SS (1997) A commotion in the blood: life, death, and the immune system. New York: Henry Holt & Company Inc. 544 pp.

Hall WH (1948) The role of initiating and promoting factors in the pathogenesis of tumors of the thyroid. Br J Cancer 2: 273-280.

Hammarberg C, Slezak P, Tribukait B (1984) Early detection of malignancy in ulcerative colitis. A flow-cytometric DNA study. Cancer 53: 291-295.

Hanahan D, Weinberg RA (2000) The hallmarks of cancer. Cell 100: 57-70.

Hanahan D, Weinberg RA (2011) Hallmarks of cancer: the next generation. Cell 144: 646-674.

Hand R, German J (1975) A retarded rate of DNA chain growth in Bloom's syndrome. Proc Natl Acad Sci U S A 72: 758-762.

Hanks S, Coleman K, Reid S, Plaja A, Firth H, Fitzpatrick D, Kidd A, Mehes K, Nash R, Robin N, Shannon N, Tolmie J, Swansbury J, Irrthum A, Douglas J, Rahman N (2004) Constitutional aneuploidy and cancer predisposition caused by biallelic mutations in BUB1B. Nat Genet 36: 1159-1161.

Hanna WM, Kahn HJ, Pienkowska M, Blondal J, Seth A, Marks A (2001) Defining a test for HER-2/neu evaluation in breast cancer in the diagnostic setting. Mod Pathol 14: 677-685.

Hansemann D (1890) Ueber asymmetrische Zelltheilung in epithel Krebsen und deren biologische Bedeutung. Virschows Arch Pathol Anat 119: 299-326.

Hansemann D (1897) Die mikroskopische Diagnose der boesartigen Geschwuelste. Berlin: August Hirschwald.

Hariharan IK, Harris AW, Crawford M, Abud H, Webb E, Cory S, Adams J (1989) A bcr-v-abl oncogene induces lymphomas in transgenic mice. Mol Cell Biol 9: 2798-2805.

Hariu H, Matsuta M (1996) Cervical cytology by means of fluorescence in situ hybridization with a set of chromosome-specific DNA probes. Journal of Obstetrics and Gynaecology Research 22: 163-170.

Harris CC (1991) Chemical and physical carcinogenesis: Advances and perspective for the 1990s. Cancer Res 51: 5023s-5044s.

Harris H (1995) The cells of the body: a history of somatic cell genetics. Plainview, NY: Cold Spring Harbor Lab Press.

Harris H (2005) A long view of fashions in cancer research. Bioessays 27: 833-838.

Harris H (2007) Concerning the origin of malignant tumours, Theodor Boveri (1914). Harris H, translator. Woodbury: Cold Spring Harbor Laboratory Press.

Harris JF, Chambers AF, Hill RP, Ling V (1982) Metastatic variants are generated spontaneously at a high rate in mouse KHT tumor. Proc Natl Acad Sci U S A 79: 5547-5551.

Hartman JL, Garvik B, Hartwell L (2001) Principles for the buffering of genetic variation. Science 291: 1001-1004.

Harvey JJ (1964) An unidentified virus which causes the rapid production of tumours in mice. Nature 204: 1104-1105.

Hasle H, Clemmensen IH, Mikkelsen M (2000) Risks of leukaemia and solid tumours in individuals with Down's syndrome. Lancet 355: 165-169.

Hassold TJ (1986) Chromosome abnormalities in human reproductive wastage. Trends Genet 2: 105-110.

Hauschka TS, Levan A (1953) Inverse Relationship Between Chromosome Ploidy and Host-Specificity of Sixteen Transplantable Tumors. Exp Cell Res 4: 457-467.

Hauschka TS, Levan A (1958) Cytologic and functional characterization of single cell clones isolated from the Krebs-2 and Ehrlich ascites tumors. J Natl Cancer Inst 21: 77-135.

Hauschka TS (1961) The chromosomes in ontogeny and oncogeny. Cancer Res 21: 957-981.

Hauschka TS (1963) Chromosome Patterns in Primary Neoplasia. Exp Cell Res 24: SUPPL9:86-98.

Hauser G (1903) Giebt es eine primaere zur Geschwulstbildung fuehrende Epithelerkrankung? Ein Beitrag zur Geschwulstlehre. Beitr path Anat allg Path 33: 1-31.

Hawkins NJ, Gorman P, Tomlinson IP, Bullpitt P, Ward RL (2000) Colorectal carcinomas arising in the hyperplastic polyposis syndrome progress through the chromosomal instability pathway. Am J Pathol 157: 385-392.

Hayflick L, Moorhead PS (1961) The serial cultivation of human diploid cell strains. Exp Cell Res 25: 585-621.

Hayflick L (1965) The limited in vitro lifetime of human diploid cell strains. Exp Cell Res 37: 614-636.

Hede K (2005) Which came first? Studies clarify role of aneuploidy in cancer. J Natl Cancer Inst 97: 87-89.

Heim S, Mitelman F (1995a) Tumors of the female genital organs. Cancer cytogenetics. 2nd ed. New York: Wiley-Liss. pp. 389-407.

Heim S, Mitelman F (1995b) Cancer cytogenetics. New York: Wiley-Liss.

Heim S, Mitelman F (2009) A New Approach to an Old Problem. In: Heim S, Mitelman F, editors. Cancer Cytogenetics: Chromosomal and Molecular Genetic Aberrations of Tumor Cells. 3rd ed. Hoboken: Wiley-Blackwell.

Heinrich R, Rapoport TA (1973) Linear theory of enzymatic chains; its application for the analysis of the crossover theorem and of the glycolysis of human erythrocytes. Acta Biol Med Ger 31: 479-494.

Heinrich R, Rapoport TA (1974) A linear steady-state treatment of enzymatic chains. General properties, control and effector strength. Eur J Biochem 42: 89-95.

Heisterkamp N, Stam K, Groffen J, de Klein A, Grosveld G (1985) Structural organization of the *bcr* gene and its role in the Ph' translocation. Nature 315: 758-761.

Hellstroem KE, Hellstroem I, Sjoegren HO (1963) Further Studies on Karyotypes of a Variety of Primary and Transplanted Mouse Polyoma Tumors. J Natl Cancer Inst 31: 1239-1253.

Helmbold P, Altrichter D, Klapperstuck T, Marsch W (2005) Intratumoral DNA stem-line heterogeneity in superficial spreading melanoma. J Am Acad Dermatol 52: 803-809.

Heng HH, Bremer SW, Stevens J, Ye KJ, Miller F, Liu G, Ye CJ (2006a) Cancer progression by non-clonal chromosome aberrations. J Cell Biochem 98: 1424-1435.

Heng HH, Stevens JB, Liu G, Bremer SW, Ye KJ, Reddy PV, Wu GS, Wang YA, Tainsky MA, Ye CJ (2006b) Stochastic cancer progression driven by non-clonal chromosome aberrations. J Cell Physiol 208: 461-472.

Heppner G, Miller FR (1998) The cellular basis of tumor progression. Int Rev Cytol 177: 1-56.

Herberman RB (1984) Possible role of natural killer cells and other effector cells in immune surveillance against cancer. J Invest Dermatol 83: 137s-140s.

Hermsen M, Postma C, Baak J, Weiss M, Rapallo A, Sciutto A, Roemen G, Arends JW, Williams R, Giaretti W, De Goeij A, Meijer G (2002) Colorectal adenoma to carcinoma progression follows multiple pathways of chromosomal instability. Gastroenterology 123: 1109-1119.

Hernandez D, Fisher EM (1999) Mouse autosomal trisomy: two's company, three's a crowd. Trends Genet 15: 241-247.

Hertzberg L, Betts DR, Raimondi SC, Schafer BW, Notterman DA, Domany E, Izraeli S (2007) Prediction of chromosomal aneuploidy from gene expression data. Gene Chromosome Canc 46: 75-86.

Heselmeyer-Haddad K, Macville M, Schrock E, Blegen H, Hellstrom AC, Shah K, Auer G, Ried T (1997) Advanced-stage cervical carcinomas are defined by a recurrent pattern of chromosomal aberrations revealing high genetic instability and a consistent gain of chromosome arm 3q. Gene Chromosome Canc 19: 233-240.

Heselmeyer-Haddad K, Sommerfeld K, White NM, Chaudhri N, Morrison LE, Palanisamy N, Wang ZY, Auer G, Steinberg W, Ried T (2005) Genomic amplification of the human telomerase gene (TERC) in pap smears predicts the development of cervical cancer. Am J Pathol 166: 1229-1238.

Heslop-Harrison JS, Bennett MD (1984) Chromosome order—possible implications for development. J Embryol Exp Morphol 83 Suppl: 51-73.

Hewitt HB, Blake ER, Walder AS (1976) A critique of the evidence for active host defence against cancer, based on personal studies of 27 murine tumours of spontaneous origin. Br J Cancer 33: 241-259.

Heylighen F (1999a) The evolution of complexity. In: Heylighen F, Bollen J, Riegler A, editors. Einstein Meets Magritte. Boston: Kluwer. 374 pp.

Heylighen F (1999b) The science of self-organization and adaptivity. Brussels: Center "Leo Apostel", Free University of Brussels, Belgium. pages 1-26, http://www.redfish.com/research/EOLSS-Self-Organiz.pdf

Hieter P, Griffiths T (1999) Polyploidy — more is more or less. Science 285: 210-211.

Hildebrandt B, Wust P, Ahlers O, Dieing A, Sreenivasa G, Kerner T, Felix R, Riess H (2002) The cellular and molecular basis of hyperthermia. Crit Rev Oncol Hematol 43: 33-56.

Hill DA, Gridley G, Cnattingius S, Mellemkjaer L, Linet M, Adami HO, Olsen JH, Nyren O, Fraumeni JF, Jr (2003) Mortality and cancer incidence among individuals with Down syndrome. Arch Intern Med 163: 705-711.

Hill J (1761) Cautions Against the Immoderate Use of Snuff. Founded on the Known Qualities of the Tobacco Plant; And the Effects it Must Produce when this Way Taken into the Body: And Enforced by Instances of Persons who have Perished Miserably of Diseases, Occasioned, or Rendered Incurable by its Use. London: Baldwin, R. & Jackson, J.

Hittelman WN (2001) Genetic instability in epithelial tissues at risk for cancer. Ann N Y Acad Sci 952: 1-12.

Hobohm U (2009) Healing heat: Harnessing infection to fight cancer. Amer Sci 97: 34-41.

Hodge LD, Barrett JM, Welter DA (1995) Computer graphics of SEM images facilitate recognition of chromosome position in isolated human metaphase plates. Microsc Res Tech 30: 408-418.

Hodgkin J (2005) Karyotype, ploidy, and gene dosage. WormBook: 1-9.

Hoeijmakers JH (2001) Genome maintenance mechanisms for preventing cancer. Nature 411: 366-374.

Hoglund M, Gisselsson D, Mandahl N, Johansson B, Mertens F, Mitelman F, Sall T (2001a) Multivariate analyses of genomic imbalances in solid tumors reveal distinct and converging pathways of karyotypic evolution. Gene Chromosome Canc 31: 156-171.

Hoglund M, Sall T, Heim S, Mitelman F, Mandahl N, Fadl-Elmula I (2001b) Identification of cytogenetic subgroups and karyotypic pathways in transitional cell carcinoma. Cancer Res 61: 8241-8246.

Hoglund M, Gisselsson D, Hansen GB, Sall T, Mitelman F, Nilbert M (2002) Dissecting karyotypic patterns in colorectal tumors: two distinct but overlapping pathways in the adenoma-carcinoma transition. Cancer Res 62: 5939-5946.

Hoglund M, Jin C, Gisselsson D, Hansen GB, Mitelman F, Mertens F (2004) Statistical analyses of karyotypic complexity in head and neck squamous cell carcinoma. Cancer Genet Cytogenet 150: 1-8.

Holland AJ, Cleveland DW (2009) Boveri revisited: chromosomal instability, aneuploidy and tumorigenesis. Nat Rev Mol Cell Biol 10: 478-487.

Holliday R (1989) Chromosome error propagation *and* cancer. Trends Genet 5: 42-45.

Holliday R (1996) Neoplastic transformation: the contrasting stability of human and mouse cells. In: Lindhal T, Tooze J, editors. Genetic Instability in Cancer. Plainview, NY: Cold Spring Harbor Lab Press. pp. 103-115.

Hollingsworth RE, Lee WH (1991) Tumor suppressor genes: new prospects for cancer research. J Natl Cancer Inst 83: 91-96.

Hollstein M, Rice K, Greenblatt MS, Soussi R, Fuchs R, Sorlie T, Hovig E, Smith-Sorensen B, Montesano R, Harris CC (1994) Database of p53 gene somatic mutations in human tumors and cell lines. Nucleic Acids Res 22: 3551-3555.

Holmberg K, Falt S, Johansson A, Lambert B (1993) Clonal chromosome aberrations and genomic instability in X-irradiated human T-lymphocyte cultures. Mutat Res 286: 321-330.

Hook EB (1985) The impact of aneuploidy upon public health: mortality and morbidity associated with human chromosome abnormalities. In: Dellarco VL, Voytek PE, Hollaender A, editors. Aneuploidy: Etiology and Mechanisms. New York: Plenum Press. pp. 7-33.

Hoption Cann SA, van Netten JP, van Netten C, Glover DW (2002) Spontaneous regression: a hidden treasure buried in time. Med Hypotheses 58: 115-119.

Hoption Cann SA, van Netten JP, van Netten C (2003) Dr William Coley and tumour regression: a place in history or in the future. Postgrad Med J 79: 672-680.

Horn HF, Vousden KH (2004) Cancer: guarding the guardian? Nature 427: 110-111.

Horowitz JM, Yandell DW, Park SH, Canning S, Whyte P, Buchkovich K, Harlow E, Weinberg RA, Dryja TP (1989) Point mutational inactivation of the retinoblastoma antioncogene. Science 243: 937-940.

Horrobin DF (2003) Modern biomedical research: an internally self-consistent universe with little contact with medical reality? Nat Rev Drug Discov 2: 151-154.

Hortobagyi GN (2005) Trastuzumab in the treatment of breast cancer. N Engl J Med 353: 1734-1736.

Howard RO (1981) Classification of chromosomal eye syndromes. Int Ophthalmol 4: 77-91.

Hsu TC (1960) Reduction of transplantability of Novikoff hepatoma cells grown in vitro and the consequent protecting effect to the host against their malignant progenitor. J Natl Cancer Inst 25: 927-935.

Hua VY, Wang WK, Duesberg PH (1997) Dominant transformation by mutated human ras genes in vitro requires more than 100 times higher expression than is observed in cancers. Proc Natl Acad Sci U S A 94: 9614-9619.

Huang Y, Malone KE, Cushing-Haugen KL, Daling JR, Li CI (2011) Relationship between menopausal symptoms and risk of postmenopausal breast cancer. Cancer Epidemiol Biomarkers Prev 20: 379-388.

Huebner RJ, Todaro G (1969) Oncogenes of RNA tumor viruses as determinants of cancer. Proc Natl Acad Sci U S A 64: 1087-1094.

Hueper WC (1952) Environmental Cancers: A Review. Cancer Res 12: 691-697.

Huettel B, Kreil DP, Matzke M, Matzke AJ (2008) Effects of aneuploidy on genome structure, expression, and interphase organization in *Arabidopsis thaliana*. PLoS Genet 4: e1000226.

Hughes TR, Roberts CJ, Dai H, Jones AR, Meyer MR, Slade D, Burchard J, Dow S, Ward TR, Kidd MJ, Friend SH, Marton MJ (2000) Widespread aneuploidy revealed by DNA microarray expression profiling. Nat Genet 25: 333-337.

Huna A, Salmina K, Jascenko E, Duburs G, Inashkina I, Erenpreisa J (2011) Self-renewal signalling in presenescent tetraploid IMR90 cells. 14 pages, http://www.sage-hindawi.com/journals/jar/2011/103253/cta/

Hunter T (1987) A tail of two src's: mutatis mutandis. Cell 49: 1-4.

Hunter WS (1968) Unexpected regressed retinoblastoma. Can J Ophthalmol 3: 376-380.

Hutchinson J (1888) On some examples of arsenic-keratosis of the skin and of arsenic-cancer. Trans Pathol Soc London 39: 352-393.

Huxley J (1956) Cancer biology: comparative and genetic. Biol Rev 31: 474-514.

Iacobuzio-Donahue CA, Maitra A, Olsen M, Lowe AW, van Heek NT, Rosty C, Walter K, Sato N, Parker A, Ashfaq R, Jaffee E, Ryu B, Jones J, Eshleman JR, Yeo CJ, Cameron JL, Kern SE, Hruban RH, Brown PO, Goggins M (2003) Exploration of global gene expression patterns in pancreatic adenocarcinoma using cDNA microarrays. Am J Pathol 162: 1151-1162.

Ifrah N, James JM, Viguie F, Marie JP, Zittoun R (1985) Spontaneous remission in adult acute leukemia. Cancer 56: 1187-1190.

Isaacs SN, Axelrod PI, Lorber B (1990) Antipyretic orders in a university hospital. Am J Med 88: 31-35.

Ising U, Levan A (1957) The chromosomes of two highly malignant human tumours. Acta Pathol Microbiol Scand 40: 13-24.

Ivanov X, Mladenov Z, Nedyalkov S, Todorov T, Yakimov M (1964) Experimental investigations into avian leukoses. V. Transmission, haematology and morphology of avian myelocytomatosis. Bulletin de l'Institute de Pathologie Comparee des Animaux 10: 5-38.

Iversen OH (1991a) The skin tumorigenic and carcinogenic effects of different doses, numbers of dose fractions and concentrations of 7,12-dimethylbenz[a] anthracene in acetone applied on hairless mouse epidermis. Possible implications for human carcinogenesis. Carcinogenesis 12: 493-502.

Iversen OH (1991b) Urethan (ethyl carbamate) is an effective promoter of 7,12-dimethylbenz[a]anthracene-induced carcinogenesis in mouse skin two-stage experiments. Carcinogenesis 12: 901-903.

Ivshina AV, George J, Senko O, Mow B, Putti TC, Smeds J, Lindahl T, Pawitan Y, Hall P, Nordgren H, Wong JEL, Liu ET, Bergh J, Kuznetsov VA, Miller LD (2006) Genetic reclassification of histologic grade delineates new clinical subtypes of breast cancer. Cancer Res 66: 10292-10301.

Jackson AL, Loeb LA (1998) On the origin of multiple mutations in human cancers. Seminars in Cancer Biology 8: 421-429.

Jain IS, Singh K (1968) Retinoblastoma in phthisis bulbi. J All India Ophthalmol Soc 16: 76-78.

Jakubezak RJ, Merlino G, French JE, Muller WJ, Paul B, Adhya S, Garges S (1996) Analysis of genetic instability during mammary tumor progression using a novel selection-based assay for in vivo mutations for a bacterial transgene target. Proc Natl Acad Sci U S A 93: 9073-9078.

Jenkins JR, Rudge K, Chumakov P, Currie GA (1985) The cellular oncogene p53 can be activated by mutagenesis. Nature 317: 816-818.

Jensen CG, Jensen LC, Rieder CL, Cole RW, Ault JG (1996) Long crocidolite asbestos fibers cause polyploidy by sterically blocking cytokinesis. Carcinogenesis 17: 2013-2021.

Jensen RV (1987) Classical chaos. American Scientist 75: 168-181.

Jiang F, Richter J, Schraml P, Bubendorf L, Gasser T, Sauter G, Mihatsch MJ, Moch H (1998) Chromosomal imbalances in papillary renal cell carcinoma: genetic differences between histological subtypes. Am J Pathol 153: 1467-1473.

Jin Y, Jin C, Lv M, Tsao SW, Zhu J, Wennerberg J, Mertens F, Kwong YL (2005) Karyotypic evolution and tumor progression in head and neck squamous cell carcinomas. Cancer Genet Cytogenet 156: 1-7.

Johansson B, Bardi G, Pandis N, Gorunova L, Backman PL, Mandahl N, Dawiskiba S, Andren-Sandberg A, Heim S, Mitelman F (1994) Karyotypic pattern of pancreatic adenomas correlates with survival and tumor grade. Int J Cancer 58: 8-13.

Johansson B, Mertens F, Mitelman F (1996) Primary vs. secondary neoplasia-associated chromosomal abnormalities—balanced rearrangements vs. genomic imbalances? Gene Chromosome Canc 16: 155-163.

Jono K, Ikebe Y, Inada K, Tsuda H (1994) A case of spontaneous remission in chronic B-cell leukemia with virus infection. Nippon Naika Gakkai Zasshi 83: 2159-2160.

Kacser H, Burns JA (1973) The control of flux. Sym Soc Exp Biol 27: 65-104.

Kacser H, Burns JA (1979) Molecular democracy: who shares the controls? Biochem Soc Trans 7: 1149-1160.

Kacser H, Burns JA (1981) The molecular basis of dominance. Genetics 97: 639-666.

Kacser H (1995) Recent developments beyond Metabolic Control Analysis. Biochem Soc Trans 23: 387-391.

Kahlem P, Sultan M, Herwig R, Steinfath M, Balzereit D, Eppens B, Saran NG, Pletcher MT, South ST, Stetten G, Lehrach H, Reeves RH, Yaspo ML (2004) Transcript level alterations reflect gene dosage effects across multiple tissues in a mouse model of down syndrome. Genome Res 14: 1258-1267.

Kahn D, Westerhoff HV (1991) Control theory of regulatory cascades. J Theor Biol 153: 255-285.

Kajii T, Kawai T, Takumi T, Misu H, Mabuchi O, Takahashi Y, Tachino M, Nihei F, Ikeuchi T (1998) Mosaic variegated aneuploidy with multiple congenital abnormalities: homozygosity for total premature chromatid separation trait. Am J Med Genet 78: 245-249.

Kallioniemi A, Kallioniemi OP, Piper J, Tanner M, Stokke T, Chen L, Smith HS, Pinkel D, Gray JW, Waldman FM (1994a) Detection and mapping of amplified DNA sequences in breast cancer by comparative genomic hybridization. Proc Natl Acad Sci U S A 91: 2156-2160.

Kallioniemi OP, Kallioniemi A, Kurisu W, Thor A, Chen LC, Smith HS, Waldman FM, Pinkel D, Gray JW (1992) ERBB2 amplification in breast cancer analyzed by fluorescence in situ hybridization. Proc Natl Acad Sci U S A 89: 5321-5325.

Kallioniemi OP, Kallioniemi A, Piper J, Isola J, Waldman FM, Gray JW, Pinkel D (1994b) Optimizing comparative genomic hybridization for analysis of DNA sequence copy number changes in solid tumors. Gene Chromosome Canc 10: 231-243.

Kammerer S, Roth RB, Hoyal CR, Reneland R, Marnellos G, Kiechle M, Schwarz-Boeger U, Griffiths LR, Ebner F, Rehbock J, Cantor CR, Nelson MR, Braun A (2005) Association of the NuMA region on chromosome 11q13 with breast cancer susceptibility. Proc Natl Acad Sci U S A 102: 2004-2009.

Kan Z, Jaiswal BS, Stinson J, Janakiraman V, Bhatt D, Stern HM, Yue P, Haverty PM, Bourgon R, Zheng J, Moorhead M, Chaudhuri S, Tomsho LP, Peters BA, Pujara K, Cordes S, Davis DP, Carlton VE, Yuan W, Li L, et al. (2010) Diverse somatic mutation patterns and pathway alterations in human cancers. Nature 466: 869-873.

Kaneko Y, Rowley JD, Variakojis D, Chilcote RR, Moohr JW, Patel D (1981) Chromosome abnormalities in Down's syndrome patients with acute leukemia. Blood 58: 459-466.

Kapp JP (1983) Microorganisms as antineoplastic agents in CNS tumors. Arch Neurol 40: 637-642.

Kappauf H, Gallmeier WM, Wunsch PH, Mittelmeier HO, Birkmann J, Buschel G, Kaiser G, Kraus J (1997) Complete spontaneous remission in a patient with metastatic non-small-cell lung cancer. Case report, review of the literature, and discussion of possible biological pathways involved. Ann Oncol 8: 1031-1039.

Kartner N, Riordan JR, Ling V (1983) Cell surface P-glycoprotein associated with multidrug resistance in mammalian cell lines. Science 221: 1285-1288.

Kato A, Kubo K, Kurokawa F, Okita K, Oga A, Murakami T (1998) Numerical aberrations of chromosomes 16, 17, and 18 in hepatocullular carcinoma. Digest Dis Sci 43: 1-7.

Kato R (1967) Localization of "spontaneous" and Rous sarcoma virus-induced breakage in specific regions of the chromosomes of the Chinese hamster. Hereditas 58: 221-247.

Kato R (1968) The chromosomes of forty-two primary Rous sarcomas of the Chinese hamster. Hereditas 59: 63-119.

Katsios C, Roukos DH (2011) Missing heritability, next-generation genome-wide association studies and primary cancer prevention: an Atlantean illusion? Future Oncol 7: 477-480.

Katsura K, Sugihara H, Nakai S, Fujita S (1996) Alteration of numerical chromosomal aberrations during progression of colorectal tumors revealed by a combined fluorescence in situ hybridization and DNA ploidy analysis of intratumoral heterogeneity. Cancer Genet Cytogenet 90: 146-153.

Katz SE, Schapira HE (1982) Spontaneous regression of genitourinary cancer—an update. J Urol 128: 1-4.

Kaye FJ, Kratzke RA, Gerster JL, Horowitz JM (1990) A single amino acid substitution results in a retinoblastoma protein defective in phosphorylation and oncoprotein binding. Proc Natl Acad Sci U S A 87: 6922-6926.

Keech ML, Wills ED (1979) The effect of hyperthermia on activation of lysosomal enzymes in HeLa cells. Eur J Cancer 15: 1025-1031.

Kendall SD, Linardic CM, Adam SJ, Counter CM (2005) A network of genetic events sufficient to convert normal human cells to a tumorigenic state. Cancer Res 65: 9824-9828.

Kilpivaara O, Rantanen M, Tamminen A, Aittomaki K, Blomqvist C, Nevanlinna H (2008) Comprehensive analysis of NuMA variation in breast cancer. BMC Cancer 8: 71.

Kim SH, Roth KA, Moser AR, Gordon JI (1993) Transgenic mouse models that explore the multistep hypothesis of intestinal neoplasia. J Cell Biol 123: 877-893.

Kimmel M, Axelrod DE (1990) Mathematical models of gene amplification with applications to cellular drug resistance and tumorigenicity. Genetics 125: 633-644.

King M (1993) Species evolution: the role of chromosome change. Cambridge: Cambridge University Press.

Kinzler K, Vogelstein B (1996) Lessons from hereditary colorectal cancer. Cell 87: 159-170.

Kinzler KW, Vogelstein B (1997) Cancer-susceptibility genes: gatekeepers and caretakers. Nature 386: 761,763.

Kirsten WH, Mayer LA (1967) Morphologic responses to a murine erythroblastosis virus. J Natl Cancer Inst 39: 311-335.

Kizaki M, Ogawa T, Watanabe Y, Toyama K (1988) Spontaneous remission in hypoplastic acute leukemia. Keio J Med 37: 299-307.

Kleef R, Jonas WB, Knogler W, Stenzinger W (2001) Fever, cancer incidence and spontaneous remissions. Neuroimmunomodulation 9: 55-64.

Klein A, Li N, Nicholson JM, McCormack AA, Graessmann A, Duesberg P (2010) Transgenic oncogenes induce oncogene-independent cancers with individual karyotypes and phenotypes. Cancer Genet Cytogenet 200: 79-99.

Klein CA, Blankenstein TJ, Schmidt-Kittler O, Petronio M, Polzer B, Stoecklein NH, Riethmuller G (2002) Genetic heterogeneity of single disseminated tumour cells in minimal residual cancer. Lancet 360: 683-689.

Klein G (1979) Lymphoma development in mice and humans: diversity of initiation is followed by convergent cytogenetic evolution. Proc Natl Acad Sci U S A 76: 2442-2446.

Knosel T, Schluns K, Stein U, Schwabe H, Schlag PM, Dietel M, Petersen I (2004) Chromosomal alterations during lymphatic and liver metastasis formation of colorectal cancer. Neoplasia 6: 23-28.

Knudson AG (1971) Mutation and cancer: statistical study of retinoblastoma. Proc Natl Acad Sci U S A 68: 820-823.

Knudson AG (1985) Hereditary cancer, oncogenes, and antioncogenes. Cancer Res 45: 1437-1443.

Knudson AG (2000) Chasing the cancer demon. Annu Rev Genet 34: 1-19.

Knudson AG (2001) Two genetic hits (more or less) to cancer. Nat Rev Cancer 1: 157-162.

Knuutila S, Siimes M, Vuopio P (1981) Chromosome pulverization in blood diseases. Hereditas 95: 15-24.

Ko JM, Yau WL, Chan PL, Lung HL, Yang L, Lo PH, Tang JC, Srivastava G, Stanbridge EJ, Lung ML (2005) Functional evidence of decreased tumorigenicity associated with monochromosome transfer of chromosome 14 in esophageal cancer and the mapping of tumor-suppressive regions to 14q32. Gene Chromosome Canc 43: 284-293.

Kobayashi K, Okamoto T, Takayama S, Akiyama M, Ohno T, Yamada H (2000) Genetic instability in intestinal metaplasia is a frequent event leading to well-differentiated early adenocarcinoma of the stomach. Eur J Cancer 36: 1113-1119.

Koeffler HP, Golde DW (1981a) Chronic Myelogenous Leukemia—New Concepts—Part 1. N Engl J Med 304: 1201-1209.

Koeffler HP, Golde DW (1981b) Chronic Myelogenous Leukemia—New Concepts—Part 2. N Engl J Med 304: 1269-1274.

Koller PC (1964) Chromosomes in neoplasia. Amsterdam: Elsevier Publishing Company.

Koller PC (1972) The role of chromosomes in cancer biology. Recent Results Cancer Res 38: 1-122.

Konishi N, Hiasa Y, Matsuda H, Tao M, Tsuzuki T, Hayashi I, Kitahori Y, Shiraishi T, Yatani R, Shimazaki J, Lin IC (1995) Intratumor cellular heterogeneity and alterations in ras oncogene and p53 tumor suppressor gene in human prostate carcinoma. Am J Pathol 147: 1112-1122.

Koprowski H, Ponten JA, Jensen F, Ravdin RG, Moorhead P, Saksela E (1962) Transformation of cultures of human tissue infected with Simian virus SV40. Journal of Cellular Comparative Physiology 59: 281-286.

Kosak ST, Scalzo D, Alworth SV, Li F, Palmer S, Enver T, Lee JS, Groudine M (2007) Coordinate gene regulation during hematopoiesis is related to genomic organization. PLoS Biol 5: e309.

Koscielny S (2008) Critical review of microarray-based prognostic tests and trials in breast cancer. Current Opinion in Obstetrics & Gynecology 20: 47-50.

Kost-Alimova M, Fedorova L, Yang Y, Klein G, Imreh S (2004) Microcell-mediated chromosome transfer provides evidence that polysomy promotes structural instability in tumor cell chromosomes through asynchronous replication and breakage within late-replicating regions. Gene Chromosome Canc 40: 316-324.

Kraemer PM, Petersen DF, Van Dilla MA (1971) DNA constancy in heteroploidy and the stem line theory of tumors. Science 174: 714-717.

Kraemer PM, Travis GL, Ray FA, Cram LS (1983) Spontaneous neoplastic evolution of Chinese hamster cells in culture: multistep progression of phenotype. Cancer Res 43: 4822-4827.

Krebs H (1981) Otto Warburg: Cell physiologist, biochemist and eccentric. New York: Clarendon Press. 141 pp.

Krontiris TG, Cooper GM (1981) Transforming activity of human tumor DNAs. Proc Natl Acad Sci U S A 78: 1181-1184.

Kuroki T, Huh NH (1993) Why are human cells resistant to malignant cell transformation in vitro? Jpn J Cancer Res 84: 1091-1100.

Kutschera U, Niklas KJ (2004) The modern theory of biological evolution: an expanded synthesis. Naturwissenschaften 91: 255-276.

Kuukasjarvi T, Karhu R, Tanner M, Kahkonen M, Schaffer A, Nupponen N, Pennanen S, Kallioniemi A, Kallioniemi OP, Isola J (1997) Genetic heterogeneity and clonal evolution underlying development of asynchronous metastasis in human breast cancer. Cancer Res 57: 1597-1604.

Kuwabara S, Ajioka Y, Watanabe H, Hitomi J, Nishikura K, Hatakeyama K (1998) Heterogeneity of p53 mutational status in esophageal cell carcinoma. Jap J Cancer Res 89: 405-410.

Kuznetsova NN, Shevliagin V, Biriulina TI (1970) Virus particle antigens in virus-free tumors of adult rats induced by the Rous virus. Biull Eksp Biol Med (Biulleten Eksperimentalnoi Biologii I Meditsiny, Moskva) 70: 66-69.

Laffaire J, Rivals I, Dauphinot L, Pasteau F, Wehrle R, Larrat B, Vitalis T, Moldrich RX, Rossier J, Sinkus R, Herault Y, Dusart I, Potier MC (2009) Gene expression signature of cerebellar hypoplasia in a mouse model of Down syndrome during postnatal development. BMC Genomics 10: 138.

Lamb P, Crawford L (1986) Characterization of the human p53 gene. Mol Cell Biol 6: 1379-1385.

Land H, Parada LF, Weinberg RA (1983a) Cellular oncogenes and multistep carcinogenesis. Science 222: 771-778.

Land H, Parada LF, Weinberg RA (1983b) Tumorigenic conversion of primary embryo fibroblasts requires at least two cooperating oncogenes. Nature 304: 596-602.

Langenegger EE (2009) Correlation between histological grade and ploidy status in potentially malignant disorders of the oral mucosa. Pretoria: University of Pretoria. 71 pp., http://upetd.up.ac.za/thesis/available/etd-08112010-143516/unrestricted/dissertation.pdf.

Lania L, Gandini-Attardi D, Griffiths M, Cooke B, De Cicco D, Fried M (1980) The polyoma virus 100K large T-antigen is not required for the maintenance of transformation. Virology 101: 217-232.

Lanza A, Lagomarsini P, Casati A, Ghetti P, Stefanini M (1997) Chromosomal fragility in the cancer-prone disease xeroderma pigmentosum preferentially involves bands relevant for cutaneous carcinogenesis. Int J Cancer 74: 654-663.

Larsen SU, Rose C (1999) Spontaneous remission of breast cancer. A literature review. Ugeskr Laeger 161: 4001-4004.

Latarjet R, Duplan JF (1962) Experiment and discussion on leukaemogenesis by cell-free extracts of radiation-induced leukaemia in mice. Int J Radiat Biol 5: 339-344.

Leaf C (2004) Why we're losing the war on cancer [and how to win it]. Fortune. pp. 42-59.

Ledford H (2010) Big science: The cancer genome challenge. Nature 464: 972-974.

Lee AJX, Endesfelder D, Rowan AJ, Walther A, Birkbak NJ, Futreal PA, Downward J, Szallasi Z, Tomlinson IPM, Kschischo M, Swanton C (2011) Chromosomal instability confers intrinsic multidrug resistance. Cancer Res 71: 1-13.

Lee ML, Kuo FC, Whitmore GA, Sklar J (2000) Importance of replication in microarray gene expression studies: statistical methods and evidence from repetitive cDNA hybridizations. Proc Natl Acad Sci U S A 97: 9834-9839.

Lee WM, Schwab M, Westaway D, Varmus HE (1985) Augmented expression of normal c-myc is sufficient for cotransformation of rat embryo cells with a mutant ras gene. Mol Cell Biol 5: 3345-3356.

Lefrere F, Hermine O, Radford-Weiss I, Veil A, Picard F, Dreyfus F, Flandrin G, Varet B (1994) A spontaneous remission of lymphoid blast crisis in chronic myelogenous leukaemia following blood transfusion and infection. Br J Haematol 88: 621-622.

Lei KI, Gwi E, Ma L, Liang EY, Johnson PJ (1997) 'Spontaneous' regression of advanced leiomyosarcoma of the urinary bladder. Oncology 54: 19-22.

Leitch IJ, Bennett MD (1997) Polyploidy in angiosperms. Trends in Plant Science 2: 470-475.

Lengauer C, Kinzler KW, Vogelstein B (1997) Genetic instability in colorectal cancers. Nature 386: 623-627.

Lengauer C, Kinzler KW, Vogelstein B (1998) Genetic instabilities in human cancers. Nature 396: 643-649.

Lengauer C, Wang Z (2004) From spindle checkpoint to cancer. Nat Genet 36: 1144-1145.

Leprohon P, Legare D, Raymond F, Madore E, Hardiman G, Corbeil J, Ouellette M (2009) Gene expression modulation is associated with gene amplification, supernumerary chromosomes and chromosome loss in antimony-resistant *Leishmania infantum*. Nucleic Acids Research 37: 1387-1399.

Levan A (1956) Chromosomes in cancer tissue. Annals of the New York Academy of Sciences 63: 774-792.

Levan A, Biesele JJ (1958) Role of chromosomes in cancerogenesis, as studied in serial tissue culture of mammalian cells. Annals of New York Academy of Sciences 71: 1022-1053.

Levan A (1959) Relation of chromosome status to the origin and progression of tumours: The evidence of chromosome number. Genetics and Cancer. Austin: University of Texas Press. pp. 151-182.

Levan A (1969) Chromosome abnormalities and carcinogenesis. In: Lima-de-Faria A, editor. Handbook of Molecular Cytology. New York: American Elsevier Publishing Co. pp. 716-731.

Levan A, Levan G, Mitelman F (1977) Chromosomes and cancer. Hereditas 86: 15-30.

Levin EJ (1957) Spontaneous regression (cure?) of a malignant tumor of bone. Cancer 10: 377-381.

Lewin B (1994) Genes V. New York: Oxford University Press

Lewin B (1997) Genes VI. Oxford: Oxford University Press

Li CI, Malone KE, Weiss NS, Daling JR (2001) Tamoxifen therapy for primary breast cancer and risk of contralateral breast cancer. J Natl Cancer Inst 93: 1008-1013.

Li R, Yerganian G, Duesberg P, Kraemer A, Willer A, Rausch C, Hehlmann R (1997) Aneuploidy 100% correlated with chemical transformation of Chinese hamster cells. Proc Natl Acad Sci U S A 94: 14506-14511.

Li R, Sonik A, Stindl R, Rasnick D, Duesberg P (2000) Aneuploidy versus gene mutation hypothesis: recent study claims mutation, but is found to support aneuploidy. Proc Natl Acad Sci U S A 97: 3236-3241.

Li R, Rasnick D, Duesberg P (2002) Correspondence re: D. Zimonjic et al., Derivation of human tumor cells in vitro without widespread genomic instability, Cancer Research 61: 8838-8844, 2001. Cancer Res 62: 6345-6349.

Li R, Hehlman R, Sachs R, Duesberg P (2005) Chromosomal alterations cause the high rates and wide ranges of drug resistance in cancer cells. Cancer Genet Cytogenet 163: 44-56.

Li R, McCormack AA, Nicholson JM, Fabarius A, Hehlmann R, Sachs RK, Duesberg PH (2009) Cancer-causing karyotypes: chromosomal equilibria between destabilizing aneuploidy and stabilizing selection for oncogenic function. Cancer Genet Cytogenet 188: 1-25.

Li X, Nicklas RB (1995) Mitotic forces control a cell-cycle checkpoint. Nature 373: 630-632.

Liberman L, Drotman M, Morris EA, LaTrenta LR, Abramson AF, Zakowski MF, Dershaw DD (2000) Imaging-histologic discordance at percutaneous breast biopsy. Cancer 89: 2538-2546.

Lieberman M, Kaplan HS (1959) Leukemogenic activity of filtrates from radiation-induced lymphoid tumors of mice. Science 130: 387-388.

Lijinsky W (1989) A view of the Relation Between Carcinogenesis and Mutagenesis. Environ Mol Mutagen 14: 78-84.

Lindsley DL, Sandler L, Baker BS, Carpenter AT, Denell RE, Hall JC, Jacobs PA, Miklos GL, Davis BK, Gethmann RC, Hardy RW, Steven AH, Miller M, Nozawa H, Parry DM, Gould-Somero M (1972) Segmental aneuploidy and the genetic gross structure of the Drosophila genome. Genetics 71: 157-184.

Lingle WL, Lutz WH, Ingle JN, Maihle NJ, Salisbury JL (1998) Centrosome hypertrophy in human breast tumors: implications for genomic stability and cell polarity. Proc Natl Acad Sci U S A 95: 2950-2955.

Lingle WL, Barrett SL, Negron VC, D'Assoro AB, Boeneman K, Liu W, Whitehead CM, Reynolds C, Salisbury JL (2002) Centrosome amplification drives chromosomal instability in breast tumor development. Proc Natl Acad Sci U S A 99: 1978-1983.

Lister J (1906) The collected papers of Joseph Lister (digitized by the Internet Archive in 2008 with funding from Microsoft Corporation). Oxford: Clarenden Press http://www.archive.org/details/collectedpaperso02listuoft.

Litmanovitch T, Altaras MM, Dotan A, Avivi L (1998) Asynchronous replication of homologous alpha-satellite DNA loci in man is associated with nondisjunction. Cytogenet Cell Genet 81: 26-35.

Little JB (2000) Radiation carcinogenesis. Carcinogenesis 21: 397-404.

Littlefield LG, Joiner EE, Sayer AM (1985) Premature separation of centromeres in marrow chromosomes from an untreated patient with acute myelogenous leukemia. Cancer Genet Cytogenet 16: 109-116.

Liu P, Zhang H, McLellan A, Vogel H, Bradley A (1998) Embryonic lethality and tumorigenesis caused by segmental aneuploidy on mouse chromosome 11. Genetics 150: 1155-1168.

Liu Y (2006) Fatty acid oxidation is a dominant bioenergetic pathway in prostate cancer. Prostate Cancer Prostatic Dis 9: 230-234.

Lodish H, Berk A, Zipursky SL, Matsudaira P, Baltimore D, Darnell J (1999) Molecular Cell Biology. New York and Basingstoke UK: W. H. Freeman & Co.

Lodish H, Berk P, Matsudaira P, Kaiser CA, Krieger M, Scott MP, et al. (2004) Molecular Cell Biology. New York and Basingstoke, UK: W. H. Freeman & Co.

Loeb LA (1991) Mutator phenotype may be required for multistage carcinogenesis. Cancer Res 51: 3075-3079.

Loeb LA (1997) Transient Expression of a Mutator Phenotype in Cancer Cells. Science 277: 1449-1450.

Loeb LA (2001) A mutator phenotype in cancer. Cancer Res 61: 3230-3239.

Loeb LA, Loeb KR, Anderson JP (2003) Multiple mutations and cancer. Proc Natl Acad Sci U S A 100: 776-781.

Loeper S, Romeike BF, Heckmann N, Jung V, Henn W, Feiden W, Zang KD, Urbschat S (2001) Frequent mitotic errors in tumor cells of genetically micro-heterogeneous glioblastomas. Cytogenet Cell Genet 94: 1-8.

Logan J, Cairns J (1982) The secrets of cancer. Nature 300: 104-105.

London RE (1955) Multiple myeloma: report of a case showing unusual remission lasting two years following severe hepatitis. Ann Intern Med 43: 191-201.

Looijenga LH, de Munnik H, Oosterhuis JW (1999) A molecular model for the development of germ cell cancer. Int J Cancer 83: 809-814.

Lucas C, editor (1999) "Complexity philosophy as a computing paradigm," Manchester, UK: UMIST Workshop.

Lucito R, Healy J, Alexander J, Reiner A, Esposito D, Chi M, Rodgers L, Brady A, Sebat J, Troge J, West JA, Rostan S, Nguyen KC, Powers S, Ye KQ, Olshen A, Venkatraman E, Norton L, Wigler M (2003) Representational oligonucleotide microarray analysis: a high-resolution method to detect genome copy number variation. Genome Res 13: 2291-2305.

Lupski JR (2007) Structural variation in the human genome. N Engl J Med 356: 1169-1171.

Luttges J, Galehdari H, Brocker V, Schwarte-Waldhoff I, Henne-Bruns D, Kloppel G, Schmiegel W, Hahn SA (2001) Allelic loss is often the first hit in the biallelic inactivation of the p53 and DPC4 genes during pancreatic carcinogenesis. Am J Pathol 158: 1677-1683.

Lyle R, Gehrig C, Neergaard-Henrichsen C, Deutsch S, Antonarakis SE (2004) Gene expression from the aneuploid chromosome in a trisomy mouse model of Down syndrome. Genome Res 14: 1268-1274.

MacCorkle RA, Slattery SD, Nash DR, Brinkley BR (2006) Intracellular protein binding to asbestos induces aneuploidy in human lung fibroblasts. Cell Motility and the Cytoskeleton 63: 646-657.

Macgregor PF, Squire JA (2002) Application of microarrays to the analysis of gene expression in cancer. Clin Chem 48: 1170-1177.

Mackey MA, Anolik SL, Roti Roti JL (1992) Cellular mechanisms associated with the lack of chronic thermotolerance expression in HeLa S3 cells. Cancer Res 52: 1101-1106.

Macville M, Schrock E, Padilla-Nash H, Keck C, Ghadimi BM, Zimonjic D, Popescu N, Ried T (1999) Comprehensive and definitive molecular cytogenetic characterization of HeLa cells by spectral karyotyping. Cancer Res 59: 141-150.

Maekawa T, Fujii H, Horiike S, Okuda T, Yokota S, Ueda K, Urata Y (1989) Spontaneous remission of four months' duration in hypoplastic leukemia with tetraploid chromosome after blood transfusions and infection. Nippon Ketsueki Gakkai Zasshi 52: 849-857.

Maher B (2008) Personal genomes: The case of the missing heritability. Nature 456: 18-21.

Mahlamaki EH, Hoglund M, Gorunova L, Karhu R, Dawiskiba S, Andren-Sandberg A, Kallioniemi OP, Johansson B (1997) Comparative genomic hybridization reveals frequent gains of 20q, 8q, 11q, 12p, and 17q, and losses of 18q, 9p, and 15q in pancreatic cancer. Gene Chromosome Canc 20: 383-391.

Maia AT, van der Velden VH, Harrison CJ, Szczepanski T, Williams MD, Griffiths MJ, van Dongen JJ, Greaves MF (2003) Prenatal origin of hyperdiploid acute lymphoblastic leukemia in identical twins. Leukemia 17: 2202-2206.

Maia AT, Tussiwand R, Cazzaniga G, Rebulla P, Colman S, Biondi A, Greaves M (2004) Identification of preleukemic precursors of hyperdiploid acute lymphoblastic leukemia in cord blood. Gene Chromosome Canc 40: 38-43.

Mailhes JB, Young D, London SN (1998) Postovulatory ageing of mouse oocytes in vivo and premature centromere separation and aneuploidy. Biol Reprod 58: 1206-1210.

Makarevitch I, Harris C (2010) Aneuploidy causes tissue-specific qualitative changes in global gene expression patterns in maize. Plant Physiol 152: 927-938.

Makino S (1974) Cytogenetics of canine veneral tumors; worldwide distribution and a common karyotype. In: German J, editor. Chromosomes and Cancer. New York: John Wiley & Sons. pp. 335-372.

Mamaeva SE (1998) Karyotypic evolution of cells in culture: a new concept. Int Rev Cytol 178: 1-40.

Mangiapan G, Guigay J, Milleron B (1994) [A new case of spontaneous regression of metastasis of kidney cancer]. Rev Pneumol Clin 50: 139-140.

Manolio TA, Collins FS, Cox NJ, Goldstein DB, Hindorff LA, Hunter DJ, McCarthy MI, Ramos EM, Cardon LR, Chakravarti A, Cho JH, Guttmacher AE, Kong A, Kruglyak L, Mardis E, Rotimi CN, Slatkin M, Valle D, Whittemore AS, Boehnke M, et al. (2009) Finding the missing heritability of complex diseases. Nature 461: 747-753.

Mao R, Zielke CL, Zielke HR, Pevsner J (2003) Global up-regulation of chromosome 21 gene expression in the developing Down syndrome brain. Genomics 81: 457-467.

Marcos Sanchez F, Juarez Ucelay F, Bru Espino IM, Duran-Perez Navarro A (1991) A new case of spontaneous tumor regression. An Med Interna 8: 468.

Marczynska B, Massey RJ (1986) Transplantable primate tumors induced by Rous sarcoma virus. I. Induction of tumors transplantable into young marmosets. J Natl Cancer Inst 77: 537-547.

Margolis J, West D (1967) Spontaneous regression of malignant disease: report of three cases. J Am Geriatr Soc 15: 251-253.

Mark J (1967) Chromosomal analysis of ninety-one primary Rous sarcomas in the mouse. Hereditas 57: 23-82.

Mark J (1969) Rous sarcomas in mice: the chromosomal progression in primary tumours. Eur J Cancer 5: 307-315.

Markovic S, Ferlan-Marolt V, Hlebanja Z (1996) Spontaneous regression of hepatocellular carcinoma. Am J Gastroenterol 91: 392-393.

Marquardt H, Glaess E (1957) Die Veraenderungen der Haeufigkeit euploider und aneuploider Chromosomenzahlen in der hepatektomierten Rattenleber bei Buttergelb-Verfuetterung. Naturwissenschaften 44: 640.

Martin GS (1970) Rous sarcoma virus: A function required for the maintenance of the transformed state. Nature 227: 1021-1023.

Marx J (1991) Possible new colon cancer gene found. Science 251: 1317.

Marx J (2002) Debate surges over the origins of genomic defects in cancer. Science 297: 544-546.

Masayesva BG, Ha P, Garrett-Mayer E, Pilkington T, Mao R, Pevsner J, Speed T, Benoit N, Moon CS, Sidransky D, Westra WH, Califano J (2004) Gene expression alterations over large chromosomal regions in cancers include multiple genes unrelated to malignant progression. Proc Natl Acad Sci U S A 101: 8715-8720.

Massague J (2007) Sorting out breast-cancer gene signatures. N Engl J Med 356: 294-297.

Masuda A, Takahashi T (2002) Chromosome instability in human lung cancers: possible underlying mechanisms and potential consequences in the pathogenesis. Oncogene 21: 6884-6897.

Matagne JHJ (1953) Vers la guerison du cancer [Towards the cure of cancer]. Le Scalpel 104 & 106: 504-544 & 1387-1395.

Matthey R (1951) The chromosomes of the vertebrates. Adv Genet 4: 159-180.

Matzke MA, Mittelsten-Scheid O, Matzke AJM (1999) Rapid structural and epigenetic changes in polyploid and aneuploid genomes. BioEssays 21: 761-767.

Matzke MA, Mette MF, Kanno T, Matzke AJ (2003) Does the intrinsic instability of aneuploid genomes have a causal role in cancer? Trends Genet 19: 253-256.

Matzker J, Steinberg A (1976) Tonsillectomy and leukemia in adults (author's transl). Laryngol Rhinol Otol (Stuttg) 55: 721-725.

Mayer VW, Aguilera A (1990) High levels of chromosome instability in polyploids of *Saccharomyces cerevisiae*. Mut Res 231: 177-186.

McCormick F (2001) New-age drug meets resistance. Nature 412: 281-282.

McDonough SK, Larsen S, Brodey RS, Stock ND, Hardy WD, Jr. (1971) A transmissible feline fibrosarcoma of viral origin. Cancer Res 31: 953-956.

McIntosh JR (1991) Structural and mechanical control of mitotic progression. Cold Spring Harb Symp Quant Biol 56: 613-619.

McLemore MR (2006) Gardasil: Introducing the new human papillomavirus vaccine. Clin J Oncol Nurs 10: 559-560.

Medzhitov R (2001) Toll-like receptors and innate immunity. Nat Rev Immunol 1: 135-145.

Meijer GA, Hermsen MA, Baak JP, van Diest PJ, Meuwissen SG, Belien JA, Hoovers JM, Joenje H, Snijders PJ, Walboomers JM (1998) Progression from colorectal adenoma to carcinoma is associated with non-random chromosomal gains as detected by comparative genomic hybridisation. J Clin Pathol 51: 901-909.

Melicow MM (1975) Percivall Pott (1713-1788): 200th anniversary of first report of occupation-induced cancer scrotum in chimmey sweepers (1775). Urology 6: 745-749.

Mellors RC, Keane JF, Jr., Papanicolaou GN (1952) Nucleic acid content of the squamous cancer cell. Science 116: 265-269.

Menke-Pluymers MB, Mulder AH, Hop WC, van Blankenstein M, Tilanus HW (1994) Dysplasia and aneuploidy as markers of malignant degeneration in Barrett's oesophagus. The Rotterdam Oesophageal Tumour Study Group. Gut 35: 1348-1351.

Mentzer SJ (1995) Immunoreactivity in lung cancer. Chest Surg Clin N Am 5: 57-71.

Merlo LM, Pepper JW, Reid BJ, Maley CC (2006) Cancer as an evolutionary and ecological process. Nat Rev Cancer 6: 924-935.

Mertens F, Johansson B, Hoglund M, Mitelman F (1997) Chromosomal imbalance maps of malignant solid tumors: a cytogenetic survey of 3185 neoplasms. Cancer Res 57: 2765-2780.

Micale MA, Visscher DW, Gulino SE, Wolman SR (1994) Chromosomal aneuploidy in proliferative breast disease. Hum Pathol 25: 29-35.

Michiels S, Koscielny S, Hill C (2005) Prediction of cancer outcome with microarrays: a multiple random validation strategy. Lancet 365: 488-492.

Michiels S, Koscielny S, Hill C (2007) Interpretation of microarray data in cancer. Br J Cancer 96: 1155-1158.

Michor F, Iwasa Y, Nowak MA (2004) Dynamics of cancer progression. Nat Rev Cancer 4: 197-205.

Michor F, Iwasa Y, Lengauer C, Nowak MA (2005) Dynamics of colorectal cancer. Seminars in Cancer Biology 15: 484-493.

Miklos GL, Maleszka R (2004a) The clinical consequences of massive genomic imbalances in cancers. Cell Oncol 26: 237-240.

Miklos GL, Maleszka R (2004b) Microarray reality checks in the context of a complex disease. Nature Biotechnol 22: 615-621.

Miller JA, Miller EC (1971) Chemical carcinogenesis: Mechanisms and approaches to its control. J Natl Cancer Inst 47: v-xiv.

Minton AP (1994) Influence of macromolecular crowding on intracellular association reactions: possible role in volume regulation. In: Strange K, editor. Cellular and Molecular Physiology of Cell Volume Regulation. Ann Arbor: CRC Press. pp. 181-189.

Mitelman F (1974) The Rous sarcoma virus story: cytogenetics of tumors induced by RSV. In: German J, editor. Chromosomes and Cancer. John Wiley & Sons, Inc. pp. 675-693.

Mitelman F (1994) Catalogue of chromosome aberrations in cancer. New York: Wiley-Liss.

Mitelman F, Mertens F, Johansson B (1997a) A breakpoint map of recurrent chromosomal rearrangements in human neoplasia. Nature Genet Special issue: 417-474.

Mitelman F, Johansson B, Mandahl N, Mertens F (1997b) Clinical significance of cytogenetic findings in solid tumors. Cancer Genet Cytogenet 95: 1-8.

Mitelman F (2011) Mitelman Database of Chromosome Aberrations in Cancer (2009). http://cgap.nci.nih.gov/Chromosomes/Mitelman

Mitterbauer M, Fritzer-Szekeres M, Mitterbauer G, Simonitsch I, Knobl P, Rintelen C, Schwarzinger I, Haas OA, Silberbauer K, Frey K, Bibus B, Pabinger I, Radaszkiewicz T, Lechner K, Jaeger U (1996) Spontaneous remission of acute myeloid leukemia after infection and blood transfusion associated with hypergammaglobulinaemia. Ann Hematol 73: 189-193.

Moloney JB (1966) A virus-induced rhabdomyosarcoma of mice. Natl Cancer Inst Monogr 22: 139-142.

Moorhead PS, Saksela E (1965) The sequence of chromosome aberrations during SV40 transformation of a human diploid cell strain. Hereditas 52: 271-284.

Mora PT, Chang C, Couvillion L, Kuster JM, McFarland VW (1977) Immunological selection of tumour cells which have lost SV40 antigen expression. Nature 269: 36-40.

Mora PT, Parrott CL, Baksi K, McFarland V (1986) Immunologic selection of simian virus 40 (SV40) T-antigen-negative tumor cells which arise by excision of early SV40 DNA. J Virol 59: 628-634.

Morgan TH (1910) Sex-linked inheritance in Drosophila. Science 32: 120-122.

Morgan TH, Bridges CB (1919) The origin of gynandromorphs. In: None, editor. Contributions to the genetics of *Drosophila melanogaster*. Washington: Carnegie Institution of Washington. pp. 108-109.

Moskovitz AH, Linford NJ, Brentnall TA, Bronner MP, Storer BE, Potter JD, Bell RH, Jr., Rabinovitch PS (2003) Chromosomal instability in pancreatic ductal cells from patients with chronic pancreatitis and pancreatic adenocarcinoma. Gene Chromosome Canc 37: 201-206.

Motofei IG (1996) Herpetic viruses and spontaneous recovery in melanoma. Med Hypotheses 47: 85-88.

Mühlbock O (1965) Note on a new inbred mouse-strain GR-A. Eur J Cancer 1: 123-124.

Mukherjee AB, Costello C (1998) Aneuploidy analysis in fibroblasts of human premature aging syndromes by FISH during in vitro cellular aging. Mech Ageing Dev 103: 209-222.

Muleris M, Chalastanis A, Meyer N, Lae M, Dutrillaux B, Sastre-Garau X, Hamelin R, Flejou JF, Duval A (2008) Chromosomal instability in near-diploid colorectal cancer: a link between numbers and structure. PLoS ONE 3: e1632.

Muller HJ (1927) Artificial transmutation of the gene. Science 66: 84-87.

Muncaster MM, Cohen BL, Phillips RA, Gallie BL (1992) Failure of RB1 to reverse the malignant phenotype of human tumor cell lines. Cancer Res 52: 654-661.

Murgia C, Pritchard JK, Kim SY, Fassati A, Weiss RA (2006) Clonal origin and evolution of a transmissible cancer. Cell 126: 477-487.

Murnane JP (1996) Role of induced genetic instability in the mutagenic effects of chemicals and radiation. Mutat Res 367: 11-23.

Murphy P, Taylor I, Alexander P (1985) Organ distribution of metastases following intracardiac injection of syngeneic rat tumour cells. In: Hellman K, Eccles S, editors. Treatment of metastasis: Problems and Prospects. London: Taylor & Francis. 195 pp.

Nakao K, Shibusawa M, Ishihara A, Yoshizawa H, Tsunoda A, Kusano M, Kurose A, Makita T, Sasaki K (2001) Genetic changes in colorectal carcinoma tumors with liver metastases analyzed by comparative genomic hybridization and DNA ploidy. Cancer 91: 721-726.

Nakao K, Mehta KR, Fridlyand J, Moore DH, Jain AN, Lafuente A, Wiencke JW, Terdiman JP, Waldman FM (2004) High-resolution analysis of DNA copy number alterations in colorectal cancer by array-based comparative genomic hybridization. Carcinogenesis 25: 1345-1357.

Nasiell M, Kato H, Auer G, Zetterberg A, Roger V, Karlen L (1978) Cytomorphological grading and Feulgen DNA-analysis of metaplastic and neoplastic bronchial cells. Cancer 41: 1511-1521.

National Cancer Institute (1989) The 1988 Bethesda System for reporting cervical/vaginal cytological diagnoses. National Cancer Institute Workshop. JAMA 262: 931-934.

National Cancer Institute (2005a) SEER Cancer Statistics Review, 1975-2002. National Cancer Institute, Bethesda, http://seer.cancer.gov/csr/1975_2002/.

National Cancer Institute (2005b) Understanding Cancer Series: Genetic Variation (SNPs). slide 41Sept 10, 2009, http://nci.nih.gov/cancertopics/understandingcancer/geneticvariation/Slide41

Nauts HC, Fowler GA, Bogatko FH (1953) A review of the influence of bacterial infection and of bacterial products (Coley's toxins) on malignant tumors in man. Acta Med Scand Suppl 276: 1-103.

Nauts HC (1969) Sarcoma of the soft tissues, other than lymphosarcoma, treated by toxin therapy; end results in 186 determinate cases with microscopic confirmation of diagnosis: 49 operable, 137 inoperable. New York: Cancer Research Institute.

Nauts HC (1980) The Beneficial Effects of Bacterial Infections on Host Resistance to Cancer: End Results in 449 Cases. New York: Cancer Research Institute.

Nauts HC (1982) Bacterial products in the treatment of cancer: past, present and future. Clogn, Federal Republic of Germany.

Navin N, Krasnitz A, Rodgers L, Cook K, Meth J, Kendall J, Riggs M, Eberling Y, Troge J, Grubor V, Levy D, Lundin P, Maner S, Zetterberg A, Hicks J, Wigler M (2010) Inferring tumor progression from genomic heterogeneity. Genome Res 20: 68-80.

Nelson-Rees WA, Hunter L, Darlington GJ, O'Brien SJ (1980) Characteristics of HeLa strains: permanent vs. variable features. Cytogenet Cell Genet 27: 216-231.

Nettesheim P, Marchok A (1983) Neoplastic development in airway epithelium. Adv Cancer Res 39: 1-70.

Neusser M, Schubel V, Koch A, Cremer T, Muller S (2007) Evolutionarily conserved, cell type and species-specific higher order chromatin arrangements in interphase nuclei of primates. Chromosoma 116: 307-320.

Newman SA (2002) Developmental mechanisms: putting genes in their place. Journal of Biosciences 27: 97-104.

Nguyen HG, Makitalo M, Yang D, Chinnappan D, St Hilaire C, Ravid K (2009) Deregulated Aurora-B induced tetraploidy promotes tumorigenesis. Faseb J 23: 2741-2748.

Niakan B (1998) A mechanism of the spontaneous remission and regression of cancer. Cancer Biother Radiopharm 13: 209-210.

Nichols WW (1963) Relationships of viruses, chromosomes, and carcinogenesis. Hereditas 50: 53-80.

Nichols WW, Levan A, Heneen WK (1967) Studies on the role of viruses in somatic mutation. Hereditas 57: 365-368.

Nicholson JM, Duesberg P (2009) On the karyotypic origin and evolution of cancer cells. Cancer Genet Cytogenet 194: 96-110.

Nicklas RB, Ward SC, Gorbsky GJ (1995) Kinetochore chemistry is sensitive to tension and may link mitotic forces to a cell cycle checkpoint. J Cell Biol 130: 929-939.

Nishisho I, Nakamura Y, Miyoshi Y, Miki Y, Ando H, Horii A, Koyama K, Utsunomiya J, Baba S, Hedge P (1991) Mutations of chromosome 5q21 genes in FAP and colorectal cancer patients. Science 253: 665-669.

Nishizaki T, DeVries S, Chew K, Goodson WH, 3rd, Ljung BM, Thor A, Waldman FM (1997) Genetic alterations in primary breast cancers and their metastases: direct comparison using modified comparative genomic hybridization. Gene Chromosome Canc 19: 267-272.

Nordling CO (1953) A new theory on the cancer-inducing mechanism. Brit J Cancer 7: 68-72.

Norrby E (1989) Award Ceremony Speech. Nobel Prize in Physiology and Medicine. Stockholm: Karolinska Institute http://nobelprize.org/nobel_prizes/medicine/laureates/1989/presentation-speech.html.

Notterman DA, Alon U, Sierk AJ, Levine AJ (2001) Transcriptional gene expression profiles of colorectal adenoma, adenocarcinoma, and normal tissue examined by oligonucleotide arrays. Cancer Res 61: 3124-3130.

Novak JP, Sladek R, Hudson TJ (2002) Characterization of variability in large-scale gene expression data: implications for study design. Genomics 79: 104-113.

Nowacki MP, Szymendera JJ (1983) The strongest prognostic factors in colorectal carcinoma. Surgicopathologic stage of disease and postoperative fever. Dis Colon Rectum 26: 263-268.

Nowell P, Hungerford D (1960) A minute chromosome in human chronic granulocytic leukemia. Science 132: 1497.

Nowell P, Rowley J, Knudson A (1998) Cancer genetics, cytogenetics—defining the enemy within. Nature Medicine 4: 1107-1111.

Nowell PC (1976) The clonal evolution of tumor cell populations. Science 194: 23-28.

Nowell PC (1982) Cytogenetics. In: Becker FF, editor. Cancer: A Comprehensive Treatise. 2nd ed. New York: Plenum Press. pp. 3-40.

Ntzani EE, Ioannidis JP (2003) Predictive ability of DNA microarrays for cancer outcomes and correlates: an empirical assessment. Lancet 362: 1439-1444.

Nupponen NN, Hyytinen ER, Kallioniemi AH, Visakorpi T (1998a) Genetic alterations in prostate cancer cell lines detected by comparative genomic hybridization. Cancer Genet Cytogenet 101: 53-57.

Nupponen NN, Kakkola L, Koivisto P, Visakorpi T (1998b) Genetic alterations in hormone-refractory recurrent prostate carcinomas. Am J Pathol 153: 141-148.

O'Brien SJ, Menotti-Raymond M, Murphy WJ, Nash WG, Wienberg J, Stanyon R, Copeland NG, Jenkins NA, Womack JE, Marshall Graves JA (1999) The promise of comparative genomics in mammals. Science 286: 458-462, 479-481.

O'Regan B, Hirschberg C (1993) Spontaneous remission: an annotated bibliography. Sausalito: Institute of Noetic Sciences. 713 p http://www.noetic. org/library/publication-books/spontaneous-remission-annotated-bibliography/.

Obama B, Burr R, Menendez R (2007) Genomics and Personalized Medicine Act of 2007. Washington, DC.

Ochs MF, Godwin AK (2003) Microarrays in cancer: research and applications. BioTechniques Suppl: 4-15.

Offner S, Schmaus W, Witter K, Baretton GB, Schlimok G, Passlick B, Riethmuller G, Pantel K (1999) p53 gene mutations are not required for early dissemination of cancer cells. Proc Natl Acad Sci U S A 96: 6942-6946.

Ohno S (1971) Genetic implication of karyological instability of malignant somatic cells. Physiol Rev 51: 496-526.

Ohno S (1974) Aneuploidy as a possible means employed by malignant cells to express recessive phenotypes. In: German J, editor. Chromosomes and Cancer. New York: John Wiley & Sons. pp. 77-94.

Oikawa T, Staubach A, Okuda M, Fukasawa K (2004) Centrosome amplification, chromosome instability, and karyotypic convergence. Cell Oncol 26: 220-222.

Oksala T, Therman E (1974) Mitotic abnormalities and cancer. In: German J, editor. Chromosomes and Cancer. New York: John Wiley & Sons. pp. 239-263.

Orr-Weaver TL, Weinberg RA (1998) A checkpoint on the road to cancer. Nature 392: 223-224.

Oshimura M, Hesterberg TW, Barrett JC (1986) An early, nonrandom karyotypic change in immortal Syrian hamster cell lines transformed by asbestos: trisomy of chromosome 11. Cancer Genet Cytogenet 22: 225-237.

Oshimura M, Barrett JC (1986) Chemically induced aneuploidy in mammalian cells: mechanisms and biological significance in cancer. Environ Mutagen 8: 129-159.

Osterheld MC, Caron L, Demierre M, Laurini R, Bosman FT (2004) DNA-ploidy in advanced gastric carcinoma is less heterogeneous than in early gastric cancer. Cell Oncol 26: 21-29.

Pacchierotti F, Eichenlaub-Ritter U (2011) Environmental Hazard in the Aetiology of Somatic and Germ Cell Aneuploidy. Cytogenetic and Genome Research 133: 254-268.

Palmieri S, Kahn P, Graf T (1983) Quail embryo fibroblasts transformed by four v-myc-containing virus isolates show enhanced proliferation but are non-tumorigenic. Embo J 2: 2385-2389.

Pan K-H, Lih C-J, Cohen SN (2005) Effects of threshold choice on biological conclusions reached during analysis of gene expression by DNA microarrays. Proc Natl Acad Sci U S A 102: 8961-8965.

Panigrahi AK, Pati D (2009) Road to the crossroads of life and death: Linking sister chromatid cohesion and separation to aneuploidy, apoptosis and cancer. Crit Rev Oncol Hematol.

Panzer-Grumayer ER, Fasching K, Panzer S, Hettinger K, Schmitt K, Stockler-Ipsiroglu S, Haas OA (2002) Nondisjunction of chromosomes leading to hyperdiploid childhood B-cell precursor acute lymphoblastic leukemia is an early event during leukemogenesis. Blood 100: 347-349.

Parada LA, McQueen PG, Misteli T (2004) Tissue-specific spatial organization of genomes. Genome Biol 5: R44.

Parada LF, Land H, Weinberg RA, Wolf D, Rotter V (1984) Cooperation between gene encoding p53 tumour antigen and ras in cellular transformation. Nature 312: 649-651.

Park JH, Wacholder S, Gail MH, Peters U, Jacobs KB, Chanock SJ, Chatterjee N (2010) Estimation of effect size distribution from genome-wide association studies and implications for future discoveries. Nat Genet 42: 570-575.

Parker SL, Tong T, Bolden S, Wingo PA (1996) Cancer statistics, 1996. CA Cancer J Clin 46: 5-27.

Parkin DM, Bray F, Ferlay J, Pisani P (2001) Estimating the world cancer burden: Globocan 2000. International Journal of Cancer 94: 153-156.

Parry JM, Sors A (1993) The detection and assessment of the aneugenic potential of environmental chemicals: the European Community Aneuploidy Project. Mutat Res 287: 3-15.

Parry JM (2000) Guidance on a strategy for testing of chemicals for mutagenicity. Department of Health (London): Committee on mutagenicity (COM): 39 pages, http://iacom.org.uk/guidstate/documents/guidance.pdf.

Parslow TG, Stites DP, Terr AI, Imboden JB (2001) Medical Immunology: McGraw-Hill. 814 pp.

Patel AS, Hawkins AL, Griffin CA (2000) Cytogenetics and cancer. Curr Opin Oncol 12: 62-67.

Pathak S, Multani AS, Furlong CL, Sohn SH (2002) Telomere dynamics, aneuploidy, stem cells, and cancer (review). Int J Oncol 20: 637-641.

Patja K, Pukkala E, Sund R, Iivanainen M, Kaski M (2006) Cancer incidence of persons with Down syndrome in Finland: a population-based study. Int J Cancer 118: 1769-1772.

Pauletti G, Lai E, Attardi G (1990) Early appearance and long-term persistence of the submicroscopic extrachromosomal elements (amplisomes) containing the amplified DHFR genes in human cell lines. Proc Natl Acad Sci U S A 87: 2955-2959.

Paulsen RD, Soni DV, Wollman R, Hahn AT, Yee MC, Guan A, Hesley JA, Miller SC, Cromwell EF, Solow-Cordero DE, Meyer T, Cimprich KA (2009) A genome-wide siRNA screen reveals diverse cellular processes and pathways that mediate genome stability. Mol Cell 35: 228-239.

Pavelka N, Rancati G, Zhu J, Bradford WD, Saraf A, Florens L, Sanderson BW, Hattem GL, Li R (2010a) Aneuploidy confers quantitative proteome changes and phenotypic variation in budding yeast. Nature 6: 368.

Pavelka N, Rancati G, Li R (2010b) Dr Jekyll and Mr Hyde: role of aneuploidy in cellular adaptation and cancer. Curr Opin Cell Biol 22: 809-815.

Pavlidis P, Li Q, Noble WS (2003) The effect of replication on gene expression microarray experiments. Bioinformatics (Oxford, England) 19: 1620-1627.

Pearse AM, Swift K (2006) Allograft theory: transmission of devil facial-tumour disease. Nature 439: 549.

Pearson H (2002) Surviving a knockout blow. Nature 415: 8-9.

Pejonic T, Heim S, Oerndal C, Jin Y, Mandahl N, Willen H, Mitelman F (1990) Simple numerical chromosome aberrations in well-defined malignant epithelial tumors. Cancer Genet Cytogenet 49: 95-101.

Pejovic T, Heim S, Mandahl N, Elmfors B, Floderus UM, Furgyik S, Helm G, Willen H, Mitelman F (1990) Trisomy 12 is a consistent chromosomal aberration in benign ovarian tumors. Gene Chromosome Canc 2: 48-52.

Pelengaris S, Littlewood T, Khan M, Elia G, Evan G (1999) Reversible activation of c-Myc in skin: induction of a complex neoplastic phenotype by a single oncogenic lesion. Mol Cell 3: 565-577.

Peltomaki P, Aaltonen LA, Sistonen P, Pylkkanen L, Mecklin J-P, Jarvinen H, Green JS, Jass JR, Weber JL, Leach FS, Petersen GM, Hamilton SR, de la Chapelle A, Vogelstein B (1993) Genetic mapping of a locus predisposing to human colorectal cancer. Science 260: 810-812.

Penner DW (1953) Spontaneous regression of a case of myosarcoma. Cancer 6: 776-779.

Pennisi E (1999) Trigger for centrosome replication found. Science 238: 770-771.

Perou CM, Sorlie T, Eisen MB, van de Rijn M, Jeffrey SS, Rees CA, Pollack JR, Ross DT, Johnsen H, Akslen LA, Fluge O, Pergamenschikov A, Williams C, Zhu SX, Lonning PE, Borresen-Dale AL, Brown PO, Botstein D (2000) Molecular portraits of human breast tumours. Nature 406: 747-752.

Perucho M, Hanahan D, Wigler M (1980) Genetic and physical linkage of exogenous sequences in transformed cells. Cell 22: 309-317.

Petersen I, Langreck H, Wolf G, Schwendel A, Psille R, Vogt P, Reichel MB, Ried T, Dietel M (1997) Small-cell lung cancer is characterized by a high incidence of deletions on chromosomes 3p, 4q, 5q, 10q, 13q and 17p. Br J Cancer 75: 79-86.

Petersen I, Petersen S (2001) Towards a genetic-based classification of human lung cancer. Anal Cell Pathol 22: 111-121.

Petithory JC, Theodorides J, Brumpt L (1997) [A challenged Nobel Prize: Johannes Fibiger, 1926]. Hist Sci Med 31: 87-95.

Pettigrew RT, Galt JM, Ludgate CM, Smith AN (1974) Clinical effects of whole-body hyperthermia in advanced malignancy. Br Med J 4: 679-682.

Phillips DH, Venitt S (1995) Environmental mutagenesis; Herndon VA, editor. Human Molecular Genetics Series. Oxford: Bios Scientific Publishers.

Pierce BA (2005) Cenetics, a Conceptual Approach. New York: W. H. Freeman & Co.

Pihan G, Doxsey SJ (2003) Mutations and aneuploidy: co-conspirators in cancer? Cancer Cell 4: 89-94.

Pihan GA, Purohit A, Wallace J, Knecht H, Woda B, Quesenberry P, Doxsey SJ (1998) Centrosome defects and genetic instability in malignant tumors. Cancer Res 58: 3974-3985.

Pihan GA, Wallace J, Zhou Y, Doxsey SJ (2003) Centrosome abnormalities and chromosome instability occur together in pre-invasive carcinomas. Cancer Res 63: 1398-1404.

Pilch H, Gunzel S, Schaffer U, Tanner B, Heine M (2000) Evaluation of DNA ploidy and degree of DNA abnormality in benign and malignant melanocytic lesions of the skin using video imaging. Cancer 88: 1370-1377.

Pitot HC (1986) Fundamentals of Oncology. New York: Marcel Dekker, Inc.

Pitot HC (2002) Fundamentals of Oncology. New York: Marcel Dekker, Inc.

Planck M (1949) Scientific Autobiography and Other Papers. pp. 33-34. New York: translated by F. Gaynor.

Plattner R, Anderson MJ, Sato KY, Fasching CL, Der CJ, Stanbridge EJ (1996) Loss of oncogenic ras expression does not correlate with loss of tumorigenicity in human cells. Proc Natl Sci Acad U S A 93: 6665-6670.

Polakova S, Blume C, Zarate JA, Mentel M, Jorck-Ramberg D, Stenderup J, Piskur J (2009) Formation of new chromosomes as a virulence mechanism in yeast *Candida glabrata*. Proc Natl Acad Sci U S A 106: 2688-2693.

Poland CA, Duffin R, Kinloch I, Maynard A, Wallace WA, Seaton A, Stone V, Brown S, Macnee W, Donaldson K (2008) Carbon nanotubes introduced into the abdominal cavity of mice show asbestos-like pathogenicity in a pilot study. Nat Nanotechnol 3: 423-428.

Pollack JR, Sorlie T, Perou CM, Rees CA, Jeffrey SS, Lonning PE, Tibshirani R, Botstein D, Borresen-Dale AL, Brown PO (2002) Microarray analysis reveals a major direct role of DNA copy number alteration in the transcriptional program of human breast tumors. Proc Natl Acad Sci U S A 99: 12963-12968.

Pongsaensook P, Ritter LE, Parks KK, Grosovsky AJ (2004) Cis-acting transmission of genomic instability. Mutat Res 568: 49-68.

Ponten J (1976) The relationship between in vitro transformation and tumor formation in vivo. Biochim Biophys Acta 458: 397-422.

Posadas EM, Criley SR, Coffey DS (1996) Chaotic oscillations in cultured cells: rat prostate cancer. Cancer Res 56: 3682-3688.

Pothuri B, Leitao MM, Levine DA, Viale A, Olshen AB, Arroyo C, Bogomolniy F, Olvera N, Lin O, Soslow RA, Robson ME, Offit K, Barakat RR, Boyd J (2010) Genetic analysis of the early natural history of epithelial ovarian carcinoma. PLoS ONE 5: e10358.

Pradhan M, Risberg BA, Trope CG, van de Rijn M, Gilks CB, Lee CH (2010) Gross genomic alterations and gene expression profiles of high-grade serous carcinoma of the ovary with and without BRCA1 inactivation. BMC Cancer 10: 493.

Prehn RT (1994) Cancers beget mutations *versus* mutations beget cancers. Cancer Res 54: 5296-5300.

Prehn RT (2005) On the nature of cancer and why anticancer vaccines don't work. Cancer Cell Int 5: 25.

Preston DL, Ron E, Tokuoka S, Funamoto S, Nishi N, Soda M, Mabuchi K, Kodama K (2007) Solid cancer incidence in atomic bomb survivors: 1958-1998. Radiation Research 168: 1-64.

Preussman R (1990) Mechanisms of chemical carcinogenesis. In: Garner RC, Hradec J, editors. Biochemistry of Chemical Carcinogenesis. New York: Plenum Press. pp. 25-35.

Purdie CA, Harrison DJ, Peter A, Dobbie L, White S, Howie SE, Salter DM, Bird CC, Wyllie AH, Hooper ML, et al. (1994) Tumour incidence, spectrum and ploidy in mice with a large deletion in the p53 gene. Oncogene 9: 603-609.

Rabbitts TH, Hamlyn PH, Baer R (1983) Altered nucleotide sequences of a translocated c-myc gene in Burkitt lymphoma. Nature 306: 760-765.

Rabinovitch PS, Longton G, Blount PL, Levine DS, Reid BJ (2001) Predictors of progression in Barrett's esophagus III: baseline flow cytometric variables. Am J Gastroenterol 96: 3071-3083.

Rabinowitz Z, Sachs L (1972) The formation of variants with a reversion of properties of transformed cells. VI. Stability of the reverted state. Int J Cancer 9: 334-343.

Rachko D, Brand KG (1983) Chromosomal aberrations in foreign body tumorigenesis of mice. Proc Soc Exp Biol Med 172: 382-388.

Radford IR (2004) Chromosomal rearrangement as the basis for human tumorigenesis. Int J Radiat Biol 80: 543-557.

Rajagopalan H, Nowak MA, Vogelstein B, Lengauer C (2003) The significance of unstable chromosomes in colorectal cancer. Nat Rev Cancer 3: 695-701.

Rajagopalan H, Jallepalli PV, Rago C, Velculescu VE, Kinzler KW, Vogelstein B, Lengauer C (2004) Inactivation of hCDC4 can cause chromosomal instability. Nature 428: 77-81.

Rajagopalan H, Lengauer C (2004) Aneuploidy and cancer. Nature 432: 338-341.

Rancati G, Pavelka N, Fleharty B, Noll A, Trimble R, Walton K, Perera A, Staehling-Hampton K, Seidel CW, Li R (2008) Aneuploidy Underlies Rapid Adaptive Evolution of Yeast Cells Deprived of a Conserved Cytokinesis Motor. Cell 135: 879-893.

Rao S, Constantini S, Gomori JM, Siegal T, Epstein F (1995) Spontaneous involution of an intra-axial brain stem lesion: a case report. Pediatr Neurosurg 23: 279-281; discussion 282.

Rasnick D, Duesberg PH (1999) How aneuploidy affects metabolic control and causes cancer. Biochem J 340 (Pt 3): 621-630.

Rasnick D, Duesberg P (2000) Metabolic control analysis shows how aneuploidy causes cancer. In: Cornish-Bowden A, Cárdenas ML, editors. Technological and Medical Implications of Metabolic Control Analysis. Dordrecht: Kluwer Academic Publishers. pp. 99-107.

Rasnick D (2000) Auto-catalyzed progression of aneuploidy explains the Hayflick limit of cultured cells, carcinogen-induced tumours in mice, and the age distribution of human cancer. Biochem J 348: 497-506.

Rasnick D (2002) Aneuploidy theory explains tumor formation, the absence of immune surveillance and the failure of chemotherapy. Cancer Genet Cytogenet 136: 66-72.

Rasnick D (2009) DATE analysis: A general theory of biological change applied to microarray data. Biotechnol Prog 25: 1275-1288.

Rauscher FJ (1962) A virus-induced disease of mice characterized by erythrocytopoiesis and lymphoid leukemia. J Natl Cancer Inst 29: 515-543.

Ray FA, Peabody DS, Cooper JL, Cram LS, Kraemer PM (1990) SV40 T antigen alone drives karyotype instability that precedes neoplastic transformation of human diploid fibroblasts. J Cellular Biochemistry 42: 13-31.

Ray FA, Meyne J, Kraemer PM (1992) SV40 T antigen induced chromosomal changes reflect a process that is both clastogenic and aneuploidogenic and is ongoing throughout neoplastic progression of human fibroblasts. Mutat Res 284: 265-273.

Rebollo J, Llorente I, Yoldi A (1990) Spontaneous tumor regression in a patient with metastatic gastric cancer. Communication of an additional case. Rev Med Univ Navarra 34: 141-142.

Reddy EP, Reynolds RK, Santos E, Barbacid M (1982) A point mutation is responsible for the acquisition of transforming properties by the T24 human bladder carcinoma oncogene. Nature 300: 149-152.

Reeves BR, Nash R, Lawler SD, Fisher C, Treleaven JG, Wiltshaw E (1990) Serial cytogenetic studies showing persistence of original clone in Hodgkin's disease. Cancer Genet Cytogenet 50: 1-8.

Reeves RH (2000) Recounting a genetic story. Nature 405: 283-284.

Reid BJ, Levine DS, Longton G, Blount PL, Rabinovitch PS (2000) Predictors of progression to cancer in Barrett's esophagus: baseline histology and flow cytometry identify low- and high-risk patient subsets. Am J Gastroenterol 95: 1669-1676.

Reid JF, Lusa L, De Cecco L, Coradini D, Veneroni S, Daidone MG, Gariboldi M, Pierotti MA (2005) Limits of predictive models using microarray data for breast cancer clinical treatment outcome. J Natl Cancer Inst 97: 927-930.

Reish O, Brosh N, Gobazov R, Rosenblat M, Libman V, Mashevich M (2006) Sporadic aneuploidy in PHA-stimulated lymphocytes of Turner's syndrome patients. Chromosome Res 14: 527-534.

Reish O, Regev M, Kanesky A, Girafi S, Mashevich M (2011) Sporadic Aneuploidy in PHA-Stimulated Lymphocytes of Trisomies 21, 18, and 13. Cytogenet Genome Res 133: 184-189.

Reisman LE, Zuelzer WW, Thompson RI (1964a) Further Observation on the Role of Aneuploidy in Acute Leukemia. Cancer Res 24: 1448-1455.

Reisman LE, Mitani M, Zuelzer WW (1964b) Chromosome Studies in Leukemia. I. Evidence for the Origin of Leukemic Stem Lines from Aneuploid Mutants. N Engl J Med 270: 591-597.

Reith A (2004) Editor and Participant, Abstracts of the 1st Conference on Aneuploidy and Cancer. Oakland, California, USA, 23-26 January 2004. Cell Oncol 26: 167-269.

Reith A. Special Issue Editor (2004) Abstracts of the 1st conference on aneuploidy and cancer, Oakland, CA, USA, 23-26 January 2004. Anal Cell Pathol 26: 167-269.

Remvikos Y, Vogt N, Muleris M, Salmon RJ, Malfoy B, Dutrillaux B (1995) DNA-repeat instability is associated with colorectal cancers presenting minimal chromosome rearrangements. Gene Chromosome Canc 12: 272-276.

Renan MJ (1993) How many mutations are required for tumorigenesis? Implications from human cancer data. Mol Carcinog 7: 139-146.

Rennert G, Bisland-Naggan S, Barnett-Griness O, Bar-Joseph N, Zhang S, Rennert HS, Narod SA (2007) Clinical outcomes of breast cancer in carriers of BRCA1 and BRCA2 mutations. N Engl J Med 357: 115-123.

Rennstam K, Baldetorp B, Kytola S, Tanner M, Isola J (2001) Chromosomal rearrangements and oncogene amplification precede aneuploidization in the genetic evolution of breast cancer. Cancer Res 61: 1214-1219.

Reshmi SC, Saunders WS, Kudla DM, Ragin CR, Gollin SM (2004) Chromosomal instability and marker chromosome evolution in oral squamous cell carcinoma. Gene Chromosome Canc 41: 38-46.

Rettig RA (2006) Strategies and Tactics in the War on Cancer: A Review. Human Affairs 25: 1446-1447.

Ribas A, Butterfield LH, Glaspy JA, Economou JS (2003) Current developments in cancer vaccines and cellular immunotherapy. J Clin Oncol 21: 2415-2432.

Richter H, Slezak P, Walch A, Werner M, Braselmann H, Jaramillo E, Ost A, Hirata I, Takahama K, Zitzelsberger H (2003) Distinct chromosomal imbalances in nonpolypoid and polypoid colorectal adenomas indicate different genetic pathways in the development of colorectal neoplasms. Am J Pathol 163: 287-294.

Richter J, Jiang F, Gorog JP, Sartorius G, Egenter C, Gasser TC, Moch H, Mihatsch MJ, Sauter G (1997) Marked genetic differences between stage pTa and stage pT1 papillary bladder cancer detected by comparative genomic hybridization. Cancer Res 57: 2860-2864.

Richter J, Beffa L, Wagner U, Schraml P, Gasser TC, Moch H, Mihatsch MJ, Sauter G (1998) Patterns of Chromosomal Imbalances in Advanced Urinary

Bladder Cancer Detected by Comparative Genomic Hybridization. Am J Pathol 153: 1615-1621.

Ried T, Heselmeyer-Haddad K, Blegen H, Schrock E, Auer G (1999) Genomic changes defining the genesis, progression, and malignancy potential in solid human tumors: a phenotype/genotype correlation. Gene Chromosome Canc 25: 195-204.

Ried T (2009) Homage to Theodor Boveri (1862-1915): Boveri's theory of cancer as a disease of the chromosomes, and the landscape of genomic imbalances in human carcinomas. Environ Mol Mutagen 50: 593-601.

Rivera MC, Lake JA (2004) The ring of life provides evidence for a genome fusion origin of eukaryotes. Nature 431: 152-155.

Robins DM, Ripley S, Henderson AS, Axel R (1981) Transforming DNA integrates into the host chromosome. Cell 23: 29-39.

Rodman TC, Flehinger BJ, Squire RD (1978) Patterns of association in the human metaphase complement: ring analysis and estimation of associativity of specific chromosome regions. Hum Genet 41: 19-34.

Rokas A (2008) The origins of multicellularity and the early history of the genetic toolkit for animal development. Annu Rev Genet 42: 235-251.

Roschke AV, Stover K, Tonon G, Schaffer AA, Kirsch IR (2002) Stable karyotypes in epithelial cancer cell lines despite high rates of ongoing structural and numerical chromosomal instability. Neoplasia 4: 19-31.

Roschke AV, Tonon G, Gehlhaus KS, McTyre N, Bussey KJ, Lababidi S, Scudiero DA, Weinstein JN, Kirsch IR (2003) Karyotypic complexity of the NCI-60 drug-screening panel. Cancer Res 63: 8634-8647.

Rosen R (1991) Life Itself; Allen TFH, Roberts DW, editors. Complexity in Ecological Systems Series. New York: Columbia University Press. 285 pp.

Rosenberg HF, Gallin JL (1999) Inflammation. In: Paul WE, editor. Fundamental Immunology. 4th ed. Philadelphia: Lippincott-Raven. 1053 pp.

Rosenberg SA, Yang JC, Restifo NP (2004) Cancer immunotherapy: moving beyond current vaccines. Nat Med 10: 909-915.

Rosner F (1978) Is chemotherapy carcinogenic? CA Cancer J Clin 28: 57-59.

Rous P (1911) A Sarcoma of the Fowl Transmissible by an Agent Separable from the Tumor Cells. J Exp Med 13: 397-411.

Rous P, Kidd JG (1941) Conditional neoplasms and subthreshold neoplastic states. A study of the tar tumors of rabbits. J Exp Med 73: 365-389.

Rous P (1959) Surmise and fact on the nature of cancer. Nature 183: 1357-1361.

Rous P (1967) Nobel Prize Lecture: The challenge to man of the neoplastic cell. Cancer Res 27: 1919-1924.

Rowley JD (1973) A new consistent chromosomal abnormality in chronic myelogenous leukaemia identified by quiacrine fluorescence and Giemsa staining. Nature 243: 290-293.

Roy-Burman P, Zheng J, Miller GJ (1997) Molecular heterogeneity in prostate cancer: can TP53 mutation unravel tumorigenesis? Mol Med Today 3: 476-482.

Rubin CE, Haggitt RC, Burmer GC, Brentnall TA, Stevens AC, Levine DS, Dean PJ, Kimmey M, Perera DR, Rabinovitch PS (1992) DNA aneuploidy in colonic biopsies predicts future development of dysplasia in ulcerative colitis. Gastroenterology 103: 1611-1620.

Rubin H (1984) Mutations and oncogenes—cause or effect. Nature 309: 518.

Rubin H, Xu K (1989) Evidence for the progressive and adaptive nature of spontaneous transformation in the NIH 3T3 cell line. Proc Natl Acad Sci U S A 86: 1860-1864.

Ruckdeschel JC, Codish SD, Stranahan A, McKneally MF (1972) Postoperative empyema improves survival in lung cancer. Documentation and analysis of a natural experiment. N Engl J Med 287: 1013-1017.

Ruddon RW (1981) Cancer Biology. New York, Oxford: Oxford University Press.

Ruddon RW (1987) Cancer Biology. New York, Oxford: Oxford University Press.

Ruiz S, Santos M, Lara MF, Segrelles C, Ballestin C, Paramio JM (2005) Unexpected roles for pRb in mouse skin carcinogenesis. Cancer Res 65: 9678-9686.

Ruley HE (1983) Adenovirus early region 1A enables viral and cellular transforming genes to transform primary cells in culture. Nature 304: 602-606.

Ruley HE (1990) Transforming collaborations between ras and nuclear oncogenes. Cancer Cells 2: 258-268.

Rygaard J, Povlsen CO (1974) The mouse mutant nude does not develop spontaneous tumours. An argument against immunological surveillance. Acta Pathol Microbiol Scand B Microbiol Immunol 82: 99-106.

Sachidanandam R, Weissman D, Schmidt SC, Kakol JM, Stein LD, Marth G, Sherry S, Mullikin JC, Mortimore BJ, Willey DL, Hunt SE, Cole CG, Coggill PC, Rice CM, Ning Z, Rogers J, Bentley DR, Kwok PY, Mardis ER, Yeh RT, et al. (2001) A map of human genome sequence variation containing 1.42 million single nucleotide polymorphisms. Nature 409: 928-933.

Sadamori N, Gomez GA, Sandberg AA (1983) Therapeutic and prognostic value of initial chromosomal findings at the blastic phase of Ph-positive chronic myeloid leukemia. Blood 61: 935-939.

Sadamori N, Matsunaga M, Yao E, Ichimaru M, Sandberg AA (1985) Chromosomal Characteristics of Chronic and Blastic Phases of Ph-Positive Chronic Myeloid Leukemia. Cancer Genet Cytogen 15: 17-24.

Sager R, Tanaka K, Lau CC, Ebina Y, Anisowicz A (1983) Resistance of human cells to tumorigenesis induced by cloned transforming genes. Proc Natl Acad Sci U S A 80: 7601-7605.

Saggioro D, Szpirer J, Szpirer C (1982) The effect of ploidy and colcemid on the frequency of spontaneous transformation of cultured cells. Cell Biol Int Rep 6: 29-38.

Saito T, Yamashita T, Notani K, Fukuda H, Mizuno S, Shindoh M, Amemiya A (1995) Flow cytometric analysis of nuclear DNA content in oral leukoplakia: relation to clinicopathologic findings. Int J Oral Maxillofac Surg 24: 44-47.

Saksela E, Moorhead PS (1963) Aneuploidy in the degenerative phase of serial cultivation of human cell strains. Proc Natl Acad Sci U S A 50: 390-395.

Sanchez JH, Abernethy DJ, Boreiko CJ (1986) Reversible expression of morphological transformation in C3H/10T1/2 mouse embryo cultures exposed to 12-O-tetradecanoylphorbol-13-acetate. Carcinogenesis 7: 1793-1796.

Sandberg AA, Ishihara T, Crosswhite LH, Hauschka TS (1962) Chromosomal dichotomy in blood and marrow of acute leukemia. Cancer Res 22: 748-756.

Sandberg AA (1990) The chromosomes in human cancer and leukemia. New York: Elsevier Science Publishing.

Sandler L, Hecht F (1973) Genetic effects of aneuploidy. Am J Hum Genet 25: 332-339.

Santarelli R, Tzeng YJ, Zimmermann C, Guhl E, Graessmann A (1996) SV40 T-antigen induces breast cancer formation with a high efficiency in lactating and virgin WAP-SV-T transgenic animals but with a low efficiency in ovariectomized animals. Oncogene 12: 495-505.

Sawada M, Ohdama S, Umino T, Tachibana S, Takano S, Miyake S, Yoshizawa Y, Aoki N, Matsubara O (1994) Metastasis of an adenocarcinoma of unknown origin to mediastinal lymph nodes, and transient regression. Nihon Kyobu Shikkan Gakkai Zasshi 32: 867-872.

Saxon PJ, Srivatsan ES, Stanbridge EJ (1986) Introduction of human chromosome 11 via microcell transfer controls tumorigenic expression of HeLa cells. Embo J 5: 3461-3466.

Schafer R, Griegel S, Dubbert MA, Willecke K (1984) Unstable transformation of mouse 3T3 cells by transfection with DNA from normal human lymphocytes. Embo J 3: 659-663.

Schar P (2001) Spontaneous DNA damage, genome instability, and cancer—when DNA replication escapes control. Cell 104: 329-332.

Schimke RT (1984) Gene amplification, drug resistance, and cancer. Cancer Res 44: 1735-1742.

Schlemper RJ, Riddell RH, Kato Y, Borchard F, Cooper HS, Dawsey SM, Dixon MF, Fenoglio-Preiser CM, Flejou JF, Geboes K, Hattori T, Hirota T, Itabashi M, Iwafuchi M, Iwashita A, Kim YI, Kirchner T, Klimpfinger M, Koike M, Lauwers GY, et al. (2000) The Vienna classification of gastrointestinal epithelial neoplasia. Gut 47: 251-255.

Schneider BL, Kulesz-Martin M (2004) Destructive cycles: the role of genomic instability and adaptation in carcinogenesis. Carcinogenesis 25: 2033-2044.

Schneider EL, Epstein CJ (1972) Replication rate and lifespan of cultured fibroblasts in Down's syndrome. Proc Soc Exp Biol Med 141: 1092-1094.

Schoch C, Haferlach T, Haase D, Fonatsch C, Loffler H, Schlegelberger B, Staib P, Sauerland MC, Heinecke A, Buchner T, Hiddemann W (2001) Patients with de novo acute myeloid leukaemia and complex karyotype aberrations show a poor prognosis despite intensive treatment: a study of 90 patients. Br J Haematol 112: 118-126.

Schoenlein PV (1993) Molecular cytogenetics of multiple drug resistance. Cytotechnology 12: 63-89.

Schurmans JR, Blijenberg BG, Mickisch GH, Schroder FH (1996) Spontaneous remission of a bony metastasis in prostatic adenocarcinoma. J Urol 155: 653.

Schuttmann W (1993) Schneeberg lung disease and uranium mining in the Saxon Ore Mountains (Erzgebirge). Am J Ind Med 23: 355-368.

Schwab M, Varmus HE, Bishop JM (1985) Human N-myc gene contributes to neoplastic transformation of mammalian cells in culture. Nature 316: 160-162.

Scribner JD, Suess R (1978) Tumor initiation and promotion. Int Rev Exp Pathol 18: 137-187.

Secker-Walker LM (1985) The meaning of a clone. Cancer Genet Cytogenet 16: 187-188.

Secretary's Advisory Committee on Genetics H, and Society, (2008) U.S. System of Oversight of Genetic Testing: A Response to the Charge of the Secretary of Health and Human Services. In: Services HaH, editor. Bethesda: DHHS.

Seeburg PH, Colby WW, Capon PJ, Goeddel DV, Levinson AD (1984) Biological properties of human c-Ha-ras1 genes mutated at codon 12. Nature 312: 71-75.

Seif R, Seif I, Wantyghem J (1983) Rat cells transformed by simian virus 40 give rise to tumor cells which contain no viral proteins and often no viral DNA. Mol Cell Biol 3: 1138-1145.

Seitz S, Frege R, Jacobsen A, Weimer J, Arnold W, von Haefen C, Niederacher D, Schmutzler R, Arnold N, Scherneck S (2005) A network of clinically and functionally relevant genes is involved in the reversion of the tumorigenic phenotype of MDA-MB-231 breast cancer cells after transfer of human chromosome 8. Oncogene 24: 869-879.

Sell S, Pierce GB (1994) Biology of disease: maturation arrest of stem cell differentiation is a common pathway for the cellular origin of teratocarcinomas and epithelial cancers. Lab Invest 70: 6-22.

Selmecki AM, Dulmage K, Cowen LE, Anderson JB, Berman J (2009) Acquisition of aneuploidy provides increased fitness during the evolution of antifungal drug resistance. PLoS Genet 5: e1000705.

Sen S (2000) Aneuploidy and cancer. Curr Opin Oncol 12: 82-88.

Sennerstam R, Kato H, Auer GU (1989) Dissociation of cellular protein and DNA content in mild and moderate dysplasia as a reflection of the degree of aneuploidy in cancer. Anal Quant Cytol Histol 11: 255-260.

Serrano M, Lin AW, McCurrach ME, Beach D, Lowe SW (1997) Oncogenic ras provokes premature cell senescence associated with accumulation of p53 and p16INK4a. Cell 88: 593-602.

Shachaf CM, Kopelman AM, Arvanitis C, Karlsson A, Beer S, Mandl S, Bachmann MH, Borowsky AD, Ruebner B, Cardiff RD, Yang Q, Bishop JM, Contag CH, Felsher DW (2004) MYC inactivation uncovers pluripotent differentiation and tumour dormancy in hepatocellular cancer. Nature 431: 1112-1117.

Shackney SE, McCormack GW, Cuchural GJ, Jr. (1978) Growth rate patterns of solid tumors and their relation to responsiveness to therapy: an analytical review. Ann Intern Med 89: 107-121.

Shackney SE, Smith CA, Miller BW, Burholt DR, Murtha K, Giles HR, Ketterer DM, Pollice AA (1989) Model for the genetic evolution of human solid tumors. Cancer Res 49: 3344-3354.

Shackney SE (1993) Tumor growth, cell cycle kinetics, and cancer treatment. New York: McGraw Hill.

Shackney SE, Berg G, Simon SR, Cohen J, Amina S, Pommersheim W, Yakulis R, Wang S, Uhl M, Smith CA, Pollice AA, Hartsock RJ (1995a) Origins and clinical implications of aneuploidy in early bladder cancer. Cytometry 22: 307-316.

Shackney SE, Singh SG, Yakulis R, Smith CA, Pollice AA, Petruolo S, Waggoner A, Hartsock RJ (1995b) Aneuploidy in breast cancer: a fluorescence in situ hybridization study. Cytometry 22: 282-291.

Shafei-Benaissa E, Savage JR, Babin P, Larregue M, Papworth D, Tanzer J, Bonnetblanc JM, Huret JL (1998) The naevoid basal-cell carcinoma syndrome (Gorlin syndrome) is a chromosomal instability syndrome. Mutat Res 397: 287-292.

Shankey TV, Kallioniemi O-P, Koslowski JM, Lieber ML, Mayall BH, Miller G, Smith GJ (1993) Consensus review of the clinical utility of DNA content cytometry in prostate cancer. Cytometry 14: 497-500.

Shapiro BL (1983) Down syndrome—a disruption of homeostasis. American Journal of Medical Genetics 14: 241-269.

Shastry BS (1995) Genetic knockouts in mice: an update. Experientia 51: 1028-1039.

Shen JJ, Williams BJ, Zipursky A, Doyle J, Sherman SL, Jacobs PA, Shugar AL, Soukup SW, Hassold TJ (1995) Cytogenetic and molecular studies of Down syndrome individuals with leukemia. Am J Hum Genet 56: 915-925.

Shen Z (2011) Genomic instability and cancer: an introduction. J Mol Cell Biol 3: 1-3.

Shi L, Perkins RG, Fang H, Tong W (2008) Reproducible and reliable microarray results through quality control: good laboratory proficiency and appropriate data analysis practices are essential. Curr Opin Biotechnol 19: 10-18.

Shibata D, Schaeffer C, Li ZH, Capella G, Perucho MX (1993) Genetic heterogeneity of the c-K-ras locus in colorectal adenomas but not in adenocarcinomas. J Natl Cancer Inst 85: 1058-1063.

Shibata D (2011) Mutation and epigenetic molecular clocks in cancer. Carcinogenesis 32: 123-128.

Shih C, Shilo BZ, Goldfarb MP, Dannenberg A, Weinberg RA (1979) Passage of phenotypes of chemically transformed cells via transfection of DNA and chromatin. Proc Natl Acad Sci U S A 76: 5714-5718.

Shih C, Padhy LC, Murray M, Weinberg RA (1981) Transforming genes of carcinomas and neuroblastomas introduced into mouse fibroblasts. Nature 290: 261-264.

Shih IM, Zhou W, Goodman SN, Lengauer C, Kinzler KW, Vogelstein B (2001) Evidence that genetic instability occurs at an early stage of colorectal tumorigenesis. Cancer Res 61: 818-822.

Shilo BZ, Weinberg RA (1981) Unique transforming gene in carcinogen-transformed mouse cells. Nature 289: 607-609.

Shimkin MB (1979) Contrary to Nature. In: NIH, editor. Washington, DC: DHEW.

Shiraishi Y, Taguchi H, Niiya K, Shiomi F, Kikukawa K, Kubonishi S, Ohmura T, Hamawaki M, Ueda N (1982) Diagnostic and prognostic significance of chromosome abnormalities in marrow and mitogen response of lymphocytes of acute nonlymphocytic leukemia. Cancer Genet Cytogenet 5: 1-24.

Sick H (1902) Karzinom der Haut das auf dem Boden eines Roentgenuleus entstanden ist. Muench Med Wochenschr 50: 1445.

Sieber OM, Heinimann K, Tomlinson IP (2003) Genomic instability—the engine of tumorigenesis? Nat Rev Cancer 3: 701-708.

Sigmund CD (2000) Viewpoint: are studies in genetically altered mice out of control? Arteriosclerosis, Thrombosis, and Vascular Biology 20: 1425-1429.

Silkworth WT, Nardi IK, Scholl LM, Cimini D (2009) Multipolar spindle pole coalescence is a major source of kinetochore mis-attachment and chromosome mis-segregation in cancer cells. PLoS ONE 4: e6564.

Simi S, Musio A, Vatteroni L, Piras A, Rainaldi G (1992) Specific chromosomal aberrations correlated to transformation in Chinese hamster cells. Cancer Genet Cytogenet 62: 81-87.

Simon R, Radmacher MD, Dobbin K, McShane LM (2003) Pitfalls in the use of DNA microarray data for diagnostic and prognostic classification. J Natl Cancer Inst 95: 14-18.

Singer MJ, Mesner LD, Friedman CL, Trask BJ, Hamlin JL (2000) Amplification of the human dihydrofolate reductase gene via double minutes is initiated by chromosome breaks. Proc Natl Acad Sci U S A 97: 7921-7926.

Singh M, Mehrotra S, Kalra N, Singh U, Shukla Y (2008) Correlation of DNA ploidy with progression of cervical cancer. J Cancer Epidemiol 2008: Article ID 298495, Epub 2008 Jan 29.

Sinn E, Muller W, Pattengale P, Tepler I, Wallace R, Leder P (1987) Coexpression of MMTV/v-Ha-ras and MMTV/c-myc genes in transgenic mice: Synergistic action of oncogenes *in vivo*. Cell 49: 465-475.

Smits R, Kielman MF, Breukel C, Zurcher C, Neufeld K, Jagmohan-Changur S, Hofland N, van Dijk J, White R, Edelmann W, Kucherlapati R, Khan PM, Fodde R (1999) Apc1638T: a mouse model delineating critical domains of the adenomatous polyposis coli protein involved in tumorigenesis and development. Genes Dev 13: 1309-1321.

Soballe PW, Montone KT, Satyamoorthy K, Nesbit M, Herlyn M (1996) Carcinogenesis in human skin grafted to SCID mice. Cancer Res 56: 757-764.

Song K, Lu P, Tang K, Osborn TC (1995) Rapid Genome Change in Synthetic Polyploids of Brassica and Its Implications for Polyploid Evolution. Proc Natl Acad Sci U S A 92: 7719-7723.

Sotillo R, Hernando E, Diaz-Rodriguez E, Teruya-Feldstein J, Cordon-Cardo C, Lowe SW, Benezra R (2007) Mad2 overexpression promotes aneuploidy and tumorigenesis in mice. Cancer Cell 11: 9-23.

Spandidos DA, Wilkie NM (1984) Malignant transformation of early passage rodent cells by a single mutated human oncogene. Nature 310: 469-475.

Spriggs AI, Boddington MM, Clarke CM (1962) Carcinoma-in-situ of the cervix uteri. Some cytogenetic observations. Lancet 279: 1383-1384.

Spriggs AI, Bowey CE, Cowdell RH (1971) Chromosomes of precancerous lesions of the cervix uteri. New data and a review. Cancer 27: 1239-1254.

Spriggs AI (1974) Cytogenetics of cancer and precancerous states of the cervix uteri. In: German J, editor. Chromosomes and Cancer. New York: John Wiley and Sons. pp. 423-450.

Springer M, Weissman JS, Kirschner MW (2010) A general lack of compensation for gene dosage in yeast. Mol Syst Biol 6: 368.

Squire J, Gallie BL, Phillips RA (1985) A detailed analysis of chromosomal changes in heritable and non-heritable retinoblastoma. Hum Genet 70: 291-301.

Stanbridge EJ (1990a) Human Tumor Suppressor Genes. Annu Rev Genet 24: 615-657.

Stanbridge EJ (1990b) Identifying tumor suppressor genes in human colorectal cancer. Science 247: 12-13.

Stanley MA, Kirkland JA (1968) Cytogenetic studies of endometrial carcinoma. Am J Obstet Gynecol 102: 1070-1079.

Steck PA, Ligon AH, Cheong P, Yung WK, Pershouse MA (1995) Two tumor suppressive loci on chromosome 10 involved in human glioblastomas. Gene Chromosome Canc 12: 255-261.

Steel GG, Lamerton LF (1969) Cell population kinetics and chemotherapy. I. The kinetics of tumor cell populations. Natl Cancer Inst Monogr 30: 29-42.

Steel GG (1972) The cell cycle in tumours: an examination of data gained by the technique of labelled mitoses. Cell Tissue Kinet 5: 87-100.

Steinbeck RG (1997) Proliferation and DNA aneuploidy in mild dysplasia imply early steps of cervical carcinogenesis. Acta Oncol 36: 3-12.

Steinberg D (2003) A cell-cycle couple loses its luster. The Scientist 17: 26-27.

Steinfeld JI, Francisco JS, Hase WL (1989) Chemical Kinetics and Dynamics. Englewood Cliffs: Prentice Hall. pp. 182.

Stewart N, Bacchetti S (1991) Expression of SV40 large T antigen, but not small t antigen, is required for the induction of chromosomal aberrations in transformed human cells. Virology 180: 49-57.

Stich HF (1963) Mosaic Composition of Preneoplastic Lesions and Malignant Neoplasms. Exp Cell Res 24: SUPPL 9: 277-285.

Stirling PC, Bloom MS, Solanki-Patil T, Smith S, Sipahimalani P, Li Z, Kofoed M, Ben-Aroya S, Myung K, Hieter P (2011) The Complete Spectrum of Yeast Chromosome Instability Genes Identifies Candidate CIN Cancer Genes and Functional Roles for ASTRA Complex Components. PLoS Genet 7: e1002057.

Stock RP, Bialy H (2003) The sigmoidal curve of cancer. Nat Biotechnol 21: 13-14.

Stoler DL, Chen N, Basik M, Kahlenberg MS, Rodriguez-Bigas MA, Petrelli NJ, Anderson GR (1999) The onset and extent of genomic instability in sporadic colorectal tumor progression. Proc Natl Acad Sci U S A 96: 15121-15126.

Stolt CM, Klein G, Jansson AT (2004) An analysis of a wrong Nobel Prize-Johannes Fibiger, 1926: a study in the Nobel archives. Adv Cancer Res 92: 1-12.

Stone J, de Lange T, Ramsay G, Jakobovits E, Bishop JM, Varmus H, Lee W (1987) Definition of regions in human c-myc that are involved in transformation and nuclear localization. Molecular and cellular biology 7: 1697-1709.

Storchova Z, Pellman D (2004) From polyploidy to aneuploidy, genome instability and cancer. Nat Rev Mol Cell Biol 5: 45-54.

Storchova Z, Breneman A, Cande J, Dunn J, Burbank K, O'Toole E, Pellman D (2006) Genome-wide genetic analysis of polyploidy in yeast. Nature 443: 541-547.

Storchova Z, Kuffer C (2008) The consequences of tetraploidy and aneuploidy. J Cell Sci 121: 3859-3866.

Storz MN, van de Rijn M, Kim YH, Mraz-Gernhard S, Hoppe RT, Kohler S (2003) Gene expression profiles of cutaneous B cell lymphoma. J Invest Dermatol 120: 865-870.

Strauss BS (1992) The origin of point mutations in human tumor cells. Cancer Res 52: 249-253.

Strauss BS (1998) Our contribution to the public fear of cancer. Environ Health Perspect 106: A312-313.

Strohman RC (1997) The coming Kuhnian revolution in biology. Nat Biotechnol 15: 194-200.

Stubbs L (2010) Functional and Comparative Genomics Fact Sheet. Lawrence Livermore National Laboratory: U.S. Department of Energy Genome Programs.

Stutman O (1974) Tumor development after 3-methylcholanthrene in immunologically deficient athymic-nude mice. Science 183: 534-536.

Stutman O (1975) Immunodepression and malignancy. Adv Cancer Res 22: 261-422.

Sureda M, Subira ML, Martin Algarra S, Prieto Valtuena J, Sangro B (1990) Spontaneous tumor regression. Report of 2 cases. Med Clin (Barc) 95: 306-308.

Surveillance Epidemiology and End Results (SEER) (2009) Overview of the Cancer Statistics Review 1975-2006. Bethesda: National Cancer Institute. pp. 36.

Sutton WS (1903) The chromosomes in heredity. Biol Bull 4: 231-250.

Swanton C, Nicke B, Schuett M, Eklund AC, Ng C, Li Q, Hardcastle T, Lee A, Roy R, East P, Kschischo M, Endesfelder D, Wylie P, Kim SN, Chen JG, Howell M, Ried T, Habermann JK, Auer G, Brenton JD, et al. (2009) Chromosomal instability determines taxane response. Proc Natl Acad Sci U S A 106: 8671-8676.

Sweet M (2008) Academic freedom is at risk in dispute over Gardasil, lecturers say. BMJ 336: 741.

Szaflarski W, Nierhaus KH (2007) Question 7: optimized energy consumption for protein synthesis. Orig Life Evol Biosph 37: 423-428.

Tabin CJ, Bradley SM, Bargmann CI, Weinberg RA, Papageorge AG, Scolnick EM, Dhar R, Lowy DR, Chang EH (1982) Mechanism of activation of a human oncogene. Nature 300: 143-149.

Tabin CJ, Weinberg RA (1985) Analysis of viral and somatic activations of the cHa-ras gene. J Virol 53: 260-265.

Takahashi Y, Sinha DK, Dao TL (1977) Comparative chromosome study of normal mammary tissue, carcinogen-induced mammary tumors, and hyperplastic alveolar nodules in the rat: Brief communication. J Natl Cancer Inst 59: 1007-1012.

Takita H (1970) Effect of postoperative empyema on survival of patients with bronchogenic carcinoma. J Thorac Cardiovasc Surg 59: 642-644.

Tan BT, Park CY, Ailles LE, Weissman IL (2006) The cancer stem cell hypothesis: a work in progress. Lab Invest 86: 1203-1207.

Tannock IF (1998) Conventional cancer therapy: promise broken or promise delayed? Lancet 351 Suppl 2: SII9-16.

Taparowsky E, Suard Y, Fasano O, Shimizu K, Goldfarb M, Wigler M (1982) Activation of the T24 bladder carcinoma transforming gene is linked to a single amino acid change. Nature 300: 762-765.

Tarazov PG (1996) Spontaneous necrosis of liver cancer: one more possible cause. Am J Gastroenterol 91: 1872-1873.

Temesrekasi D (1969) Complete regression of 2 non-operated hypopharyngeal carcinomas. Arch Klin Exp Ohren Nasen Kehlkopfheilkd 194: 323-328.

Temin HM, Rubin H (1958) Characteristics of an assay for Rous sarcoma virus and Rous sarcoma cells in tissue culture. Virology 6: 669-688.

Temin HM (1971) The protovirus hypothesis: speculations on the significance of RNA-directed DNA synthesis for normal development and for carcinogenesis. J Natl Cancer Inst 46: 3-7.

Temin HM (1980) Viral oncogenes. Cold Spring Harb Symp Quant Biol 44 Pt 1: 1-7.

Terwilliger JD, Weiss KM (2003) Confounding, ascertainment bias, and the blind quest for a genetic 'fountain of youth'. Ann Med 35: 532-544.

Terwilliger JD, Hiekkalinna T (2006) An utter refutation of the "fundamental theorem of the HapMap". Eur J Hum Genet 14: 426-437.

Testa JR, Misawa S, Oguma N, Van Sloten K, Wiernik PH (1985) Chromosomal alterations in acute leukemia patients studied with improved culture methods. Cancer Res 45: 430-434.

Theile M, Hartmann S, Scherthan H, Arnold W, Deppert W, Frege R, Glaab F, Haensch W, Scherneck S (1995) Suppression of tumorigenicity of breast cancer cells by transfer of human chromosome 17 does not require transferred BRCA1 and p53 genes. Oncogene 10: 439-447.

Thiery P (1909) À propos de la fulguration dans le cancer. Bull Mem Soc Chir Paris 35: 604-698.

Thomas L (1959) Cellular and humoral aspects of the hypersensitive state (discussion); Lawrence HS, editor. New York: Harper. pp. 529-530.

Thomas V, Riegel B, Andrea J, Murray P, Gerhart A, Gocka I (1994) National survey of pediatric fever management practices among emergency department nurses. J Emerg Nurs 20: 505-510.

Thompson SL, Compton DA (2008) Examining the link between chromosomal instability and aneuploidy in human cells. J Cell Biol 180: 665-672.

Thompson SL, Compton DA (2010) Chromosomes and cancer cells. Chromosome Res 19: 433-444.

Thomson I, Gilchrist S, Bickmore WA, Chubb JR (2004) The radial positioning of chromatin is not inherited through mitosis but is established de novo in early G1. Curr Biol 14: 166-172.

Thraves P, Reynolds S, Salehi Z, Kim WK, Yang JH, Rhim JS, Dritschilo A (1991) Detection of transforming genes from radiation transformed human epidermal keratinocytes by a tumorigenicity assay. In: Rhim JS, Dritschilo A, editors. Neoplastic Transformation in Human Cell Culture. Totowa, NJ: The Humana Press Inc. pp. 93-101.

Tjio JH, Levan A (1956) The chromosome number of man. Hereditas 42: 1-6.

Tlsty TD (1990) Normal diploid human and rodent cells lack a detectable frequency of gene amplification. Proc Natl Acad Sci U S A 87: 3132-3136.

Tlsty TD (1997) Genomic instability and its role in neoplasia. Curr Top Microbiol Immunol 221: 37-46.

Todaro GJ, Wolman SR, Green H (1963) Rapid transformation of human fibroblasts with low growth potential into established cell lines by SV40. Journal of Cellular and Comparative Physiology 62: 257-265.

Todaro GT, Green H (1963) Quantitative studies of the growth of mouse embryo cells in culture and their development into established lines. J Cell Biol 17: 299-313.

Tomatis L, Aitio A, Wilbourn J, Shuker L (1989) Human carcinogens so far identified. Jpn J Cancer Res 80: 795-807.

Tomatis L, Huff J (2002) Evolution of Research in Cancer Etiology. In: Coleman WB, Tsongalis GJ, editors. The Molecular Basis of Human Cancer. Totowa, NJ: Humana Press Inc. pp. 189-201.

Tomita T (1995) DNA ploidy and proliferating cell nuclear antigen in colonic adenomas and adenocarcinomas. Dig Dis Sci 40: 996-1004.

Tomlinson I, Bodmer W (1999) Selection, the mutation rate and cancer: ensuring that the tail does not wag the dog. Nat Med 5: 11-12.

Tomlinson IP, Novelli MR, Bodmer WF (1996) The mutation rate and cancer. Proc Natl Acad Sci U S A 93: 14800-14803.

Tonon G, Wong KK, Maulik G, Brennan C, Feng B, Zhang Y, Khatry DB, Protopopov A, You MJ, Aguirre AJ, Martin ES, Yang Z, Ji H, Chin L, Depinho RA (2005) High-resolution genomic profiles of human lung cancer. Proc Natl Acad Sci U S A 102: 9625-9630.

Tooze J (1973) The Molecular Biology of Tumour Viruses. New York: Cold Spring Harbor

Torres EM, Sokolsky T, Tucker CM, Chan LY, Boselli M, Dunham MJ, Amon A (2007) Effects of aneuploidy on cellular physiology and cell division in haploid yeast. Science 317: 916-924.

Torres EM, Williams BR, Amon A (2008) Aneuploidy: cells losing their balance. Genetics 179: 737-746.

Torres EM, Dephoure N, Panneerselvam A, Tucker CM, Whittaker CA, Gygi SP, Dunham MJ, Amon A (2010) Identification of aneuploidy-tolerating mutations. Cell 143: 71-83.

Treon SP, Broitman SA (1992) Beneficial effects of post-transfusional hepatitis in acute myelogenous leukemia may be mediated by lipopolysaccharides, tumor necrosis factor alpha and interferon gamma. Leukemia 6: 1036-1042.

Trott DA, Cuthbert AP, Overell RW, Russo I, Newbold RF (1995) Mechanisms involved in the immortalization of mammalian cells by inonizing radiation and chemical carcinogens. Carcinogenesis 16: 193-204.

Tsafrir D, Bacolod M, Selvanayagam Z, Tsafrir I, Shia J, Zeng Z, Liu H, Krier C, Stengel RF, Barany F, Gerald WL, Paty PB, Domany E, Notterman DA (2006) Relationship of gene expression and chromosomal abnormalities in colorectal cancer. Cancer Res 66: 2129-2137.

Tsao JL, Yatabe Y, Markl ID, Hajyan K, Jones PA, Shibata D (2000) Bladder cancer genotype stability during clinical progression. Gene Chromosome Canc 29: 26-32.

Tsutsui T, Maizumi H, McLachlan JA, Barrett JC (1983) Aneuploidy induction and cell transformation by diethylstilbestrol: a possible chromosomal mechanism in carcinogenesis. Cancer Res 43: 3814-3821.

Tuma RS (2010) The Power of Cancer Genome Sequencing: Identifying the Drivers in the Coding Region and Beyond. Lippincott Williams & Wilkins, Inc. 24-26 January 29, 2011, http://journals.lww.com/oncology-times/Fulltext/2010/08100/The_Power_of_Cancer_Genome_Sequencing__Identifying.5.aspx

Twombly R (2004) Experts Debate Message Sent by Increased Cancer Survival Rates. J Natl Cancer Inst 96: 1412-1413.

Tzankov A, Ludescher C, Duba HC, Steinlechner M, Knapp R, Schmid T, Grunewald K, Gastl G, Stauder R (2001) Spontaneous remission in a secondary acute myelogenous leukaemia following invasive pulmonary aspergillosis. Ann Hematol 80: 423-425.

Ubeda JM, Legare D, Raymond F, Ouameur AA, Boisvert S, Rigault P, Corbeil J, Tremblay MJ, Olivier M, Papadopoulou B, Ouellette M (2008) Modulation of gene expression in drug resistant Leishmania is associated with gene amplification, gene deletion and chromosome aneuploidy. Genome Biol 9: R115.

Umayahara K, Numa F, Suehiro Y, Sakata A, Nawata S, Ogata H, Suminami Y, Sakamoto M, Sasaki K, Kato H (2002) Comparative genomic hybridization detects genetic alterations during early stages of cervical cancer progression. Gene Chromosome Canc 33: 98-102.

Upender MB, Habermann JK, McShane LM, Korn EL, Barrett JC, Difilippantonio MJ, Ried T (2004) Chromosome transfer induced aneuploidy results in complex dysregulation of the cellular transcriptome in immortalized and cancer cells. Cancer Res 64: 6941-6949.

Urano K, Katakai Y, Tokuda Y, Ueyama Y, Nomura T, Yamamoto S (1995) Failure of genotoxic carcinogens to produce tumors in human skin xenografts transplanted to SCID mice. Carcinogenesis 16: 2223-2226.

van den Berghe H (1989) Chromosomes in plasma-cell malignancies. Eur J Haematol Suppl 51: 47-51.

van der Zee J (2002) Heating the patient: a promising approach? Ann Oncol 13: 1173-1184.

Van Goethem F, de Stoppelaar J, Hoebee B, Kirsch-Volders M (1995) Identification of clastogenic and/or aneugenic events during the preneoplastic stages of experimental rat hepatocarcinogenesis by fluorescence in situ hybridization. Carcinogenesis 16: 1825-1834.

Van Valen LM, Maiorana VC (1991) HeLa, a new microbial species. Evolutionary Theory & Review 10: 71-74.

Vander Heiden MG, Cantley LC, Thompson CB (2009) Understanding the Warburg effect: the metabolic requirements of cell proliferation. Science 324: 1029-1033.

Vanderlaan M, Steele V, Nettesheim P (1983) Increased DNA content as an early marker of transformation in carcinogen-exposed rat tracheal cell cultures. Carcinogenesis 4: 721-727.

Varmus H (1989a) An historical overview of oncogenes. In: Weinberg RA, editor. Oncogenes and the Molecular Origins of Cancer. Cold Spring Harbor: Cold Spring Harbor Laboratory Press. pp. 3-44.

Varmus HE (1984) The Molecular Genetics of Cellular Oncogenes. Annu Rev Genet 18: 533-612.

Varmus HE (1989b) Retroviruses and oncogenes, I—Nobel Lecture. The Nobel Prize in Physiology or Medicine 1989accessed May 31, 2011, http://nobelprize.org/nobel_prizes/medicine/laureates/1989/varmus-lecture.pdf

Vazquez A, Beg Q, de Menezes M, Ernst J, Bar-Joseph Z, Barabasi A-L, Boros L, Oltvai Z (2008) Impact of the solvent capacity constraint on *E. coli* metabolism. BMC Systems Biol 2: 7.

Vazquez A, Oltvai ZN (2011) Molecular crowding defines a common origin for the Warburg effect in proliferating cells and the lactate threshold in muscle physiology. e19538, http://dx.doi.org/10.1038/npre.2011.5784.1

Velu TJ, Vass WC, Lowy DR, Tambourin PE (1989) Harvey murine sarcoma virus: influences of coding and noncoding sequences on cell transformation in vitro and oncogenicity in vivo. J Virol 63: 1384-1392.

Venter JC, Adams MD, Myers EW, Li PW, Mural RJ, Sutton GG, Smith HO, Yandell M, Evans CA, Holt RA, Gocayne JD, Amanatides P, Ballew RM, Huson DH, Wortman JR, Zhang Q, Kodira CD, Zheng XH, Chen L, Skupski M, et al. (2001) The sequence of the human genome. Science 291: 1304-1351.

Verhoeff FH (1966) Retinoblastoma undergoing spontaneous regression. Calcifying agent suggested in treatment of retinoblastoma. Am J Ophthalmol 62: 573-574.

Vessey CJ, Norbury CJ, Hickson ID (2000) Genetic disorders associated with cancer predisposition and genetic instability. Prog Nucleic Acid Res Mol Biol 63: 189-221.

Via S (2009) Natural selection in action during speciation. Proc Natl Acad Sci U S A 106 Suppl 1: 9939-9946.

Vig BK (1984) Sequence of centromere separation another mechanism for the origin of nondisjunction. Hum Genet 66: 239-243.

Vig BK (1987) Sequence of centromere separation: a possible role for repetitive DNA. Mutagenesis 2: 155-159.

Vincent MD (2010) The animal within: carcinogenesis and the clonal evolution of cancer cells are speciation events sensu stricto. Evolution 64: 1173-1183.

Virtaneva K, Wright FA, Tanner SM, Yuan B, Lemon WJ, Caligiuri MA, Bloomfield CD, de La Chapelle A, Krahe R (2001) Expression profiling reveals fundamental biological differences in acute myeloid leukemia with isolated trisomy 8 and normal cytogenetics. Proc Natl Acad Sci U S A 98: 1124-1129.

Visscher PM, Montgomery GW (2009) Genome-wide Association Studies and Human Disease. JAMA 302: 2028-2029.

Vladimirskaia EB (1962) A case of prolonged spontaneous remission in a patient with chronic lymphatic leukemia. Probl Gematol Pereliv Krovi 7: 51-54.

Voet D, Voet T (1995) Biochemistry. New York: John Wiley & Sons.

Vogel F, Motulsky AG (1986) Human Genetics: Problems and Approaches. Berlin, Heidelberg, New York, Tokio: Springer Verlag.

Vogelstein B, Fearon ER, Stanley BA, Hamilton R, Kern SE, Preisinger AC, Leppert M, Nakamura Y, White R, Smits AMM, Bos JL (1988) Genetic alterations during colorectal-tumor development. N Engl J Med 319: 525-532.

Vogelstein B, Kinzler KW (1993) The multistep nature of cancer. Trends Genet 9: 138-141.

Vogelstein B, Kinzler KW (1998) Introduction. In: Vogelstein B, Kinzler K, editors. The Genetic Basis of Human Cancer. New York: McGraw-Hill Health Professions Division. pp. 3-6.

Vogelstein B, Kinzler KW (2004) Cancer genes and the pathways they control. Nat Med 10: 789-799.

Vogt M (1959) A Study of the Relationship Between Karyotype and Phenotype in Cloned Lines of Strain HeLa. Genetics 44: 1257-1270.

Vogt M, Dulbecco R (1960) Virus-Cell Interaction with a Tumor-Producing Virus. Proc Natl Acad Sci U S A 46: 365-370.

Vogt M, Dulbecco R (1963) Steps in the neoplastic transformation of hamster embryo cells by polyoma virus. Proc Natl Acad Sci U S A 49: 171-179.

Voncken JW, Kaartinen V, Pattengale PK, Germeraad WT, Groffen J, Heisterkamp N (1995) BCR/ABL P210 and P190 cause distinct leukemia in transgenic mice. Blood 86: 4603-4611.

Wacholder S, Hartge P, Prentice R, Garcia-Closas M, Feigelson HS, Diver WR, Thun MJ, Cox DG, Hankinson SE, Kraft P, Rosner B, Berg CD, Brinton LA, Lissowska J, Sherman ME, Chlebowski R, Kooperberg C, Jackson RD, Buckman DW, Hui P, et al. (2010) Performance of common genetic variants in breast-cancer risk models. N Engl J Med 362: 986-993.

Wade N (2009) Genes Show Limited Value in Predicting Diseases. New York Times. New York.

Wadman M (2011) Fifty genome sequences reveal breast cancer's complexity: Decoding of ten trillion bases yields no simple patterns or silver bullets. Nature Publishing Group, http://www.nature.com/news/2011/020411/full/news.2011.203.html

Wagner RF, Jr., Nathanson L (1986) Paraneoplastic syndromes, tumor markers, and other unusual features of malignant melanoma. J Am Acad Dermatol 14: 249-256.

Wagner TM, Moslinger R, Langbauer G, Ahner R, Fleischmann E, Auterith A, Friedmann A, Helbich T, Zielinski C, Pittermann E, Seifert M, Oefner P (2000) Attitude towards prophylactic surgery and effects of genetic counselling in families with BRCA mutations. Austrian Hereditary Breast and Ovarian Cancer Group. Br J Cancer 82: 1249-1253.

Wainscoat JS, Fey MF (1990) Assessment of clonality in human tumors: a review. Cancer Res 50: 1355-1360.

Walch A, Bink K, Gais P, Stangl S, Hutzler P, Aubele M, Mueller J, Hofler H, Werner M (2000) Evaluation of c-erbB-2 overexpression and Her-2/neu gene copy number heterogeneity in Barrett's adenocarcinoma. Anal Cell Pathol 20: 25-32.

Walen KH, Stampfer MR (1989) Chromosome analyses of human mammary epithelial cells at stages of chemical-induced transformation progression to immortality. Cancer Genet Cytogenet 37: 249-261.

Walter J, Schermelleh L, Cremer M, Tashiro S, Cremer T (2003) Chromosome order in HeLa cells changes during mitosis and early G1, but is stably maintained during subsequent interphase stages. J Cell Biol 160: 685-697.

Walther A, Houlston R, Tomlinson I (2008) Association between chromosomal instability and prognosis in colorectal cancer: a meta-analysis. Gut 57: 941-950.

Wang E, Voiculescu S, Le Poole IC, El-Gamil M, Li X, Sabatino M, Robbins PF, Nickoloff BJ, Marincola FM (2006) Clonal persistence and evolution during a decade of recurrent melanoma. J Invest Dermatol 126: 1372-1377.

Wang LH, Duesberg P, Beemon K, Vogt PK (1975) Mapping RNase T1-resistant oligonucleotides of avian tumor virus RNAs: sarcoma-specific oligonucleotides are near the poly(A) end and oligonucleotides common to sarcoma and transformation-defective viruses are at the poly(A) end. J Virol 16: 1051-1070.

Wang TL, Rago C, Silliman N, Ptak J, Markowitz S, Willson JK, Parmigiani G, Kinzler KW, Vogelstein B, Velculescu VE (2002) Prevalence of somatic alterations in the colorectal cancer cell genome. Proc Natl Acad Sci U S A 99: 3076-3080.

Warburg O (1956) On the origin of cancer cells. Science 123: 309-314.

Warth A, Herpel E, Krysa S, Hoffmann H, Schnabel PA, Schirmacher P, Mechtersheimer G, Blaker H (2009) Chromosomal instability is more frequent in metastasized than in non-metastasized pulmonary carcinoids but is not a reliable predictor of metastatic potential. Exp Mol Med 41: 349-353.

Waterfield MD, Scrace GT, Whittle N, Stroobant P, Johnsson A, Wasteson A, Westermark B, Heldin CH, Huang JS, Deuel TF (1983) Platelet-derived growth factor is structurally related to the putative transforming protein p28sis of simian sarcoma virus. Nature 304: 35-39.

Waterston RH, Lindblad-Toh K, Birney E, Rogers J, Abril JF, Agarwal P, Agarwala R, Ainscough R, Alexandersson M, An P, Antonarakis SE, Attwood J, Baertsch R, Bailey J, Barlow K, Beck S, Berry E, Birren B, Bloom T, Bork P, et al. (2002) Initial sequencing and comparative analysis of the mouse genome. Nature 420: 520-562.

Watson JD, Crick FH (1953) Molecular structure of nucleic acids; a structure for deoxyribose nucleic acid. Nature 171: 737-738.

Watson JD, Hopkins NH, Jeffrey WR, Steitz JA, Weiner AM (1987) Molecular Biology of the Gene. New York: Benjamin.

Weaver BA, Cleveland DW (2006) Does aneuploidy cause cancer? Curr Opin Cell Biol 18: 658-667.

Weaver BA, Cleveland DW (2007) Aneuploidy: instigator and inhibitor of tumorigenesis. Cancer Res 67: 10103-10105.

Weaver BA, Silk AD, Montagna C, Verdier-Pinard P, Cleveland DW (2007) Aneuploidy acts both oncogenically and as a tumor suppressor. Cancer Cell 11: 25-36.

Weber RG, Scheer M, Born IA, Joos S, Cobbers JM, Hofele C, Reifenberger G, Zoller JE, Lichter P (1998) Recurrent chromosomal imbalances detected in biopsy material from oral premalignant and malignant lesions by combined tissue microdissection, universal DNA amplification, and comparative genomic hybridization. Am J Pathol 153: 295-303.

Webster MA, Cardiff RD, Muller WJ (1995) Induction of mammary epithelial hyperplasias and mammary tumors in transgenic mice expressing a murine mammary tumor virus/activated c-src fusion gene. Proc Natl Acad Sci U S A 92: 7849-7853.

Weinberg RA (1989) Introduction. In: Weinberg RA, editor. Oncogenes and the Molecular Origins of Cancer. Cold Spring Harbor: Cold Spring Harbor Laboratory Press. pp. 1-2.

Weinberg RA (1991) Tumor suppressor genes. Science 254: 1138-1146.

Weinberg RA (2007) The Biology of Cancer. New York: Garland Science.

Weinstein IB (2002) Addiction to oncogenes—the Achilles heal of cancer. Science 297: 63-64.

Weintraub LR (1969) Lymphosarcoma. JAMA 210: 1590-1591.

Weiss KM, Terwilliger JD (2000) How many diseases does it take to map a gene with SNPs? Nat Genet 26: 151-157.

Weiss R, Teich N, Varmus H, Coffin J (1985) Molecular Biology of RNA Tumor Viruses. Plainview, NY: Cold Spring Harbor Lab. Press

Weitzman JB, Yaniv M (1999) Rebuilding the road to cancer. Nature 400: 401-402.

Welter DA, Hodge LD (1985) A scanning electron microscopic technique for three-dimensional visualization of the spatial arrangement of metaphase, anaphase and telophase chromatids. Scan Electron Microsc 2: 879-888.

Westaway D, Payne G, Varmus HE (1984) Proviral deletions and oncogene base-substitutions in insertionally mutagenized c-myc alleles may contribute to the progression of avian bursal tumors. Proc Natl Acad Sci U S A 81: 843-847.

Westerhoff HV, Koster JG, van Workum M, Rudd KE (1990) On the control of gene expression. In: Cardenas AC-BML, editor. Control of Metabolic Processes. New York: Plenum Press.

White CA, Weaver RL, Grillo-Lopez AJ (2001) Antibody-targeted immunotherapy for treatment of malignancy. Annu Rev Med 52: 125-145.

White M (1978) Modes of Speciation. San Francisco: W. H. Freeman & Co.

Whitman RC (1919) Somatic mutation as a factor in the production of cancer; a critical review of v. Hansemann's theory of anaplasia in the light of modern knowledge of genetics. J Cancer Res 4: 181-202.

Wiernik PH (1976) Spontaneous regression of hematologic cancers. Natl Cancer Inst Monogr 44: 35-38.

Wilkens L, Flemming P, Gebel M, Bleck J, Terkamp C, Wingen L, Kreipe H, Schlegelberger B (2004) Induction of aneuploidy by increasing chromosomal instability during dedifferentiation of hepatocellular carcinoma. Proc Natl Acad Sci U S A 101: 1309-1314.

Wilkins AS (1996) Are there 'Kuhnian' revolutions in biology? Bioessays 18: 695-696.

Willenbucher RF, Aust DE, Chang CG, Zelman SJ, Ferrell LD, Moore DH, 2nd, Waldman FM (1999) Genomic instability is an early event during the progression pathway of ulcerative-colitis-related neoplasia. Am J Pathol 154: 1825-1830.

Williams BR, Prabhu VR, Hunter KE, Glazier CM, Whittaker CA, Housman DE, Amon A (2008) Aneuploidy affects proliferation and spontaneous immortalization in mammalian cells. Science 322: 703-709.

Williams BR, Amon A (2009) Aneuploidy: Cancer's Fatal Flaw? Cancer Res. 69: 5289-5291.

Williams GP (1997) Chaos Theory Tamed. Washington DC: Joseph Henry Press. pp. 161-167.

Willis RA (1967) Pathology of Tumours. New York: Appleton-Century-Crofts

Winge O (1930) Zytologische Untersuchungen ueber die Natur maligner Tumoren. II. Teerkarzinome bei Maeusen. Zeitschrift Zellforschung Mikroskopische Anatomie 10: 683-735.

Wolf JW (1989) Prolonged spontaneous remission of case of malignant lymphoma. Mo Med 86: 275-277.

Wolf U (1974) Theodor Boveri and his book "On the problem of the origin of malignant tumors". In: German J, editor. Chromosomes and Cancer. New York: John Wiley & Sons. pp. 3-20.

Wolman SR, Cohen TI, Becker FF (1977) Chromosome analysis of hepatocellular carcinoma 7777 and correlation with alpha-fetoprotein production. Cancer Res 37: 2624-2627.

Wolman SR, Steinberg ML, Defendi V (1980) Simian virus 40-induced chromosome changes in human epidermal cultures. Cancer Genet Cytogenet 2: 39-46.

Wolman SR (1983) Karyotypic progression in human tumors. Cancer Metastasis Rev 2: 257-293.

Wolman SR (1986) Cytogenetic heterogeneity: its role in tumor evolution. Cancer Genet Cytogenet 19: 129-140.

Woodruff MF, Ansell JD, Forbes GM, Gordon JC, Burton DI, Micklem HS (1982) Clonal interaction in tumours. Nature 299: 822-824.

Woods JE (1975) The influence of immunologic responsiveness on head and neck cancer. Therapeutic implications. Plast Reconstr Surg 56: 77-80.

Wormald RP, Harper JI (1983) Bilateral black hypopyon in a patient with self-healing cutaneous malignant melanoma. Br J Ophthalmol 67: 231-235.

Wright EG (1999) Inherited and inducible chromosomal instability: a fragile bridge between genome integrity mechanisms and tumourigenesis. J Pathol 187: 19-27.

Wu Y, Basir Z, Kajdacsy-Balla A, Strawn E, Macias V, Montgomery K, Guo SW (2003) Resolution of clonal origins for endometriotic lesions using laser capture microdissection and the human androgen receptor (HUMARA) assay. Fertil Steril 79 Suppl 1: 710-717.

Wust P, Hildebrandt B, Sreenivasa G, Rau B, Gellermann J, Riess H, Felix R, Schlag PM (2002) Hyperthermia in combined treatment of cancer. Lancet Oncol 3: 487-497.

Wyszkowski J, Armata J, Cyklis R, Halikowski B (1969) Remission in acute drug resistant leukemia complicated with steroid diabetes and severe infection. Pol Tyg Lek 24: 1974-1975.

Xu HJ, Sumegi J, Hu SX, Banerjee A, Uzvolgyi E, Klein G, Benedict WF (1991) Intraocular tumor formation of RB reconstituted retinoblastoma cells. Cancer Res 51: 4481-4485.

Yachida S, Jones S, Bozic I, Antal T, Leary R, Fu B, Kamiyama M, Hruban RH, Eshleman JR, Nowak MA, Velculescu VE, Kinzler KW, Vogelstein B, Iacobuzio-Donahue CA (2010) Distant metastasis occurs late during the genetic evolution of pancreatic cancer. Nature 467: 1114-1117.

Yamagiwa K, Ichikawa K (1915) Experimentelle Studie ueber die Pathogenese der Epithelialgeschwuelste. Mitteillungen Med Fakultaet Kaiserl Univ Tokyo 15: 295-344.

Yamamoto T, Rabinowitz Z, Sachs L (1973) Identification of the chromosomes that control malignancy. Nat New Biol 243: 247-250.

Ye CJ, Stevens JB, Liu G, Bremer SW, Jaiswal AS, Ye KJ, Lin MF, Lawrenson L, Lancaster WD, Kurkinen M, Liao JD, Gairola CG, Shekhar MP, Narayan S, Miller FR, Heng HH (2009) Genome based cell population heterogeneity promotes tumorigenicity: the evolutionary mechanism of cancer. J Cell Physiol 219: 288-300.

Yerganian G, Shein HM, Enders JF (1962) Chromosomal disturbances observed in human fetal renal cells transformed in vitro by simian virus 40 and carried in culture. Cytogenetics 1: 314-324.

Yosida T (1983) Karyotype evolution and tumor development. Cancer Genetics and Cytogenetics 8: 153-179.

Youn A, Simon R (2010) Identifying Cancer Driver Genes in Tumor Genome Sequencing Studies. Bioinformatics (Oxford, England) 27: 175-181.

Yunis JJ, Bloomfield CD, Ensrud K (1981) All patients with acute non-lymphocytic leukemia may have a chromosomal defect. N Engl J Med 305: 135-139.

Yunis JJ (1984) Recurrent chromosomal defects are found in most patients with acute nonlymphocytic leukemia. Cancer Genet Cytogenet 11: 125-137.

Zambrana Garcia JL, Torres Serrano F, Lopez Rubio F, Jimenez-Pereperez JA, Perez-Jimenez F (1996) [Spontaneous tumor regression and gastric cancer]. An Med Interna 13: 47-48.

Zanazzi C, Hersmus R, Veltman IM, Gillis AJ, van Drunen E, Beverloo HB, Hegmans JP, Verweij M, Lambrecht BN, Oosterhuis JW, Looijenga LH (2007) Gene expression profiling and gene copy-number changes in malignant mesothelioma cell lines. Genes Chromosomes Cancer 46: 895-908.

Zang KD, Singer H (1967) Chromosomal consitution of meningiomas. Nature 216: 84-85.

Zang KD (1982) Cytological and cytogenetical studies on human meningioma. Cancer Genet Cytogenet 6: 249-274.

Zaridze DG, Arkadieva MA, Day NE, Duffy SW (1993) Risk of leukaemia after chemotherapy in a case-control study in Moscow. Br J Cancer 67: 347-350.

Zhang L, Zhou W, Velculescu VE, Kern SE, Hruban RH, Hamilton SR, Vogelstein B, Kinzler KW (1997) Gene expression profiles in normal and cancer cells. Science 276: 1268-1272.

Zhao H, Langerod A, Ji Y, Nowels KW, Nesland JM, Tibshirani R, Bukholm IK, Karesen R, Botstein D, Borresen-Dale AL, Jeffrey SS (2004) Different gene expression patterns in invasive lobular and ductal carcinomas of the breast. Mol Biol Cell 15: 2523-2536.

Zhao X, Weir BA, LaFramboise T, Lin M, Beroukhim R, Garraway L, Beheshti J, Lee JC, Naoki K, Richards WG, Sugarbaker D, Chen F, Rubin MA, Janne PA, Girard L, Minna J, Christiani D, Li C, Sellers WR, Meyerson M (2005) Homozygous deletions and chromosome amplifications in human lung carcinomas revealed by single nucleotide polymorphism array analysis. Cancer Res 65: 5561-5570.

Zhou H, Duesberg PH (1990) A retroviral promoter is sufficient to convert proto-src to a transforming gene that is distinct from the src gene of Rous sarcoma virus. Proc Natl Acad Sci U S A 87: 9128-9132.

Zhou RP, Duesberg PH (1988) myc protooncogene linked to retroviral promoter, but not to enhancer, transforms embryo cells. Proc Natl Acad Sci U S A 85: 2924-2928.

Zhou RP, Duesberg PH (1989) Avian proto-myc genes promoted by defective or nondefective retroviruses are single-hit transforming genes in primary cells. Proc Natl Acad Sci U S A 86: 7721-7725.

Zhu XQ, Qian JW (1986) Remission of acute lymphoblastic leukemia of childhood following acute infectious disease. A case report. Chin Med J (Engl) 99: 433-434.

Ziegler JL (1976) Spontaneous remission in Burkitt's lymphoma. Natl Cancer Inst Monogr 44: 61-65.

Zimonjic D, Brooks MW, Popescu N, Weinberg RA, Hahn WC (2001) Derivation of human tumor cells in vitro without widespread genomic instability. Cancer Res 61: 8838-8844.

Zimonjic D, Brooks MW, Popescu N, Weinberg RA, Hahn WC (2002) Correspondence re: D. Zimonjic et al., Derivation of human tumor cells in vitro without widespread genomic instability. Cancer Res., 61: 8838-8844, 2001, authors' reply. Cancer Res 62: 6348-6349.

Zipursky A, Thorner P, De Harven E, Christensen H, Doyle J (1994) Myelodysplasia and acute megakaryoblastic leukemia in Down's syndrome. Leuk Res 18: 163-171.

Zirkle RE (1970) Ultraviolet-microbeam irradiation of newt-cell cytoplasm: spindle destruction, false anaphase, and delay of true anaphase. Radiation Research 41: 516-537.

Zygiert Z (1971) Hodgkin's disease: remissions after measles. Lancet 1: 593.

Index

Color Plate Section

CHAPTER-4

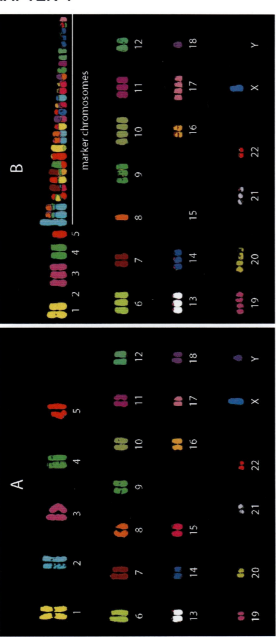

Fig. 4.1   Karyotypes of a normal and a highly aneuploid human cells.
Comparison of the two karyotypes shows that the normal cell (A) differs from the aneuploid cancer cell with numerous numerical and structural chromosomal alterations or aneusomies (B) (Duesberg et al. 2005, Rasnick 2002). Marker chromosomes (B) are structurally abnormal chromosomes, which are either rearranged intra-chromosomally or inter-chromosomally to form various hybrid chromosomes. Owing to their unique structure, marker chromosomes can serve as tracers for the origin of possibly metastatic cancer cells from primary cancers and for the origin of primary cancer cells from possibly aneuploid preneoplastic precursors (Koller 1972, Sandberg 1990).

CHAPTER-5

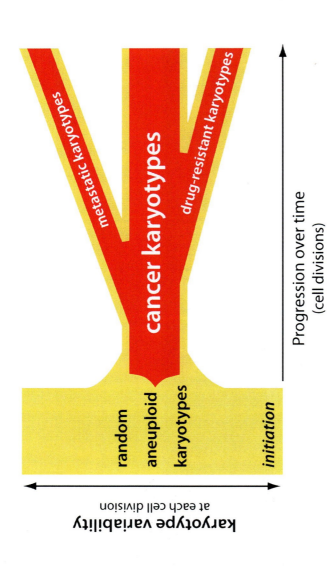

Fig. 5.1   Quasi-stable cancer and precancer karyotypes.

Cancers are initiated and maintained by individual cells with aneuploid karyotypes, much like new species. The tip at the left of the red zone signals the origin of the cancer-causing karyotype from an initiation pool of cells consisting of random aneuploid non-neoplastic karyotypes (yellow). The cancer-causing karyotypes are in flexible or dynamic equilibrium — destabilized by random aneuploidy and stabilized (within narrow limits of variation) by selection for viability and oncogenic function. Together, the two competing forces form quasi-stable average cancer-causing karyotypes (red zones) as the populations proliferate. The range of variability of the cancer-causing karyotypes is always accompanied by a range of non-neoplastic aneuploid variants (yellow zones). Occasionally, stochastic karyotypic evolutions generate new cancer-specific phenotypes, such as drug resistance and metastasis in a process termed tumor progression.

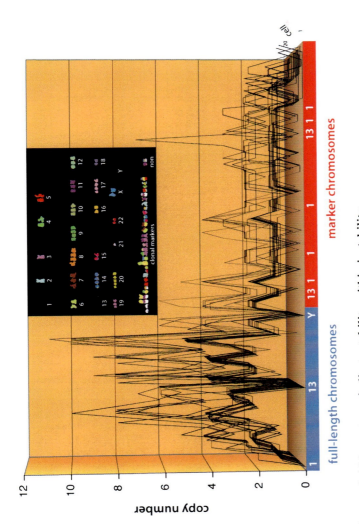

full-length chromosomes

marker chromosomes

copy number

Fig. 5.2   Clonal aneuploid karyotypes indicate stability within instability.

Twenty cells from a primary bladder cancer had unique karyotypes. Nevertheless, the karyograph reveals regions of similarity. The abscissa is the full-length chromosome number (blue region) and the marker chromosome designation (red region). The ordinate is the number of copies of each chromosome. The z axis represents the metaphases of 20 different cells. The most notable feature shared by every cell was the complete absence of intact chromosomes 1 and 13. The insert shows the karyotype of one of the cancer cells lacking chromosomes 1 and 13 (blue region). Cells cannot live without at least one copy of each chromosome. Therefore, the missing chromosomes 1 and 13 appear as part of several marker chromosomes (red region).

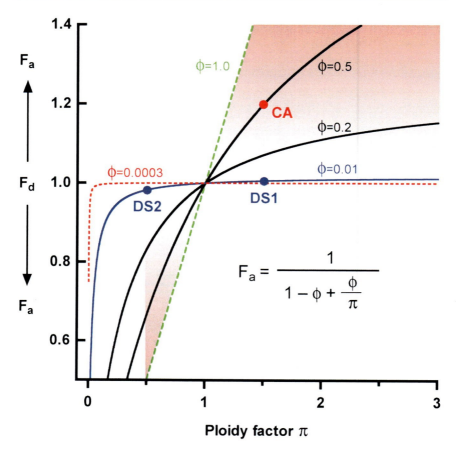

Fig. 5.3 **A graphical representation of Equation 5.7 showing how aneuploidy changes phenotypes.**

The normal diploid metabolic phenotype F_d is perturbed by varying the ploidy factor π and the aneuploid fraction ϕ to produce an ensemble of aneuploid metabolic phenotypes F_a. The phenotypes (F) of polyploid cells with balanced karyotypes fall on a straight line ($\phi=1$, broken green line), with haploids at $\pi=0.5$, diploids at $\pi=1$, triploids at $\pi=1.5$ and tetraploids with $\pi=2$, differing by equal increments of 0.5 F units. An ensemble of aneuploid metabolic phenotypes, F_a, was produced by varying the ploidy factor, π, and the fraction of the normal chromosome set ($0<\phi<1$, black and blue lines) according to Equation 5.7. $F_a>1$ represents positive aneuploidy, corresponding to gain-of-flux relative to the diploid cell, and $F_a<1$ represents negative aneuploidy, corresponding to loss of biochemical flux. Specific examples of aneuploid phenotypes are Down syndrome ($\phi=0.01$, blue line) with trisomy ($\pi=1.5$) of chromosome 21, $F_a=1.006$ (DS1),

Fig. 5.3 Contd. ...

Fig. 5.3 Contd. ...

and monosomy ($\pi=0.5$) of chromosome 21, $F_a=0.98$ (DS2). DS2 is more severe than DS1, consistent with the general principle that a lost of gene dose is more deleterious than a gain. Another example is a typical, near triploid colon cancer (CA, red dot) with an average of 69 chromosomes, corresponding to $\phi=0.5$, $\pi=1.5$, and $F_a=1.2$. The effect on the phenotype of increasing or decreasing the functional dosage from $\pi=0$ to 3 of the seven genes ($\phi=0.0003$) thought to cause colon cancer is indicated by the dotted red line. For $\pi>0.05$, the metabolic phenotype described by the dotted red line nearly coincides with that of the normal diploid cell, which is far from sufficient to generate cancer. The shaded area represents the cancer phenotypes.

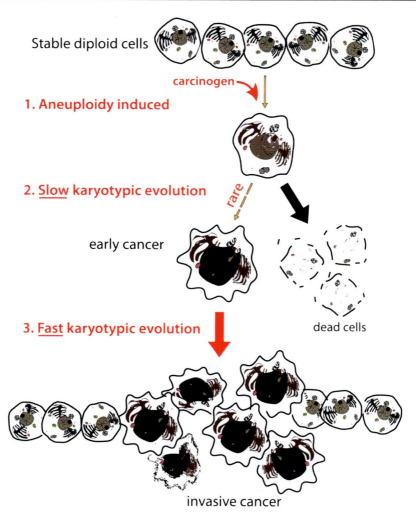

Stable diploid cells

carcinogen

1. Aneuploidy induced

2. Slow karyotypic evolution

rare

early cancer

dead cells

3. Fast karyotypic evolution

invasive cancer

Fig. 6.1   Three steps in the evolution to invasive cancer.

1) Initiation: A carcinogen or a spontaneous accident induces random aneuploidy by various mechanisms, e.g. nondisjunction, breaking and rearranging chromosomes, etc. 2) Slow pre-neoplastic chromosomal evolutions: By unbalancing thousands of genes, aneuploidy corrupts teams of proteins that segregate, synthesize and repair chromosomes. Aneuploidy is therefore a steady source of chromosomal variations, from which, in classical Darwinian terms, neoplastic karyotypes eventually evolve. The initial low level of aneuploidy catalyzes a slow progression of pre-neoplastic chromosomal evolutions. While chromosomal imbalance is necessary for progression it also retards it because many aneuploid cells die from loss of both copies of a chromosome and non-viable chromosome combinations. 3) Fast neoplastic evolutions: Once a neoplastic chromosome

Fig. 6.1 Contd. ...

Fig. 6.1 Contd. ...

combination evolves, subsequent karyotypic variations are accelerated, because neoplastic cells are generally more aneuploid and thus more adaptable than pre-neoplastic cells and can form locally large pools by outgrowing normal cells. Thus, neoplastic cells evolve independently within tumors forming ever-more heterogeneous and malignant phenotypes such as invasiveness, metastasis and drug-resistance at high rates. In sum: Malignancy can be seen as a consequence of autonomous chromosomal evolutions that increase karyotypic entropy to its biological limits, at or near a DNA index of 1.7 (Section 6.2.3).

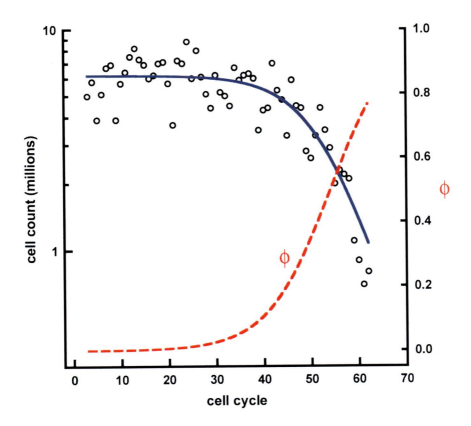

Fig. 6.3   Hayflick limit for the human cell.

The blue solid line is the best-fit curve of Equation 6.4 to the serial passaging data from Figure 3 of Hayflick for an embryonic human cell strain (Hayflick 1965). The broken red line represents the autocatalyzed progression of the aneuploid fraction ϕ for the same data using Equation 6.3. The parameters from the best fit were $k = 0.15$ cell-cycle^{-1}, $\phi_0 = 3.6 \times 10^{-4}$.

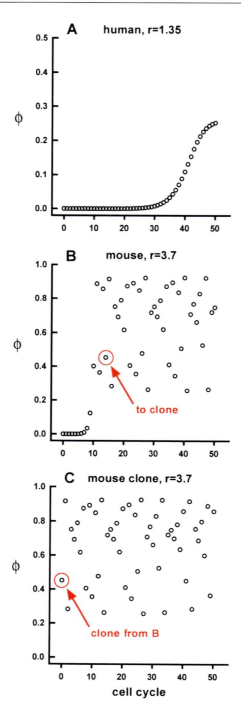

Fig. 6.4 Contd. ...

Fig. 6.4 Contd. ...

Fig. 6.4   Chaotic growth of aneuploidy drives transformation.

Equation 6.5 was used to model the autocatalyzed progression of aneuploidy for primary human (A) and mouse (B) fetal cells in culture. The control parameter r=1.35 and the initial average aneuploid fraction $\phi_0 = 10^{-6}$ modeled the 50 cell cycle Hayflick limit for primary human fetal cells (A). The data points represent the average aneuploid fraction ϕ for the population of cells at each cell cycle. In panel B the same ϕ_0 was used for the mouse fetal cells. Although values of r>1.5 were completely unrealistic for modeling the Hayflick limit of human cells, values of the control parameter greater than 3.57 could be used to model mouse cells. A control parameter greater than 3.57 for the logistic equation (Equation 6.5) produces chaotic growth patterns (Williams 1997). Therefore, the value r=3.7 produced a chaotic progression of the aneuploid fraction ϕ for all cell divisions beyond around 12 cycles (B). While the aneuploid human cells would probably die out before being transformed into an immortal cell line (because so little genome space is being explored) the chaotic redistribution of the mouse genome provides a greater opportunity for the cells to hit upon a genetic combination that leads to transformation and immortalization. Panel C shows that when the transforming genome in panel B (red arrow) is cloned, its intrinsic karyotypic instability immediately leads to a heterogeneous population of heteroploid offspring.

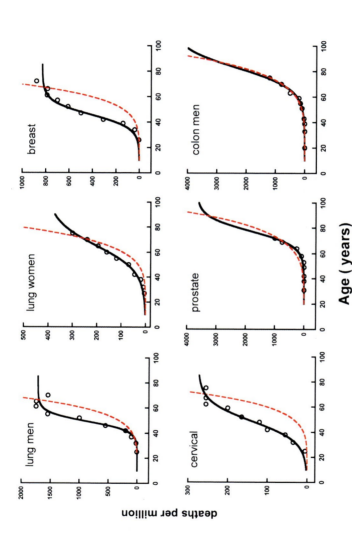

Fig. 6.5 **Autocatalyzed growth of aneuploidy explains age distribution of human cancers.**

The superiority of the sigmoidal curve of Equation 6.7 for the autocatalyzed progression of aneuploidy was best demonstrated by comparing it with Equation 6.6 for the multi-hit version of the gene mutation theory of human carcinogenesis. Equation 6.7 gives a good fit (solid black lines) to the number of deaths per million people for six typical cancers as a function of age (Armitage and Doll 1954). The broken red lines show the best-fit curves to the same data for the 7-successive mutation model (Equation 6.6). The only good fit for Equation 6.6 was with colon cancer deaths in men.

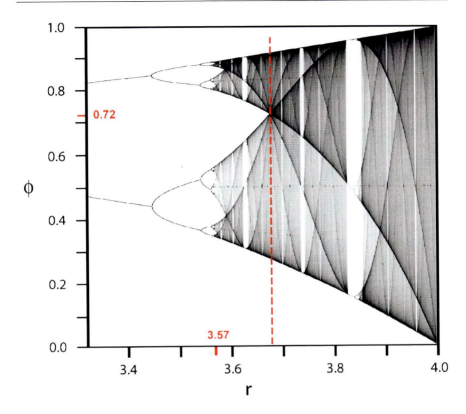

Fig. 6.6   The autocatalyzed progression of aneuploidy leads to cancer DNA indices near 1.7.

After a large number of cell divisions, DNA index = 1+ϕ (Rasnick and Duesberg 1999, Rasnick 2000). Equation 6.9 was iterated to generate a map of the probability that after numerous divisions cells will have particular values of the aneuploid fraction ϕ for various values of the growth control parameter r (Rasnick 2002). At relatively low values of r, the aneuploid states bifurcate until r reaches the critical value of 3.57 (red tick mark on abscissa), beyond which the progression of aneuploidy becomes chaotic. The denser regions of the probability map represent the more likely values of ϕ. Aneuploid cells evolve towards the attractor readily visible at r=3.68 (broken red line) and ϕ= 0.72 (DNA index = 1.7) (red tick mark on the ordinate). At values of r greater than 3.68, the aneuploid cells become less coherent as their genomes become too disorganized and chaotic to sustain viability. That is why mature cancers tend to have DNA indices near 1.7 and its overtone multiples — the point of maximum disorder of the genome that still sustains life.

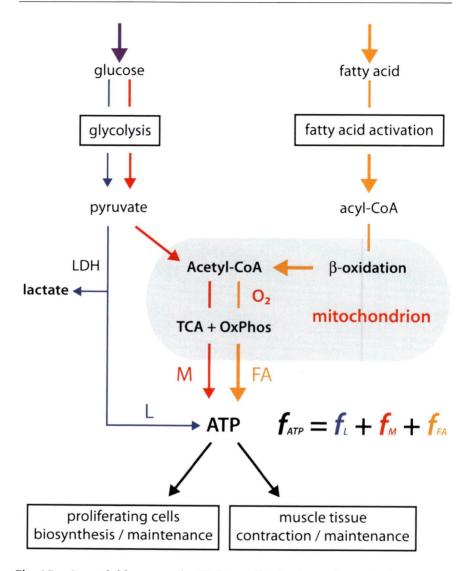

Fig. 6.7   Aneuploidy causes the Warburg effect by increasing ATP demand.

The energy requirements needed to produce the significantly increased protein levels of cancer cells necessarily places extra demands on ATP production. The production of ATP is distributed among three fundamental pathways: 1) glycolysis via pyruvate through mitochondria (f_M, red arrow), 2) aerobic cytoplasmic glycolysis with production of lactic acid (f_L, blue arrow), and 3) fatty acid oxidation through the mitochondria (f_{FA}, gold arrow). The normal distribution of metabolic flux along these three pathways changes dramatically as a consequence of the unavoidable physical

Fig. 6.7 Contd. ...

Fig. 6.7 Contd. ...

constraints imposed by the chromosomal imbalance of cancer cells (see Section 6.3). ATP is produced entirely by the mitochondrial aerobic pathway during the low ATP demand of the normal cell (f_M, red arrow and f_{FA}, gold arrow). The production of ATP increases with demand until the mitochondrial capacity is saturated. Beyond this limiting value, the excessive ATP demand of aneuploid cancer cells leads to the abrupt turning on of the cytosolic glycolytic pathway for additional ATP production with increasing secretion of lactate (f_L, blue arrow).

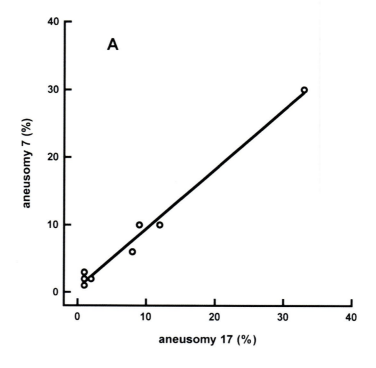

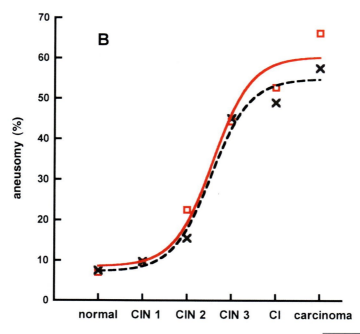

Fig. 7.2 Contd. ...

Fig. 7.2 Contd. ...

Fig. 7.2   Chromosomes are equally efficient at detecting neoplasia.

Measuring the level of aneusomy of any single chromosome (or combination of chromosomes for better statistics and reliability) could provide an objective means of determining the presence or absence of neoplasia at all the grade levels pathologists use. (A) This has been tested experimentally for similar levels of aneusomy of chromosomes 7 and 17 present in the same cervical samples (author's unpublished results). (B) Equivalent fractions of cells with aneusomy for chromosomes 1 (red) and 17 (black) in the same grades of cervical neoplasia (Hariu and Matsuta 1996).

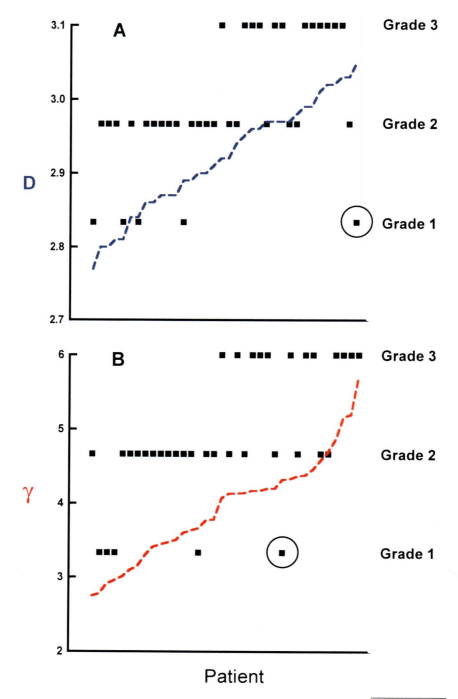

Fig. 7.4 Contd. ...

Fig. 7.4 Contd. ...

Fig. 7.4   Thirty-six invasive ductal carcinomas of the breast correlated with two measures of chromosomal instability.

The solid squares represent a patient's tumor that had been graded 1, 2, or 3, for increasing severity. It is generally accepted that the most malignant cancers are the most genetically unstable. Both D and γ are measures of the genomic instability of cancer cells (See reference (Rasnick 2009) for detailed descriptions of D (the distribution entropy of histogram data) and γ). A) Patients along the horizontal axis were sorted by increasing D (broken blue line). B) Patients were sorted by increasing γ (broken red line). Grade 3 tumors were concentrated at high values of D and γ (indicating high levels of chromosome instability). The few examples of Grade 1 favored low values of D and γ. The Grade 1 tumor circled at the lower right of both graphs was likely misclassified and probably highly malignant. Grade 2 tumors were disperse but tended to the left side of the graphs, with low and intermediate values of D and γ. It is likely intermediate Grade 2 is so uninformative as to be of little value. This was recognized some years ago for cervical cancer when the intermediate category CIN-2 was eliminated (Bollmann et al. 2001, National Cancer Institute 1989). Now there are only low and high grade cervical lesions. This simplified classification scheme has also been recommended for neoplastic lesions of esophagus, stomach, colon and rectum (Schlemper et al. 2000).